Polymers in Oil and Gas Industry

Polymers in Oil and Gas Industry

Dr. Vikas Mittal
Editor and Lead Author

𝒞𝒲𝒫
Central West Publishing

Disclaimer
Every effort has been made by the publisher, editor and authors while preparing this book, however, no warranties are made regarding the accuracy and completeness of the content. The publisher, editor and authors disclaim without any limitation all warranties as well as any implied warranties about sales, along with fitness of the content for a particular purpose. Citation of any website and other information sources does not mean any endorsement from the publisher and authors. For ascertaining the suitability of the contents contained herein for a particular lab or commercial use, consultation with the subject expert is needed. In addition, while using the information and methods contained herein, the practitioners and researchers need to be mindful for their own safety, along with the safety of others, including the professional parties and premises for whom they have professional responsibility. To the fullest extent of law, the publisher, editor and authors are not liable in all circumstances (special, incidental, and consequential) for any injury and/or damage to persons and property, along with any potential loss of profit and other commercial damages due to the use of any methods, products, guidelines, procedures contained in the material herein.

NATIONAL
LIBRARY
OF AUSTRALIA

A catalogue record for this book is available from the National Library of Australia

ISBN (print): 978-0-6482205-1-0
ISBN (e-book): 978-0-6482205-0-3

Contents

Preface

Polymers have found widespread application (or have exhibited application potential) in a wide range of areas related to oil and gas industry. Though majority of these uses are in infancy or are not yet commercialized, however, the indications obtained from the laboratory and pilot tests confirm the usefulness of polymers for entire oil and gas industry. Some of the aspects of oil and gas industry benefiting from the use of polymers include enhanced oil recovery, oil spill clean up, membranes, adsorbents and coatings, among others. The current book aims to present in detail the advances gained on these aspects in the recent years, so as to provide deep insights into the current state of the art as well as future trends. In addition, ample fundamental insights have been provided to make the readers aware about the importance of polymers and polymer nanotechnologies prevalent in oil and gas industry as well as to tune their performance to achieve certain set of properties useful for their applications.

In this respect, Chapter 1 reviews the nano-enhanced polymer applications in oil and gas industry. The use of polymer nanocomposites in the oil and gas industry is also reviewed in detail in Chapter 2. Chapters 3 and 4 specifically target the use of polymers in enhanced oil recovery processes, where the use of associative polymers for chemical injection and xanthan gum as advanced polymer EOR system has been reviewed. Various polymer systems for oil spill clean up as well as phenol removal are discussed in Chapter 5. Chapter 6 presents an experimental study to synthesize resorcinol-formaldehyde based cryogel nanocomposites for oil spill clean up. Chapter 7 reviews the membrane technology for gas separation, especially dealing with polyurethane based materials. Polymer modified/enhanced adsorbents for gas adsorption and sweetening are the focus of Chapter 8. Chapter 9 reviews the use of various polymers in pipeline coatings for oil and gas industry. The use of biopolymers for the industrial coatings is reviewed in Chapter 10, whereas Chapter 11 presents an experimental study for the development of epoxy composite coatings for achieving enhanced corrosion resistance. Finally, Chapter 12 details the important subject of achieving enhanced thermal conductivity in the polymeric materials for a large range of applications in the oil and gas industry.

Finally, the support of all the chapter contributors is greatly acknowledged towards the successful accomplishment of the book. The book is dedicated to my family for their constant support.

Vikas MITTAL
Australia

1

Nano-enhanced Polymer Applications in Oil and Gas Industry: A Review

1.1 Introduction

Nanotechnology and nanomaterials comprise a growing field that has inter-
sected with many major industries over the past decade. Specifically, the oil
industry has taken an interest, in recent years, in the numerous potential
technological advancements offered by nanoscale technologies. An extensive
array of research and development in this field has provided several concepts
and prototypes, many of which stand to improve the efficiency and reduce the
cost of upstream oil recovery operations. Nanotechnology is a field that involves materials featuring particle sizes or structural ordering at the scale of
approximately 1-100 nm in one or more dimensions. A branch of nanotechnology that has great potential for advances in oil and gas applications is polymer nanotechnology. Polymer nanotechnology encompasses the design and
manipulation of polymeric materials at the nanometer scale, as well as the incorporation of nanoparticles into polymers to form nanocomposites [1]. Enhancement of materials with nanoparticles has brought about novel materials
with properties that are far superior to the constituent materials in their bulk
phases. Properties such as strength, electrical and thermal conductivity, catalytic activity, and dispersion are only some of the many that are greatly enhanced at the nanometer scale [1].

There are several specific forms of polymer nanotechnology that are currently under development for various uses in the oil and gas industry. One of
the more traditional forms of polymer nanotechnology is polymer nanocomposites, which dates back before the 1990s [2]. These are materials that contain dispersed nanoparticles in a polymeric matrix, which results in enhanced
properties. Polymers that are used as part of machinery and in structural applications can provide great engineering and functional benefits when they
are made into polymer nanocomposites. A form of polymer nanotechnology
that is more focused on direct usage of nanoparticles and polymers for oil recovery purposes is the generation of enhanced oil recovery (EOR) agents. Pol-

Arjun Dhillon and Vikas Mittal**, The Petroleum Institute (part of Khalifa University of
Science and Technology), Abu Dhabi, UAE
*Current address: University of Waterloo, Canada; **Current address: Bletchington,
Wellington County, Australia

ymers are used in the fabrication or stabilization of nanoparticles that can be injected into oil wells to improve the oil production after conventional methods of oil recovery have been exhausted. EOR processes designed/based on these nanoparticles may result in oil recovery that is more efficient and cost-effective. A reservoir stimulation process that can also benefit from polymer nanotechnology is hydraulic fracturing, which relies on various materials such as fracturing fluids and proppants to improve production. Many materials employed in such processes can be enhanced or substituted with nanomaterials exhibiting preferred behavior. Even conventional oil drilling processes can be augmented through the inclusion of polymer nanotechnology into the range of materials. The present work reviews some of the current developments in the area of polymer nanotechnology intended for use in upstream oil and gas applications. Many potential uses of polymer nanotechnology are though in the research and development phase, however, many commercial applications have already been realized, indicating the immense importance of polymers in oil and gas industry.

1.2 Polymer Nanocomposites for Engineering Material Enhancements

The fabrication and characterization of polymer nanocomposites has taken place for decades, making it one of the better-established applications of polymer nanotechnology [1]. Polymer nanocomposites typically entail the inclusion of nanoparticles as filler in a polymeric matrix to enhance mechanical, electrical, and thermal properties of the resulting composite material. Polymer nanocomposites can possess unique properties due to the small size of nanoparticles. There have been a number of polymer nanocomposites proposed and developed in recent years for various general and specific applications within the oil and gas industry.

An important consideration to make when utilizing polymers in certain applications is their thermal behavior. In the oil and gas industry, high-temperature conditions are often present, which place engineering constraints on the types of polymers that can be used. It is, therefore, highly desirable to obtain information on the thermal degradation characteristics of polymer nanocomposites, so that their suitability can be assessed for operations at elevated temperatures. A recent comprehensive review of polymer nanocomposites comprising several condensation polymers with various nanoparticle fillers studied available literature to draw conclusions regarding the impact of nanomaterials on the thermal stability of polymer nanocomposites [3]. Findings of the review suggested that thermal stability of certain polymer nanocomposites was improved, while for others, it was worsened. Some polymer nanocomposites that included clay nanoparticles exhibited better thermal stability, due in part to the thermal characteristics of clay itself and the obstruction of diffusive transport processes by the nanoparticles. This phenomenon is illustrated in Figure 1.1 for poly(butylene terephthalate)

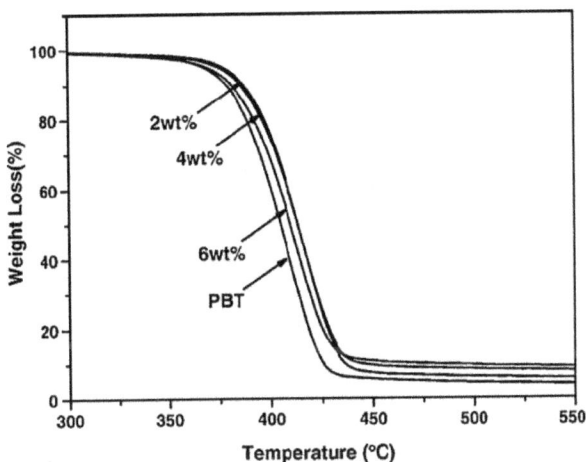

Figure 1.1 Effect of clay concentration on thermal stability of poly(butylene terephthalate). Reproduced from Reference 4 with permission from American Chemical Society.

nanocomposites [4]. Carbon nanotubes (CNTs) were found to be beneficial for thermal stability of condensation polymers, particularly when carboxylated. Boehmite (AlOOH) and fumed silica were also found to improve thermal stability in their nanocomposites. The typically optimal concentration of nanoparticles with respect to enhancement of thermal stability was found to be in the range of 2-5 wt% [3]. Surface modifications of the fillers have also been observed to enhance thermal performance of the filler, and subsequently the polymer matrices in which these modified fillers are incorporated. This behavior is shown in Figure 1.2 through thermal degradation behavior of the modified montmorillonite with poly(vinylpyrrolidone) (PVP) [5].

Among functional reinforcements for polymer nanocomposites, properties and applications of CNTs have been explored extensively in recent years. These fillers possess a tensile strength in the range of 50-200 GPa, with the elastic modulus up to 1200 GPa. Owing to these high mechanical properties, attempts

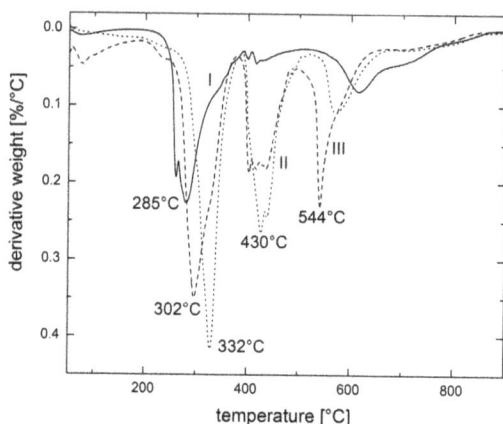

Figure 1.2 Derivative weight of organically modified montmorillonite before adsorption (I), after adsorption with 1.5 g of initially added PVP (for 2 g of organically modified montmorillonite), (II) and after adsorption with 1 g of initially added PVP (for 2 g of organically modified montmorillonite) (III). Reproduced from reference 5 with permission from Elsevier.

have been made to strengthen a number of thermoplastic and thermoset polymers with CNT fillers. The resulting mechanical properties of these nanocomposites have been both positive and negative in different cases, depending on dispersion and filler-matrix interaction characteristics. Some thermal stability and conductivity improvements have also been reported in literature, although many of these results are not yet significant enough to warrant the usage of these nanocomposites in applications where these characteristics are required. One encouraging instance of a CNT polymer nanocomposite that gained enhanced thermal stability involved the inclusion of multi-walled carbon nanotubes (MWCNTs) in a polychloroprene rubber matrix. Polychloroprene rubber has properties that enable it to be useful in oil and gas industry applications, including its resistance to combustion and degradation by heat and hydrocarbons. Researchers have identified hoses, belts, gaskets, and adhesives as some of the acceptable applications of this material [6]. Nanocomposites were prepared using MWCNTs that were modified with an ionic liquid based on imidazolium. Analysis of the prepared nanocomposite revealed that the ionic liquid modification of the MWCNTs improved their dispersion behavior, while the unmodified nanotubes did not disperse as well in the polymeric matrix. The modification of the MWCNTs also resulted in more favorable interactions at matrix-filler interfaces. The thermal stability of the unmodified MWCNT composite was greater than that of the pure polymer, while the stability of the modified MWCNT composite was greater than that of the unmodified MWCNT composite [6].

Elastomer/MWCNT nanocomposites have also proven useful for applications in challenging environments, including subsea operations and exploiting high temperature and high pressure (HTHP) reservoirs. In such environments, it is frequently necessary to seal equipment operating at high pressures while under extreme temperatures. For this purpose, a rubber/MWCNT nanocomposite was prepared using a method known as "cellulation" along with surface treatment of the CNTs, which achieved good dispersion and interfacial interactions between the MWCNTs and matrix [7]. Characterization of the resulting nanocomposites revealed that the composites were able to seal effectively at both high and low temperatures, for high pressures, and outperformed commercial sealing materials. The additional advantage of low-temperature sealing applications resulted from the fact that the glass transition temperature of the nanocomposite was relatively low, at -11 °C, meaning greater stability at lower temperatures. The nanocomposite has shown favorable results in several field trials, conducted in several different regions of the world. Another application of elastomer/CNT nanocomposites that has been evaluated recently consists of a fluoroelastomer matrix with dispersed MWCNTs that can be used for O-rings in the oil and gas industry [8]. When compared with both the fluoroelastomer itself and a composite with the carbon black, the MWCNT nanocomposite was found to be superior in terms of physical, chemical, and thermal properties.

Graphene is another functional filler which has added tremendous potential to the use of the polymer nanocomposites in challenging situations. For instance, Figure 1.3 shows the results of the increment in the mechanical properties of various polymers when incorporated with 7 wt% graphene [9].

1.3 Polymeric Nanoparticles and Polymer-Stabilized Nanomaterials as EOR Agents

Among the potential applications of polymer nanotechnology are novel polymeric and polymer-stabilized nanoparticles for EOR purposes. EOR generally refers to the processes intended to increase well production beyond the limitations of primary production and basic water flooding. There are a vast number of EOR technologies that have been explored over time, with some having

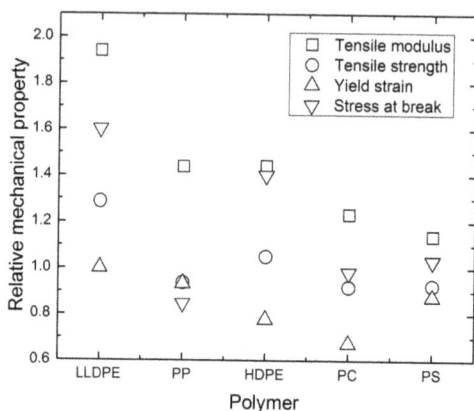

Figure 1.3 Mechanical performance of polymer-graphene nanocomposites with 7 wt% filler content. Reproduced from reference 9 with permission from Wiley.

been adopted widely throughout the oil and gas industry. As with other areas within the oil and gas industry, nanotechnology offers several novel potential EOR agents that may increase oil production to a reasonable degree.

EOR operations frequently involve measures to improve the conditions of oil recovery by modifying the behavior and flow characteristics of injected and produced fluids with respect to the reservoir rock. Conformance control research constitutes a present example of such methods. It has been established that heterogeneities in reservoir rock structure and regions where permeability is higher reduce volumetric sweep efficiency, which is a strong indicator of the performance of a given oil recovery method and overall oil production [10]. To mitigate the effects of such heterogeneities on sweep efficiency, blocking agents are sometimes deployed in formations. These particles are injected into the reservoir to reduce the problem of channelling, where the flow of injected fluid follows high-permeability paths that do not necessarily make contact with a sufficient amount of oil. By preventing fluids from following the high-permeability paths formed from the heterogeneities, blocking agents improve the sweep efficiency in oil recovery processes [11]. A nanostructured polymeric blocking agent has been developed recently to improve sweep efficiency [12]. This agent consists of polymeric microsphere particles based on acrylamide with an average size of 70 nm. These particles have favorable transport properties due to their small size, making it simple to inject them into an oil reservoir. The particles can swell slowly with water, increasing their size and enabling their blocking capability. The particles have good resistance to various reservoir conditions, such as high temperatures up to 110°C and salt concentrations of up to 200 g/L. Despite reducing permeability to water, the particles have not been found to restrict produced oil from flowing. Experiments conducted with the microspheres showed that these were able to increase oil recovery by 22.8%. Field tests conducted using a multi-step process involving microspheres and surfactant showed that the particles were able to enhance oil recovery to a significant degree. Experiments involving similar polyacrylamide nanoparticles that were 50 nm in size also showed improvement in oil recovery by 20% of oil in place (OIP) following primary oil recovery processes [13].

Certain nanoparticles can also augment the effectiveness of existing EOR chemicals, such as surfactants. This is true in the case of silica nanoparticles in the size range of 50-100 nm that were encapsulated in polymeric shells [14]. These organic/inorganic hybrid nanoparticles were combined in various ratios with surfactant solutions to evaluate their performance in EOR applications. The resulting suspensions were then tested to determine their effect on incremental oil recovery. It was observed that at quantities of 800 ppm surfactant and 200 ppm nanoparticles, viscosity of the solution was improved and the oil/water interfacial tension was reduced. The interfacial tension is an important parameter in EOR, as it plays a major role in capillary forces that act to entrap oil in pores of formation rock [15]. A great advantage of the hybrid nanoparticles is their suitability to environments where high temperatures and highly saline conditions are present. With the deployment of EOR processes under increasingly extreme conditions, it is important for potential EOR agents to perform well under such conditions. With respect to improvement to EOR performance, use of the aforementioned nanoparticle-surfactant solution managed to recover 6.2% of additional oil from a sample that had approximately 30% of its oil in place (OIP) remaining after initial water flooding [14].

Carbon dioxide foam flooding is another EOR process that can be improved with certain nanoparticles. Typical CO_2 foams are comprised of carbon dioxide gas, as well as water containing surfactants [16]. The foams are more viscous than carbon dioxide gas, allowing them to mobilize more oil than with carbon dioxide flooding. A major drawback that hinders the feasibility of foams for EOR processes under various circumstances is their inherent lack of stability. This is especially pronounced in high-temperature environments, where commonly used surfactants are unable to resist degradation. To help prolong the life of carbon dioxide foams for EOR purposes, nanoparticles have been developed and tested. These nanoparticles are based on 5 nm-diameter silica spheres, with surface modifications of short chains of polyethylene glycol (PEG). The polymer-modified silica nanoparticles were used to stabilize the foams by adsorption at gas/liquid interfaces of the foam [17]. The inclusion of nanoparticles in concentrations starting at 0.5 wt% resulted in foams that compared favorably to those generated through the use of surfactants. Noteworthy and beneficial characteristics of the nanoparticle-stabilized foams included stability at high temperatures and relatively significant salt concentrations. It has also been shown in a different study that similar silica nanoparticles with PEG surface modification have effective transport properties that make them highly suited to this EOR application [18]. Findings suggest that the nanoparticles can move through reservoir rock with relative ease, regardless of permeability. Additionally, the relative quantity of nanoparticles retained by the rock is very low, which is a highly desirable trait.

1.4 Improvements to Hydraulic Fracturing Processes

Hydraulic fracturing is a process for well stimulation that is commonly employed throughout the oil and gas industry to improve production. It is of particular importance in production improvements for unconventional reservoirs [19]. The hydraulic fracturing process entails the injection of pressurized fracturing fluids into a well to create numerous fractures in the formation rock, which are then kept open with proppants. Proppants are particles such as silica sand that are suspended in the fracturing fluid [20]. Fracturing process can cause significant impact on the stability of additives present in fracturing fluid due to high pressure, high temperature, and salinity. In addition, presence of organic matter in the oil formations would accelerate the abiotic transformations, thus, leading to the degradation of organic additives, thus, requiring a very stable hydraulic fracturing liquid formulation [21].

A consideration that is usually significant during the design of effective gelled fracturing fluids is achieving a sufficiently high viscosity. It has been found that a higher fluid viscosity can result in increased fracturing effectiveness [22]. To reach this objective, a typical additive in fracturing fluids is a polymeric thickening agent that acts to increase the viscosity of the fluid. Guar gum is consistently used as such a thickening agent in gelled fracturing fluids [19]. Guar is a water-soluble polymer whose viscosity in solution generally increases with increasing concentration [23]. For this reason, a desirable area of improvement for guar-based fracturing fluids is the capability to reach required solution viscosities at lower polymer concentrations. The field of nano-technology offers some promising solutions to this challenge. Silica nanoparti-cles have been explored for enhancement of rheology and viscosity of fractur-ing fluids [24]. In the investigation, researchers dispersed 20 nm silica nano-particles in both guar-based fluids and fluids that contained blends of guar and surfactant-based fluid (SBF). It was found that guar-based fluids experi-enced an increase in apparent viscosity from increased concentration of the silica nanoparticles, up to the highest tested concentration of 0.4 wt%. The reason for this was attributed to adsorption of polymer chains on the nano-particles, resulting in structured aggregation. The addition of silica nanoparti-cles was also found to benefit the guar-SBF blends, but optimal concentration and conditions for their use varied from the strictly guar-based fluids.

Cross-linking constitutes another means for increasing the viscosity of guar-based fracturing fluids. Cross-linking of guar gum in fracturing fluids is not possible if the concentration of polymer is below the value of the critical overlap concentration [23]. A novel nanostructured polymer has been devel-oped to promote cross-linking of guar in fracturing fluids, thereby reducing the critical overlap concentration [25]. The 15 nm nanolatex particles were functionalized with boronic acid and used to cross-link guar at different con-centrations and conditions. The boron concentration required to reach an op-timum viscosity at room temperature was 2 ppm. The functionalized nanola-tex particles compared favorably to the more traditional borate cross-linker, showing the ability to efficiently cross-link lower concentrations of guar. Re-searchers identified the size of the nanolatex as a likely contributing factor, as the particles are larger than borate ions by a factor of 100. This size advantage facilitates the interaction of the nanolatex with guar molecules in dilute solu-tion. A unique advantage of the nanolatex cross-linker is its behavior at differ-ent pressures. The guar gel cross-linked with nanolatex containing 3 ppm bo-ron, unlike with typical borate cross-linkers, exhibited insignificant depend-ence on pressure up to 20,000 psi.

Since nanotechnology has a number of potential future applications in hy-draulic fracturing processes, the rheological side effects of nanoparticle addi-tion to guar gels have also been studied [26]. Vapor grown carbon fibers with silica coating and octadecyltrichlorosilane group functionalization (ODS-SiO$_2$-VGCF) were studied as part of this analysis. The nanomaterials were added to guar gel containing borate cross-linkers. The findings showed that the nano-particles did not have a significant adverse impact on the guar viscosity under typical fracturing operating conditions. Some interactions, however, were ob-served at low pH values outside the range of regular fracturing applications, as shown in Figure 1.4. The results indicated that nanoparticles with a high degree of compatibility with the base fracturing fluids should not negatively affect the rheology of these fluids.

An alternative to polymeric fracturing fluids that has been explored in great depth is the use of viscoelastic surfactant (VES) fluids. VES fluids are based on certain amphiphilic surfactants that associate in structured micellar formations, increasing the viscosity of the fracturing fluid. As suggested by the name, VES fluids possess viscoelastic properties that result from the contained self-assembled structures [27,28]. Experimentation has been conducted to improve the proppant transport characteristics in VES fracturing fluids using polymeric fiber nanocomposites. These nanocomposites resulted from inorganic nanoparticles that were dispersed in polyester. The nanocomposite fibers were found to reduce flowback of proppant, as well as surfactant loss, when tested in coal seam samples with quantities between 0.4% and 0.7% of the weights of proppant in the fluid. It is of particular significance to fracturing applications pertaining to extraction of the coalbed methane [29].

Following the pumping stage of fracturing fluids into the well, it is necessary to remove the fluids to enable efficient production of hydrocarbons from the well. For removal of gelled fracturing fluids to be feasible, the polymeric thickening agents must be degraded chemically to lower the fluid viscosity. This is achieved through the utilization of agents known as gel breakers, including enzymes, acids, and oxidizers. These act to reduce the polymeric chain length of the guar macromolecules, reducing their average molecular weight as well as the gel viscosity [30,31]. Oxidizers are currently the most widely deployed class of breakers in the oil and gas industry. However, there are some severe drawbacks to the usage of oxidative breakers, including environmental and equipment damage. Enzymatic breakers are much safer, as these do not share these drawbacks [32]. A fundamental disadvantage of enzyme breakers is the fact that their activity is typically very strongly dependent on temperature and pH, which gets worsened by the high temperature and high pH conditions incurred as part of numerous fracturing applications [33]. It is this challenge that has motivated research into the usage of nanoparticles for the entrapment of enzyme breakers. Polyelectrolyte complex (PEC) nanoparticles were shown to encapsulate and protect both pectinase and EL2X enzymes, maintaining their activity for longer periods of time when compared to their activity in the absence of the nanoparticles [34,35]. In addition to preventing the enzymes from denaturing under unfavorable conditions, the nanoparticles managed to delay the enzyme-catalyzed guar degradations by as much as 12 hours - another potentially beneficial feature. It was

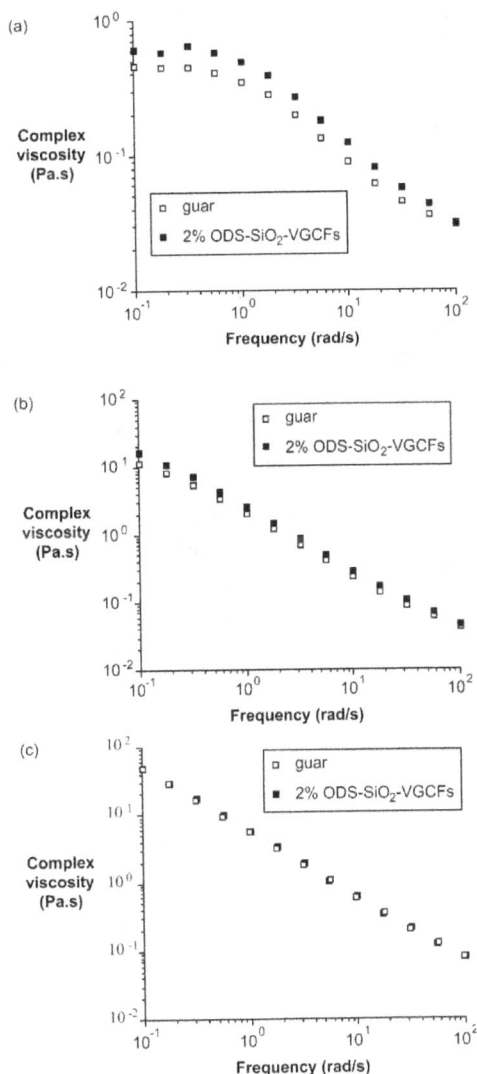

Figure 1.4 Viscosity with respect to frequency at pH = 8.6 (a), 9.3 (b) and 10.3 (c) for boric acid crosslinked guar gel in comparison with boric acid crosslinked guar gel with 2 wt% of ODS-SiO$_2$-VGCF. Reproduced from reference 26 with permission from American Chemical Society.

also discovered that these same PEC nanoparticles, when dispersed in fracturing fluids, cause a significant decrease in fluid loss, which is of great importance to fracturing applications [36].

1.5 Improvements to Oil Drilling Processes

Nanotechnology is able to offer substantial improvements to conventional processes in the oil and gas industry, even from the initial stages of drilling an oil well. Popular applications involve the dispersal of nanomaterials in drilling fluids to achieve various objectives. Drilling fluids are a necessary component of all drilling operations, as they perform several necessary functions. The cuttings, which are the fragments of the drilled formation rock, are brought to the surface via the flow of drilling fluid. The pressurization of the drilling fluid prevents oil and gas from flowing into the well as it is being drilled [37]. The drilling fluid also has a major role in keeping the terminal end of the drilling apparatus functioning properly by acting as both a heat transfer fluid and a lubricant for the drill string and bit [37]. Drilling fluids can contain several additives, including various appropriate polymers. Research in recent years has been conducted into nanomaterial enhancements to drilling fluids, including polymeric nanoparticles and inorganic nanoparticles in the presence of polymer additives. Water-based drilling fluids containing CuO and ZnO nanoparticles, as well as xanathan gum polymer as a stabilizing dispersal agent, have been studied. In one experiment, thermal conductivity was improved greatly by both nanoparticles, although CuO proved more effective and more stable than ZnO [38]. A water-based fluid containing 0.4 wt% xanathan gum and 0.5 wt% CuO nanoparticles showed an increase in thermal conductivity greater than 50% over that of the water/polymer mixture, while a fluid containing 0.5 wt% of ZnO yielded an increase in thermal conductivity of nearly 25%. The thermal conductivity improvement afforded by these polymer-stabilized nanoparticles makes them ideal candidates for inclusion in water-based drilling fluids, as they would be capable of transferring heat from the drill bit much more efficiently than conventional drilling fluids. In another experiment, a drilling fluid was prepared with the addition of 1 vol% of the aforementioned CuO and ZnO fluid [39]. The addition of the 0.5 wt% CuO fluid resulted in a thermal conductivity increase of 38.8%, while the 0.5 wt% ZnO fluid provided an increase of 34.4%. It was also found that the fluid containing CuO was more stable than the one containing ZnO under high temperature and pressure conditions, further establishing its superior feasibility as a drilling fluid additive.

During drilling operations, a portion of the drilling fluid is lost to the formation as a result of its high hydrostatic pressure. This can cause the fluid pressure to drop, which compromises its ability to carry out necessary functions such as preventing oil and gas from entering the drilled well, potentially resulting in severe negative consequences. Fluid loss additives are present in drilling muds to mitigate the effects of fluid loss during drilling operations. Nanoparticle-based fluid loss additives have been developed and analyzed as well. Silica nanoparticles were incorporated into drilling fluids with varying proportions of xanthan gum polymer and viscoelastic surfactants, then evaluated with respect to fluid loss properties [40]. The performance of the nanoparticle-enhanced fluids was also compared with that of similar fluids containing a commercially available polymeric fluid loss additive. The results showed that the silica nanoparticles drastically outperformed the polymer fluid loss additive. Due to the impressive performance of silica nanoparticles as fluid loss additives, these have also been studied with respect to applications in unconventional formations such as shale [41]. Li *et al.* [42] also developed cellulose nanoparticles as modifiers for rheology and fluid loss in bentonite water-based drilling fluids (Figure 1.5). In another study, a novel fluid loss ad-

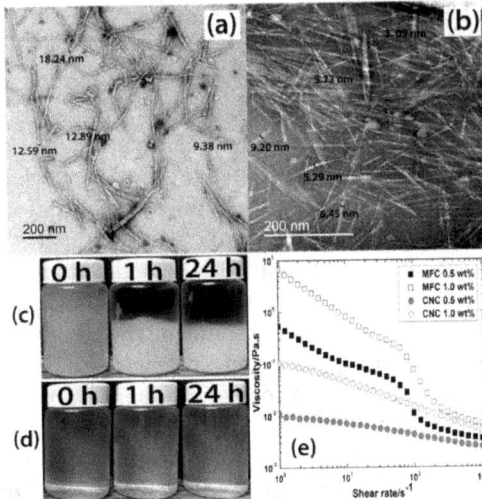

Figure 1.5 (a and b) TEM micrographs of microfibrillated cellulose (MFC) and cellulose nanocrystals (CNC), (c and d) dispersion of MFC and CNCs in aqueous solution at the concentration of 1.0 wt% and (e) shear-thinning behavior of MFC and CNC suspensions. Reproduced from reference 42 with permission from American Chemical Society.

ditive was also developed from a polymer-based nanocomposite gel. The gel consisted of colloidal particles in addition to cross-linked polymeric components. The gel is able to reduce permeability of reservoir rock to fluids, and is able to perform well for long periods of time under high temperature conditions [43].

1.6 Conclusion

Research and development in the oil and gas industry has been changed dramatically by the progression of nanotechnology as a field in engineering. The applications of systems involving particles smaller than 100 nm in at least one dimension are numerous in oil and gas industry research. Several polymer-nanoparticle systems with potential or realized applications in the oil and gas industry were reviewed. Polymer nanocomposites are among the easiest forms of polymer nanotechnology to implement, as they consist of materials that serve traditional functions but possess improved properties. High temperature resistance can be improved in certain polymers with the appropriate nanoparticles, including clay, carbon nanotubes, AlOOH, and fumed silica in low concentrations. Physical features such as tensile strength and elastic modulus can also be augmented with nanoscale fillers such as carbon nanotubes. Enhanced oil recovery processes can be based around nanoparticle agents, and existing processes can be improved by nanoparticles. Acrylamide-based polymer nanoparticles have the capability to act as effective blocking agents by swelling. Silica/polymer hybrid nanoparticles can be added to surfactant flooding EOR processes to improve effectiveness. Polymer-modified silica nanoparticles have been used to stabilize carbon dioxide foams for enhanced oil recovery processes. Guar-based hydraulic fracturing fluids can achieve improved viscosity from silica nanoparticles and crosslinking from nanolatex particles. Polyelectrolyte complex nanoparticles can protect enzymatic breakers in fracturing applications from the unfavorable conditions of the reservoir. In drilling fluids, CuO and ZnO inorganic nanoparticles can be added to polymer-containing fluids to promote heat transfer from the drill bit. Drilling fluid loss can also be managed using silica nanoparticles or nanocomposite gels. The vast array of improvements offered to the oil and gas industry by the field of nanotechnology continues to grow, as less traditional options are explored in greater detail and at a larger scale. Polymer nanotechnology will likely produce great benefits for the oil and gas industry in the coming years, even though many challenges still persist for the large scale application of these technologies.

References

1. Paul, D. R., and Robeson, L. M. (2008) Polymer nanotechnology: Nanocomposites. *Polymer*, **49**(15), 3187-3204.
2. Gao, F. (2004) Clay/polymer composites: the story. *Materials Today*, **7**(11), 50-55.
3. Bikiaris, D. (2011) Can nanoparticles really enhance thermal stability of polymers? Part II: An overview on thermal decomposition of polycondensation polymers. *Thermochimica Acta*, **523**(1-2), 25-45.
4. Tripathy, A. R., Burgaz, E., Kukureka, S. N., and MacKnight, W. J. (2003) Poly(butylene terephthalate) nanocomposites prepared by in-situ polymerization. *Macromolecules*, **36**, 8593-8595.
5. Mittal, V., and Herle, V. (2008) Physical adsorption of organic molecules on the surface of layered silicate clay platelets: A thermogravimetric study. *Journal of Colloid and Interface Science*, **327**, 295-301.
6. Subramaniam, K., Das, A., Haussler, L., Harnisch, C., Stockelhuber, K. W., and Heinrich, G. (2012) Enhanced thermal stability of polychloroprene rubber composites with ionic liquid modified MWCNTs. *Polymer Degradation and Stability*, **97**(5), 776-785.
7. Ito, M., Madhavan, R., Osawa, O., Noguchi, T., Ueki, H., Takeuchi, K., and Endo, M. (2012) Game Changing Technology with MWNT Nanocomposites for HTHP and Hostile Environment Sealing in Enhancing Oil Recovery. *SPE International Oilfield Nanotechnology Conference*. Online: https://www.onepetro.org/download/conference-paper/SPE-156347-MS?id=conference-paper%2FSPE-156347-MS (assessed 19.01.2017).
8. Heidarian, J. and Hassan, A. (2014) Microstructural and thermal properties of fluoroelastomer/carbon nanotube composites. *Composites Part B: Engineering*, **58**, 166-174.
9. Mittal, V., and Chaudhry, A. U. (2015) Polymer - graphene nanocomposites: Effect of polymer matrix and filler amount on properties. *Macromolecular Materials and Engineering*, **300**, 510-521.
10. Glasbergen, G., Abu-Shiekah, I., Balushi, S., and van Wunnik, J. (2014) Conformance Control Treatments for Water and Chemical Flooding: Material Screening and Design. *SPE EOR Conference at Oil and Gas West Asia*. Online: https://www.onepetro.org/download/conference-paper/SPE-169664-MS?id=conference-paper%2FSPE-169664-MS (assessed 19.01.2017).
11. Seright, R. S., and Liang, J. (1995) A Comparison of Different Types of Blocking Agents. *SPE European Formation Damage Conference*. Online: https://www.onepetro.org/download/conference-paper/SPE-30120-MS?id=conference-paper%2FSPE-30120-MS (assessed 11.01.2017).
12. Tian, Q. Y., Wang, L., Tang, Y., Liu, C., Ma, C., and Wang, T. (2012) Research and Application of Nano Polymer Microspheres Diversion Technique of Deep Fluid. *SPE International Oilfield Nanotechnology Conference and Exhibition*. Online: https://www.onepetro.org/download/conference-paper/SPE-156999-MS?id=conference-paper%2FSPE-156999-MS (assessed 12.02.2017).

13. Wang, L., Zhang, G., Ge, J., Li, G., Zhang, J., and Ding, B. (2010) Preparation of Microgel Nanospheres and Their Application in EOR. *International Oil and Gas Conference and Exhibition in China*. Online: https://www.onepetro.org/download/conference-paper/SPE-130357-MS?id=conference-paper%2FSPE-130357-MS (assessed 17.02.2017).
14. Nguyen, P.-T., Do, B.-P. H., Pham, D.-K., Nguyen, Q.-T., Dao, D.-Q. P., and Nguyen, H.-A. (2012) Evaluation on the EOR Potential Capacity of the Synthesized Composite Silica-Core/ Polymer-Shell Nanoparticles Blended with Surfactant Systems for the HPHT Offshore Reservoir Conditions. *SPE International Oilfield Nanotechnology Conference and Exhibition*. Online: https://www.onepetro.org/download/conference-paper/SPE-157127-MS?id=conference-paper%2FSPE-157127-MS (assessed 21.01.2017).
15. Najurieta, H. L., Galacho, N., Chimienti, M. E., and Illiano, S. N. (2001) Effects of Temperature and Interfacial Tension in Different Production Mechanisms. *SPE Latin American and Caribbean Petroleum Engineering Conference*. Online: https://www.onepetro.org/download/conference-paper/SPE-69398-MS?id=conference-paper%2FSPE-69398-MS (assessed 22.01.2017).
16. Thomas, S. (2007) Enhanced oil recovery - an overview. *Oil & Gas Science and Technology - Revue de l'IFP*, **63**(1), 9-19.
17. Espinosa, D. A., Caldelas, F. M., Johnston, K. P., Bryant, S. L., and Huh, C. (2010) Nanoparticle-Stabilized Supercritical CO_2 Foams for Potential Mobility Control Applications. *SPE Improved Oil Recovery Symposium*. Online: https://www.onepetro.org/download/conference-paper/SPE-129925-MS?id=conference-paper%2FSPE-129925-MS (assessed 30.01.2017).
18. Rodriguez Pin, E., Roberts, M., Yu, H., Huh, C., and Bryant, S. L. (2009) Enhanced Migration of Surface-Treated Nanoparticles in Sedimentary Rocks. *SPE Annual Technical Conference and Exhibition*. Online: https://www.onepetro.org/download/conference-paper/SPE-124418-MS?id=conference-paper%2FSPE-124418-MS (assessed 21.01.2017).
19. Trabelsi, S., and Kakadjian, S. (2013) Comparative Study Between Guar and Carboxymethylcellulose Used as Gelling Systems in Hydraulic Fracturing Application. *SPE Production and Operations Symposium*. Online: https://www.onepetro.org/download/conference-paper/SPE-164486-MS?id=conference-paper%2FSPE-164486-MS (assessed 26.01.2017).
20. Veatch, R. W. (2013) Overview of current hydraulic fracturing design and treatment technology: Part 2. *Journal of Petroleum Technology*, **35**(05), 853-864.
21. Kahrilas, G. A., Blotevogel, J., Stewart, P. S. and Borch, T. (2015) Biocides in hydraulic fracturing fluids: A critical review of their usage, mobility, degradation, and toxicity. *Environmental Science and Technology*, **49**(1), 16-32.
22. Howard, G. C., and Fast, C. R. (1957) Optimum Fluid Characteristics for Fracture Extension. In *Drilling and Production Practice*. Online: https://www.onepetro.org/download/conference-paper/API-57-261?id=conference-paper%2FAPI-57-261 (assessed 31.01.2017).
23. Lei, C., and Clark, P. E. (2013) Crosslinking of guar and guar Derivatives. *SPE Journal*, **12**(03), 316-321.

24. Fakoya, M. F., and Shah, S. N. (2013) Rheological Properties of Surfactant-Based and Polymeric Nano-Fluids. *SPE/ICoTA Coiled Tubing & Well Intervention Conference & Exhibition.* Online: https://www.onepetro.org/download/conference-paper/SPE-163921-MS?id=conference-paper%2FSPE-163921-MS (assessed 28.01.2017).

25. Lafitte, V., Tustin, G. J., Drochon, B., and Parris, M. D. (2012) Nanomaterials in Fracturing Applications. *SPE International Oilfield Nanotechnology Conference and Exhibition.* Online: https://www.onepetro.org/download/conference-paper/SPE-155533-MS?id=conference-paper%2FSPE-155533-MS (assessed 21.01.2017).

26. Jafry, H. R., Pasquali, M., and Barron, A. R. (2011) Effect of functionalized nanomaterials on the rheology of borate cross-linked guar gum. *Industrial & Engineering Chemistry Research*, **50**(6), 3259-3264.

27. Samuel, M., Card, R. J., Nelson, E. B., Brown, J. E., Vinod, P. S., Temple, H. L., Qu, Q., and Fu, D. K. (1997) Polymer-Free Fluid for Hydraulic Fracturing. *SPE Annual Technical Conference and Exhibition.* Online: https://www.onepetro.org/download/conference-paper/SPE-38622-MS?id=conference-paper%2FSPE-38622-MS (assessed 01.02.2017).

28. Sullivan, P. F., Gadiyar, B. R., Morales, R. H., Holicek, R. A., Sorrells, D. C., Lee, J., and Fischer, D. D. (2006) Optimization of a Visco-Elastic Surfactant (VES) Fracturing Fluid for Application in High-Permeability Formations. *SPE International Symposium and Exhibition on Formation Damage Control.* Online: https://www.onepetro.org/download/conference-paper/SPE-98338-MS?id=conference-paper%2FSPE-98338-MS (assessed 03.02.2017).

29. Xiao, B., Zhang, S., and Zhang, J. (2013) A Novel Nano-Composite Fiber Laden Viscoelastic Fracturing Fluid for Coal-Bed Methane (CBM) Reservoir Stimulation: Laboratory Study and Test. *SPE Asia Pacific Oil and Gas Conference and Exhibition.* Online: https://www.onepetro.org/download/conference-paper/SPE-165862-MS?id=conference-paper%2FSPE-165862-MS (assessed 04.02.2017).

30. Terracina, J. M., McCabe, M. A., Shuchart, C. E., and Walker, M. L. (2013) Novel oxidizing breaker for high-temperature fracturing. *SPE Production & Facilities*, **14**(02), 139-143.

31. Weaver, J., Schmelzl, E., Jamieson, M., and Schiffner, G. (2002) New Fluid Technology Allows Fracturing Without Internal Breakers. *SPE Gas Technology Symposium.* Online: https://www.onepetro.org/download/conference-paper/SPE-75690-MS?id=conference-paper%2FSPE-75690-MS (assessed 03.02.2017).

32. Barati, R., and Liang, J.-T. (2014) A review of fracturing fluid systems used for hydraulic fracturing of oil and gas wells. *Journal of Applied Polymer Science*, **131**(16), doi:10.1002/app.40735.

33. Zhang, B., Davenport, A. H., Whipple, L., Urbina, H., Barrett, K., Wall, M. and Mirakyan, A. (2013) A superior, high-performance enzyme for breaking borate crosslinked fracturing fluids under extreme well conditions. *SPE Production & Operations*, **28**(02), 210-216.

34. Barati, R., Johnson, S. J., McCool, S., Green, D. W., Willhite, G. P., and Liang, J.-T. (2011) Fracturing fluid cleanup by controlled release of enzymes from polyelectrolyte complex nanoparticles. *Journal of Applied Polymer Science*, **121**(3), 1292-1298.

35. Barati, R., Johnson, S. J., McCool, S., Green, D. W., Willhite, G. P., and Liang, J.-T. (2012) Polyelectrolyte complex nanoparticles for protection and delayed release of

enzymes in alkaline pH and at elevated temperature during hydraulic fracturing of oil wells. *Journal of Applied Polymer Science*, **126**(2), 587-592.

36. Bose, C. C., Alshatti, B., Swartz, L., Gupta, A., and Barati, R. (2014) Dual Application of Polyelectrolyte Complex Nanoparticles as Enzyme Breaker Carriers and Fluid Loss Additives for Fracturing Fluids. *SPE/CSUR Unconventional Resources Conference – Canada*. Online: https://www.onepetro.org/download/conference-paper/SPE-171571-MS?id=conference-paper%2FSPE-171571-MS (assessed 19.02.2017).

37. Abraham, W. (1933) The Functions of Mud Fluids used in Rotary Drilling. *1st World Petroleum Congress*. Online: https://www.onepetro.org/download/conference-paper/WPC-1093?id=conference-paper%2FWPC-1093 (assessed 12.02.2017).

38. Ponmani, S., William, J. K. M., Samuel, R., Nagarajan, R., and Sangwai, J. S. (2014) Formation and characterization of thermal and electrical properties of CuO and ZnO nanofluids in xanthan gum. *Colloids and Surfaces A: Physicochemical and Engineering Aspects*, **443**, 37-43.

39. William, J. K. M., Ponmani, S., Samuel, R., Nagarajan, R., and Sangwai, J. S. (2014) Effect of CuO and ZnO nanofluids in xanthan gum on thermal, electrical and high pressure rheology of water-based drilling fluids. *Journal of Petroleum Science and Engineering*, **117**, 15-27.

40. Srivatsa, J. T., and Ziaja, M. B. (2011) An Experimental Investigation on Use of Nanoparticles as Fluid Loss Additives in a Surfactant - Polymer Based Drilling Fluids. *International Petroleum Technology Conference*. Online: https://www.onepetro.org/download/conference-paper/IPTC-14952-MS?id=conference-paper%2FIPTC-14952-MS (assessed 21.02.2017).

41. Riley, M., Stamatakis, E., Young, S., Price, K., and De Stefano, G. (2012) Novel Water-Based Mud for Shale Gas Part II: Mud Formulation and Performance. *SPE Americas Unconventional Resources Conference*. Online: https://www.onepetro.org/download/conference-paper/SPE-152945-MS?id=conference-paper%2FSPE-152945-MS (assessed 21.02.2017).

42. Li, M.-C., Wu, Q., Song, K., Qing, Y., and Wu, Y. (2015) Cellulose nanoparticles as modifiers for rheology and fluid loss in bentonite water-based fluids. *ACS Applied Materials and Interfaces*, **7**, 5006-5016.

43. Lecolier, E., Herzhaft, B., Rousseau, L., Neau, L., Quillien, B., and Kieffer, J. (2005) Development of a Nanocomposite Gel for Lost Circulation Treatment. *SPE European Formation Damage Conference*. Online: https://www.onepetro.org/download/conference-paper/SPE-94686-MS?id=conference-paper%2FSPE-94686-MS (assessed 21.02.2017).

2

Application of Polymer Nanocomposites in Oil and Gas Industry

2.1 Introduction

Polymer nanocomposites are composed of polymer matrices with different types of filler materials in nanoscale range [1]. Owing to their superior properties, nanocomposites exhibit strong potential for various industrial applications such as marine coatings, engineering materials, automobile components, sensor materials, etc. [2-5]. A large number of thermoplastic and thermosetting polymers are utilized in oil and gas industries. However, pure polymeric materials have limitation of usage in high temperature, high pressure and other harsh environments [6]. In order to enhance the properties of the polymers in the petroleum industry, nanotechnology has been implemented, i.e., various types of nanomaterials have been incorporated into polymer matrices, thus, resulting in superior properties in upstream and downstream processes [7].

The interfacial and colloidal chemistry of nanomaterials exerts significant influence on the polymeric materials, which thus impacts productivity of various oil and gas processes. The oil and gas production processes are generally divided into three categories as upstream, midstream and downstream processes. Though polymer nanocomposites provide high temperature and pressure resistivity in both upstream and downstream processes as compared to pure polymeric materials, these nanocomposites have found major utilization in upstream processes in the form of drilling fluids, dispersants, thermoset coatings, anti-scaling agents, etc. For instance, polymer nanocomposites possessing superior rheological properties have been observed to result in improved efficiency in well drilling and enhanced oil recovery (EOR) processes [8]. In addition, the polymer nanocomposites exhibiting excellent adhesive and anti-corrosion properties have been utilized as coating materials for pipelines and storage tanks [9-11]. In downstream industry, nanocomposites have been used as adsorbents, viscosity modifiers, anti-scaling agents, etc. [7]. Overall, polymer nanocomposites possessing improved thermal, mechanical, rheological and permeation properties have helped to minimize many frequent problems in the petroleum industry [12-14].

Muthukumaraswamy R. Vengatesan and Vikas Mittal*, The Petroleum Institute (part of Khalifa University of Science and Technology), Abu Dhabi, UAE
*Current address: Bletchington, Wellington County, Australia

This review is focused on the application of polymer nanocomposites in oil and gas industry. The utilization of polymer nanocomposites in both upstream and downstream processes has been summarized and particularly the application of nanocomposites in enhanced oil recovery, oil spill cleanup, acid gas removal, anti-corrosion coatings for pipelines and oily wastewater treatment are discussed.

2.2 Polymer Nanocomposites in Upstream Processes

As mentioned earlier, polymeric nanocomposites have been utilized as additives, adhesives and coating materials to improve the upstream processes. For example, polymeric nanocomposites are used in upstream processes as floating materials for EOR, mud additives in well drilling process and anti-corrosion agents in pipeline coatings, which are detailed below.

2.2.1 EOR

EOR corresponds to various techniques applied to increase the recovery of crude oil from the oil fields. Three methods such as gas injection, thermal injection and chemical injection methods are generally employed for EOR [15]. For the chemical injection method, polymers have been used to increase the water sweep efficiency in oil reservoirs and to decrease the water mobility, thus, resulting in enhanced oil recovery [16]. Generally, polyacrylamide and polyethylene oxides are widely used polymers as the flooding materials in the oil reservoirs [17,18]. High temperature, high pressure, salinity as well as bacteria affect the stability of polymer solutions in the oil reservoirs [19,20]. Also, the polymer adsorption on the rock surface blocks the pores, which is one of the main drawbacks in the EOR process [8]. In order to improve the efficiency of polymers in EOR process, polymer nanomaterials have been utilized. In many instances, the addition of nanomaterials into polymer solution improves the oil recovery significantly. For instance, nanomaterials in the polymer nanocomposites reduce the interfacial tension and improve the mobility of the capillary trapped oil in the reservoirs [21]. Silica (SiO_2), alumina (Al_2O_3), magnesium oxide (MgO) and clay particles are widely used nanomaterials for the preparation of polymer nanocomposites for the EOR process. Clay based minerals have an advantage over other nanofillers due to the presence of cationic, anionic or neutral sites which alter the rheological behavior and control the polymer adsorption/retention mechanism in porous media [21]. Cheraghian [22] studied the effect of nanoclay on polymer flooding in heavy oil recovery and compared it with the polymer solution for degradation temperature and salinity [22]. It was observed that the nanoclay reduced the temperature rise of polymer fluids in the recovery process which resulted in a 5.8% increase in oil recovery. The effect of clay concentration (0.8, 0.9 and 1.0 wt%) on heavy oil recovery during polymer flooding was also investigated and 0.9 wt% was observed as the optimum level which resulted in a 5% increase in oil recovery [23].

Rheological properties of the polymer solution strongly influence the polymer flooding in oil recovery process due to adsorption and precipitation on the reservoir rocks. The changes in rheological properties result mainly due to the surface properties of the rock, composition of the oil and reservoir fluids, nature of the polymer and solution conditions (salinity, pH and temperature). The addition of nanoclay has been reported to strongly alter the solution viscosity and salinity. The salinity and viscosity also varied with the content of nanoclay in the polymer nanocomposites [24]. Maghzi *et al.* [16] studied the effect of silica nanoparticles on the performance of polymer solutions with different salinity during polymer flooding in heavy oil recovery process. The authors reported that the increase in silica content increased the pore to pore displacement efficiency and increased the oil recovery efficiency. Ye *et al.* [25] developed polyacrylamide silica nanocomposites for the EOR application. The authors observed that the nanocomposite exhibited higher viscosity with 43.7% retention rate at 95 °C. The nanocomposite exhibited an improved oil recovery of 20.1% in core flooding experiments at 65 °C. Maurya and Mandal [26] also generated polyacrylamide/silica hybrid nanocomposite, where the presence of silica in the polymer nanocomposite resulted in higher viscosity and viscosity retention in the presence of salt at high temperature and the nanocomposite was concluded to be suitable for potential EOR process. Thus, these nanocomposites are functional materials which are able to generate effective solutions in oil and gas industry by surpassing the properties and performance of the existing materials by a large extent.

Recently, graphene based polymer nanocomposites have exhibited great potential for a variety of applications in different fields. For instance, graphene oxide (GO) has been used as a viscosity enhancer in EOR processes [27]. Nguyen *et al.* [28] studied the impact of graphene oxide particles on viscosity stabilization for diluted polymer solutions. The addition of GO improved the viscosity stability of the polymer solution in sea water conditions and also increased the thermal stability in reservoir conditions for long period of time. Zuniga *et al.* [29] studied the dispersability and long-term high temperature stability of partially reduced GO/polyzwitterionic nanocomposite in the presence of both salinity brine and deep oil reservoir conditions. It was observed in the study that the nanocomposite solution was stable even after 140 days at 90 °C (as shown in Figure 2.1).

Magnetic polymer nanocomposites have also been used in the EOR processes because of easy recovery and reuse. In a magnetic polymer nanocomposite, the magnetic nanoparticles are embedded in polymer layer which enables effective dispersion in injected brine and compatibility with oil. The magnetic nanocomposites are also stable at high temperatures which avoids adsorption on the surface of the rock reservoirs [30]. Nguyen *et al.* [30] prepared polymer magnetic nanocomposite from red waste mud. The nanocomposite was blended with carboxylate surfactant which reduced the interfacial tension between the oil and water and the nanocomposite was thermally stable even after 31 days at 100 °C, thus, confirming the potential of utilization in EOR process at offshore reservoirs.

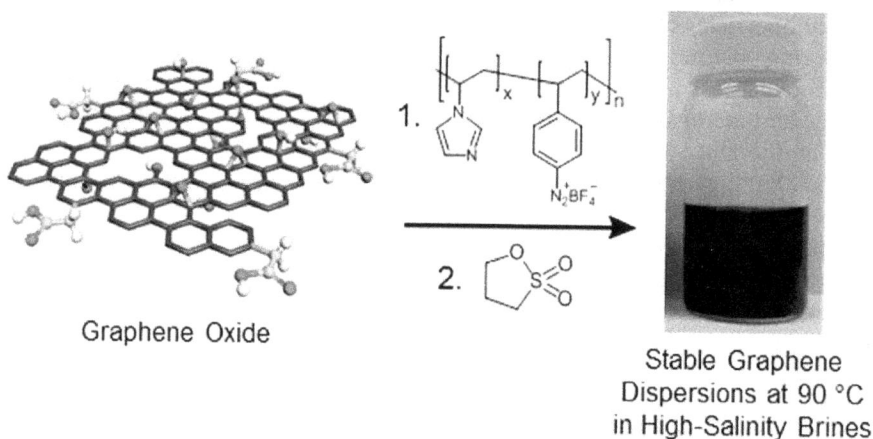

Graphene Oxide

Stable Graphene
Dispersions at 90 °C
in High-Salinity Brines

Figure 2.1 Image of dispersed graphene at 90 °C in high-salinity brine condition. Reproduced from reference 29 with permission from American Chemical Society.

2.2.2 Drilling Fluids

In rotary drilling processes, drilling fluids or drilling muds are used to drill the oil and gas on land and offshore. Drilling fluids play an important role during the drilling process such as transporting cutting to the surface; balancing the subsurface; cooling, lubricating and supporting the drill bit and drill pipe [31-34]. Drilling muds are classified as oil based muds (OBM), water based muds

(WBM) and synthetic muds. OBM are formulated based on crude oil or diesel, whereas the WBM are formulated using different types of water soluble polymers and biopolymers. OBM though exhibit better drilling properties compared to WBM, but also cause toxicity in the environment [35,36]. During the drilling process, the drilling fluid performance and stability is also important over the temperature range of 200 °C. Normally, water soluble polymers such as partially hydrolyzed polyacrylamide and polyethylene glycol (PEG) are used as polymers for WBM to support the rheological properties, filtration and shale inhibition characteristics [37]. Though polymers have been used as viscosifying agents and filtrate reducers to control rheological properties and fluid loss, however, the gelation of drilling polymers makes it difficult to control the drilling process under ultra-high temperature conditions [38]. Therefore, the current research has been focused on the development of water based drilling fluids with improved thermal, mechanical and rheological properties which enhance the drilling performance. The major problem of the polymer additives is shear thinning nature, known as thixotropy [39]. The addition of nanofiller reduces the thixotropic nature of the drilling fluid which results in an improved stability. The presence of nanomaterials in the polymer solution helps in sealing of the pores and micro-cracks near the wellbore surface during the drilling process [37]. Normally, inorganic nanoparticles and clay materials are widely used as the filler materials with polymers for the mud additives of drilling fluids. Jain *et al.* [37] generated a polyacrylamide-grafted-polyethylene glycol/silica nanocomposite and studied the rheological and shale inhibition properties. The authors observed that the nanocomposite had good rheological properties, low formation damage and high thermal stability which resulted in a potential advanced additive for water based drilling mud [37]. Similarly, Sadeghalvaad and Sabbaghi [40] also developed polyacrylamide/TiO_2 nanocomposite as an effective mud additive for water based drilling fluids. Mao *et al.* [41] prepared nano-silica based hydrophobic polymer nanocomposite with core-shell structure. The nano-silica particles cross-linked and bridged with the polymer matrix, then rapidly paddled the micro-pores and micro-cracks larger than their own size through a self-assembling process. The nanocomposite exhibited good thermal stability, fluid loss lubricity properties and was concluded as an effective micro-nano drilling fluid additive under ultra-high temperature/pressure conditions. Madkour *et al.* [42] developed biopolymer nanocomposites using multi walled carbon nanotube and graphene with polylactic acid. The nanocomposites exhibited good rheological, electrical and filtration properties. The composites were environmental friendly and could be utilized as OBM viscosifier. William *et al.* [43] prepared nanofluid-enhanced WBM (NWBM) using the nanofluids consisting of copper oxide (CuO) and zinc oxide (ZnO) (size <50 nm) in a xanthan gum aqueous solution. The addition of CuO and ZnO nanoparticles into the xanthan resulted in improved thermal, electrical and rheological properties of the water-based drilling fluids. The increase in concentration of nanoparticles enhanced rheological stability of NWBM. The high pressure rheological studies of NWBM were conducted at different pressures (0.1 and 10 MPa) and temperature ranges (25, 70, 90 and 110 °C). The authors observed that the effect of pressure on the rheology of NWBM was more significant at higher temperatures, and the NWBM could be used as well viscosity stabilizer at higher temperature.

2.2.3 Anti-corrosion Coatings

Corrosion is the major problem in oil and gas industries which reduces the life of pipelines and also causes major incidents. Pipelines are one of the most important oil and gas transportation means. The pipes used are made up of steel alloys, which get corroded due to interaction with oil products internally as well as high temperature and humidity conditions externally. Thus, corrosion may occur as either internal or external corrosion. As mentioned above, internal corrosion occurs due to the contact between inner side of the pipeline and chemicals in oil and gas products during the flow. The acid gases such as hydrogen sulfide (H_2S), carbon dioxide (CO_2) and sulfur dioxide (SO_2) are the major sources which contribute to chemical corrosion inside the pipelines. External corrosion occurs due to chemical, electrochemical, physical and environmental factors [44]. Coatings are commonly used method to safeguard the pipeline surfaces against corrosion, where coating materials are required to possess good surface adhesion, corrosion resistance, wear resistance, high thermal stability and mechanical properties. Organic polymer coatings are widely accepted pipeline coatings in oil and gas industries due to good corrosion resistance, durability, resistant to salinity, immunity to large variation in pH, chemical and physical stabilities at moderately high temperatures. Epoxy and 2-layer/3-layer polyethylene coatings are predominantly used organic coatings in the oil and gas industries [45]. Currently, the research is driven to enhance the properties of organic coatings further in order to meet the continuously increasing need to perform in harsh conditions [46]. Similar to other fields, organic-inorganic hybrid nanocomposite coatings have received recent attention due to their improved corrosion resistance, adhesion, chemical resistance, etc. [47-49]. Different studies have been carried out to investigate the effectiveness of polymer nanocomposite coatings in the industrial environments. For this, different types of polymer nanocomposites have been investigated, along with corrosion inhibition mechanisms of the coated substrates in different environments [50-54]. Figure 2.2 exhibits the corrosion protection achieved by using polyaniline/partially phosphorylated poly(vinyl alcohol) (PANI/P-PVA) spherical nanoparticles based composites [54]. A variety of other nanomaterials such as

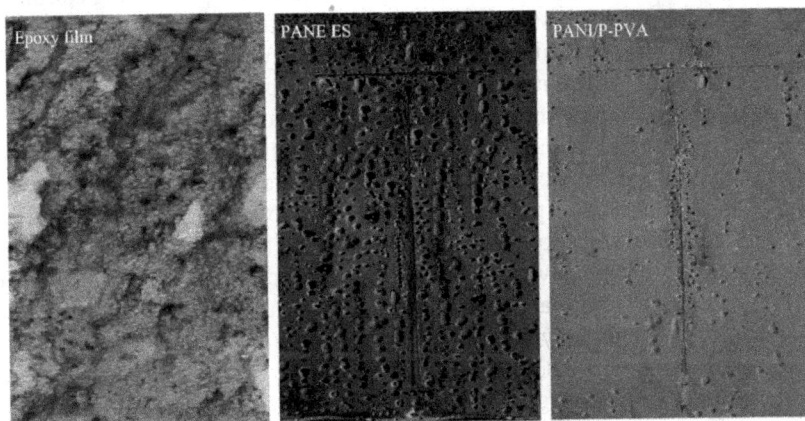

Figure 2.2 Images of pure epoxy-coated panel, 2.5 wt% PANI ES-coated panel, and 2.5 wt% PANI/P-PVA-coated panel after 30 days exposure to salt spray test. Reproduced from Reference 54 with permission from American Chemical Society.

clay, SiO_2, titanium dioxide (TiO_2), Al_2O_3, ZnO, iron oxide (Fe_3O_4), zirconia (ZrO_2), polyhedral oligomeric silsequioxane (POSS), boron nitride, carbon nanotubes (CNT), graphene, etc., have contributed to enhance the anti-corrosion properties of the organic polymers.

Zaarei *et al.* [55] coated a cold rolled steel (CRS) substrate using isophoronediamine-modified montmorillonite clay-epoxy nanocomposite and studied the corrosion behavior in presence of 3.5% salt solution medium for more than 210 days. It was observed that the addition of clay in coating compositions led to superior anti-corrosion properties compared with pure resin. Yeh et al. [56] developed siloxane modified epoxy-clay nanocomposites and studied the anti-corrosion behavior in 5 wt% aqueous sodium chloride (NaCl) electrolyte medium. The nanocomposite exhibited advanced protection against corrosion on CRS. In another study, Yu *et al.* [57] prepared organo-soluble polyimide-clay nanocomposite and studied the anti-corrosion behavior. The incorporation of clay platelets into soluble polyimide membrane was observed to result in an enhancement of oxygen (O_2) and water (H_2O) molecular barrier properties and excellent anti-corrosion properties.

Nearly 20 to 30% of external corrosion on underground pipelines occurs due to microbial influenced corrosion [58]. Commonly, anti-microbial agents are added in polymers to protect against microbial growth, however, these agents might not withstand the commonly used polymer processing conditions. Silver (Ag) nanoparticles, which have anti-microbial properties, have been utilized as more effective filler material for variety of nanocomposites [59]. Manjumeena *et al.* [60] developed anti-microbial coating on metal steel plate using epoxy silver nanocomposites and studied the anti-corrosion properties. Epoxy/silver nanocomposite coatings offered manifold anti-microbial protection to the metal surfaces by inhibiting the growth of microbes. The incorporation of nanoparticles into polymeric resin has also been reported to enhance the integrity and durability of coatings and, therefore, the nanoparticles dispersed coatings can fill cavities as well as bridge, deflect and bow cracks [61-64]. The addition of nanoparticle also tend to fill small hole defects formed from local shrinkage during polymer curing and can also act as a bridge for interconnecting molecules [65]. Bakhshandeh *et al.* [66] studied the anti-corrosion properties of organic–inorganic hybrid coatings based on diglycidylether of bisphenol A (DGEBA) epoxy resin and hydrolyzation of tetraethoxysilane (TEOS) through a sol-gel process. It was observed that the corrosion resistance of the hybrid nanocomposite coatings improved with increasing silica content. Ramezanzadeh *et al.* [67] studied the anti-corrosion behavior of ZnO nanoparticles filled epoxy nanocomposites. The addition of nanoparticles in the epoxy matrix resulted in improved barrier properties and coating resistance against hydrolytic degradation. In a similar study, Zhang *et al.* [68] fabricated silica coated iron oxide for the enhancement of anti-corrosion performance of epoxy coatings. The incorporation of silica coated iron oxide nanoparticles enhanced the anti-corrosion behavior of the epoxy coatings [68].

Conducting polymer nanocomposites (CPN) are known to improve the anti-corrosion behavior of metal substrates significantly through different corrosion protection mechanisms, such as (i) anodic protection and passivation of the metal substrate, (ii) formation of a protective metal/polymer complex, (iii) absorption of OH- and inhibition of cathodic disbondment and (iv) inhibition of the cathodic reactions [69]. Ganash [70] developed poly(o-phenylenediamine)/ZnO (PoPd/ZnO) nanocomposites and coated on type-304 austenitic stainless steel (SS). The corrosion behavior was studied by E_{ocp}-time measurement, anodic and cathodic potentiodynamic polarization and impedance techniques in the presence of 3.5% NaCl as corrosive solution. The ZnO nanoparticle were concluded to significantly improve the barrier and electrochemical anti-corrosion properties of poly(o-phenylenediamine) [70]. Chang *et al.* [71] demonstrated the application of polyaniline (PANI)/graphene composites (PAGCs) for corrosion protection of steel. The nanocomposite coatings were shown to effectively protect the steel substrate due to the good O_2 and H_2O barrier property of graphene [71]. In other studies reporting the use of graphene, Yu *et al.* [72] successfully prepared polystyrene (PS)/modified-GO nanocomposites and studied the anti-corrosion properties. It was observed that the addition of 2% of GO improved the corrosion protection efficiency from 37.90% to 99.53%. Li *et al.* [73] also developed graphene reinforced waterborne polyurethane (PU) composite and studied the barrier and anti-corrosion properties. It was opined that the graphene layers in the nanocomposites were self-aligned parallel to the substrate surface, and the high surface area of graphene helped it to interact with the electrolyte and prevent the electrolyte penetration to the substrate. In another recent study using biopolymer based anti-corrosion coatings, Luckachan and Mittal [74] fabricated functional coatings by layer-by-layer (lbl) addition of chitosan (Ch) and poly vinyl butyral (PVB) on mild carbon steel substrate (Figure 2.3).

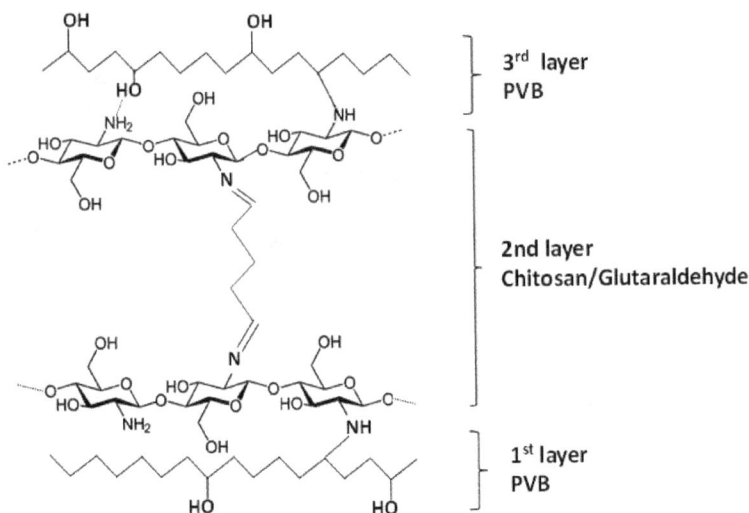

Figure 2.3 Schematic representation of crosslinking of chitosan with PVB and glutaraldehyde in the coatings. Reproduced from Reference 74 with permission from Springer.

Self-healing coatings are the alternative method for efficient external anti-corrosion protection in oil and gas industries. Self-healing coatings have been developed for metal substrates under cathodic protection of specific-film-formers which are sensitive to the electrical field and pH. The incorporation of micro- or nano-capsules into the polymer coatings formulations containing film-formers is a common method to achieve self-healing properties. The nano-capsules automatically heal/repair the coating damage, even if the coating gets damaged from external electrical field or environment. Fan *et al.* [75] developed self-healing coatings using cerium based conversion layer, a GO layer, and a branched poly(ethylene imine) (PEI)/poly(acrylic acid) (PAA) multilayer. The GO layer acted as corrosion inhibitor and PEI/PAA multilayers provided self-healing ability to the coating systems. GO also acted as a barrier layer by stopping the penetration of corrosive electrolytes. Niratiwongkorn *et al.* [76] prepared self-healing polyvinyl butyral (PVB) based organic coating with the incorporation of polypyrrole-carbon black (PPyCB) composite as an inhibiting pigment. It was concluded that the addition of more conducting particles enhanced the protective nature of the PVB/PPyCB composite coatings (Figure 2.4).

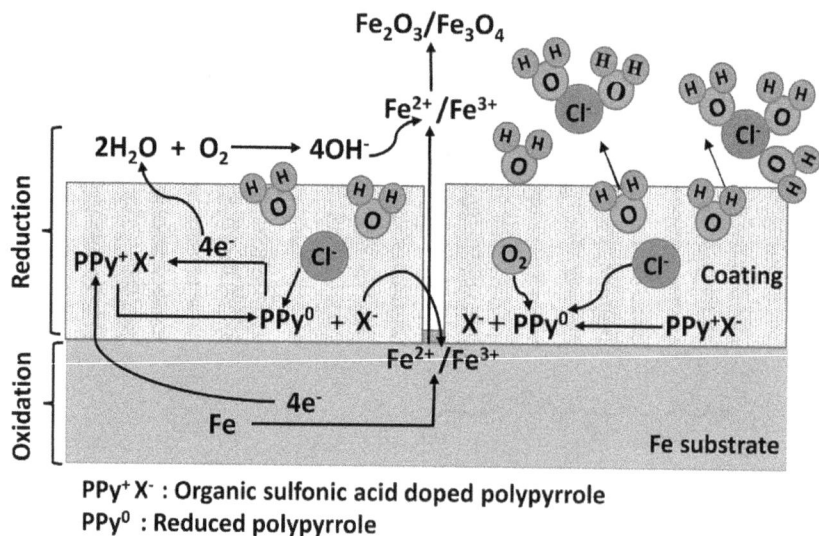

PPy⁺ X⁻ : Organic sulfonic acid doped polypyrrole
PPy⁰ : Reduced polypyrrole

Figure 2.4 Schematic representation of corrosion protection imparted by PPy pigments in PVB/PPyCB composite coatings. Reproduced from reference 76 with permission from Royal Society of Chemistry.

2.2.4 Oil Spill Cleanup

Oil spill is the major issue in the upstream oil and gas industries, which causes a great environmental pollution. Lot of precautions have been put in place to avoid the oil spills, however, it is impossible to completely avoid the accidents involving oil spills. A variety of methods have been developed to clean up the oil spills such as chemical treatments, mechanical and biological methods [77]. Among these, chemical treatment using sorbents are extensively used for the oil spill treatment and considered to be effective method for the complete recovery of the spilled oil [78]. Hydrophobic and lipophilic polymers, exhibiting higher oil retention and prompt oil pick up from the spilled area, have been used either to adsorb or absorb the spilled oil. In recent years, advanced nanotechnology methods have been adopted to address the oil spill remediation [78]. A large variety of nanomaterials such as carbon, clay, CNT, graphene, POSS, etc., have been used for separation and removal of organics or oils from water [79]. The inclusion of nano-sized fillers in the polymers significantly alters the properties of the corresponding polymers leading to enhanced potential for oil spill cleanup applications. Zhang *et al.* [80] fabricated polymer of intrinsic micro-porosity (PIM-1)/POSS micro-fibrous membranes using electrospinning technology. The hydrophobic nature of POSS and an increase in the fiber surface roughness led to the superhydrophobicity and superoleophilicity of the nanocomposites. It was observed that the PIM-1 fibrous membrane nanocomposite exhibited the ability to adsorb a large amount of oil from the contaminants (Figure 2.5). Nikkhah *et al.* [81] developed PU nanocomposite foam with modified clay and studied the oil removal from the water-oil mixture. It was observed that the addition of 3 wt % of nanoclay enhanced the sorption capacity up to 16% and oil removal efficiency up to 56% as compared to the capacity of the pure PU foam in water-oil system [81]. Carbon nanomaterial (carbon black, CNT and graphene) filled polymer nanocomposites possess superhydrophobicity and can be effectively used in oil-water separation. Maphutha *et al.* [82] prepared CNT doped polymer nanocomposite membrane and studied the oil separation efficiency for oil-water mixture. The authors observed the membrane nanocomposite oil rejection of over 95% from the oil-water mixture. Liu *et al.* [83] developed PU-reduced GO nanocomposite foam and conducted the oil absorption studies. It was observed that the reduced GO enhanced the hydrophobicity and oleophilicity of the composite, which resulted in an enhanced oil absorption. Tran *et al.* [84] prepared porous and green three-dimensional (3-D) polydimethylsiloxane (PDMS)–graphene sponge and studied the selective adsorption of oil-water mixtures. The nanocomposite foam was observed selectively adsorbed the petroleum products with adsorption capacity of 4.5 L of hexane in 30 min.

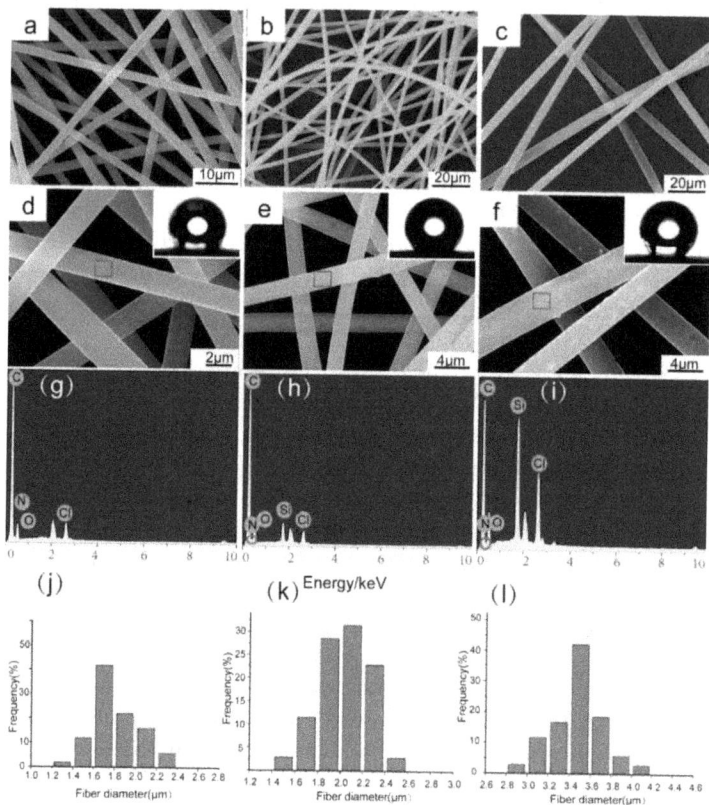

Figure 2.5 FE-SEM images of (a,d) pristine PIM-1, (b,e) 20 wt% and (c,f) 40 wt% PIM-1/POSS fibers. EDS spectra of (g) pristine PIM-1, (h) 20 wt% and (i) 40 wt% PIM-1/POSS fibers taken from the rectangles in panels d, e, and f, respectively; and (j-l) fiber diameter distributions of these materials respectively. Reproduced from Reference 80 with permission from American Chemical Society.

Among other categories, magnetic polymer nanocomposites have also been developed for oil spill cleanup application. Saber *et al.* [85] proposed the mechanism of oil spill using magnetic nanocomposites. The mechanism consists of three stages: (i) nano-size and low density of the nanocomposite which facilitated the process of penetration of magnetic nanomaterial through floatation of the nanoparticles with oil at the water surface, (ii) high surface area of the magnetic nanocomposites which increased the surface contact of the magnetic materials with the oil, and (iii) the superparamagnetic behavior of the magnetic nanocomposites generated the magnetic behavior for the oil spill. Finally, oil spill had magnetic property and the oil could be easily recovered by the external magnetic field. Zhang *et al.* [86] prepared magnetic poly(styrene-divinylbenzene) (poly(St-DVB)) monolith with highly opened porous structure and lipophilicity. The porous monolith had hydrophobicity and adsorbed oilsfrom water selectively with an oil intake capacity of approximately 23 times. Souza *et al.* [87] prepared biopolymer magnetic nanocomposites and carried out the oil absorption analyses. It was concluded that one part of the nanocomposite could be used to remove more than eight parts of oil from water. Yu *et al.* [88] developed durable and modified magnetic polystyrene foam (DMMPF) with superhydrophobicity, superoleophilicity and fast magnetic response. The DMMPF exhibited absorption capacity of 40.1 times as compared to its own weight and the absorbed oils/organic solvents could be collected by simple mechanical extrusion. Liu *et al.* [89] fabricated magnetic polymer-based graphene foam for the applications in oil-water separation. Graphene sheets were assembled on the surface of PU chains with ferrous ions, which improved the adhesion of graphene on PU matrix. The resulting magnetic polymer graphene (MPG) nanocomposite exhibited superior hydrophobicity and oleophilicity. The MPG nanocomposite were observed to endure high absorption capacity for oil-water separation even after several absorption/desorption cycles.

2.3 Polymer Nanocomposites in Downstream Processes

As mentioned earlier, in downstream operations, polymeric nanocomposites have been used in the form of additives, sorbents and membranes which enhance the performance of gas separation and sorption processes, among other benefits. Similar to earlier examples, the incorporation of nano-reinforcements in the polymer matrices has evolved more efficient use of polymers for these applications. Some of these downstream applications of polymer nanocomposites, such as natural gas processing as well as oily wastewater treatment, are reviewed in the below sections.

2.3.1 Natural Gas Processing

Raw natural gas from the reservoirs generally consists of methane (CH_4) (30-90%), other light hydrocarbons, such as ethane and propane, along with many heavier hydrocarbons. The raw gas also contains acid gases (CO_2, H_2S and SO_2), H_2O, helium and nitrogen (N_2) in different concentration levels. It is essential to purify the raw natural gas to avoid the pipeline corrosion and meet the regulatory standards on calorific value [90].

Acid Gas Removal

Acidic gases are the components in the raw natural gas and must be removed (also called natural gas sweetening) to meet specifications, to increase heating value and to reduce corrosion of pipelines. Amine absorption is the most widely used technology for natural gas sweetening. Efforts are being made to find alternatives to amine absorption, because of its high capital cost, high energy consumption for absorbent regeneration and potential environmental pollution. Polymer membrane technology has been suggested as an alternative for the natural gas purification due to low-cost, energy savings and environment friendly nature [91]. The structure and orientations in polymeric membrane are required to be controlled in order to achieve a good balance of permeability and selectivity. Currently, much research attention has been focused to improve the performance of the polymeric membranes by achieving high permeability and selectivity. It has also been suggested that organic and inorganic hybrid nanocomposite membranes are superior to pristine polymers with enhanced properties and performance [92-94]. The presence of inorganic nanomaterials may increase the mean distance between the polymer chains which results in an increase in the permeability of the membranes. The surface OH groups of inorganic nanomaterials embedded in the polymer matrix interact with the polar gases like CO_2 and SO_2, thus, increasing the solubility by increasing Henry's law coefficient [95]. Various types of metal and metal oxide nanoparticles, clay minerals, metal organic frameworks (MOF), CNTs and graphene have been used as filler materials for the polymer matrix membranes to enhance the permeability and selectivity characteristics.

Rafiq *et al.* [96] prepared polysulfone (PSf)/polyimide-silica nanocomposite membrane and studied the CO_2 removal from methane. The authors observed that an increase in silica content increased the permeability of CO_2 from the gas mixture. Xin *et al.* [97] developed sulfonated poly (ether ether ketone) membrane nanocomposite with aminated titania nanotubes. The amine groups in nanotubes increased the CO_2-facilitated transport sites in the nanocomposite membrane. The authors concluded that titania polymer nanocomposite exhibited the highest ideal selectivity of CO_2 from the CO_2/CH_4 and CO_2/N_2 mixtures, with a CO_2 permeability of 2090 Barrer. In another study, Rahman *et al.* [98]

studied CO_2 separation from the gas mixtures using two types PEBAX® polymer membranes with the incorporation of PEG functionalized POSS. The addition of 30 wt% of POSS into the polymer resulted an improved selectivity and permeability of CO_2. Liu *et al.* [99] prepared polyzwitterion coated CNT (SBMA@CNT) incorporated polyimide nanocomposites and studied the CO_2 permeability. 5 wt% SBMA@CNT doped hybrid membrane exhibited the maximum CO_2 permeability of 103 Barrer with a CO_2/CH_4 selectivity of 36. The laminar structures and the controlled structural defects on GO sheets have the potential to demonstrate efficient gas separation performance. In addition, the hydrophilic nature of GO might be able to hold the water molecules inside the membranes, thus, providing an alternative method for stable gas separation process [100-102]. Dong *et al.* [102] prepared fixed carrier composite membrane using GO nanosheets, hyperbranched polyethylenimine (HPEI) and trimesoyl chloride (TMC) coating on a polysulfone membrane via interfacial polymerization method. It was observed that the addition of GO significantly improved the CO_2 permeability to 9.7 gas permeance units (GPU) and CO_2/N_2 selectivity of over 80 (Figure 2.6). Wang *et al.* [103] studied CO_2 separation performance of PU-GO nanocomposites. Well-distributed GO in the PU membrane hybrid was observed to lead to higher permeability and selectivity for CO_2. Polymeric membranes also exhibit the capability to separate H_2S together with CO_2 from CH_4 gas mixtures due to solution-diffusion mechanism. The separation and removal

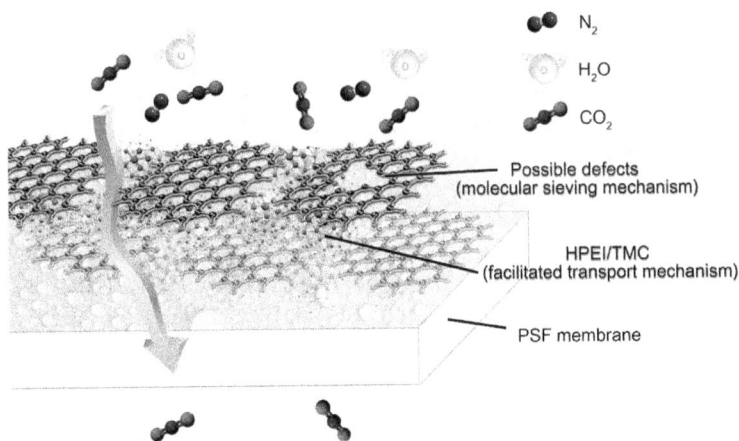

Figure 2.6 Schematic representation of transport mechanism of mixed gas through the HPEI/GO-TMC composite membrane. Reproduced from reference 102 with permission from American Chemical Society.

ability of these membranes generally depend on the selectivity of H_2S and CO_2 over CH_4. In order to achieve this, a dual process has been implemented i.e. the acid gases initially separated together from the natural gas using membranes and subsequently separated one by one using various techniques [92]. Catalytic polymeric membrane reactors (CPMRs) is another facile technique which is capable to catalytically react with chosen gas species in single step method [104]. Nour *et al.* [105] developed silver nanoparticle doped PDMS nanocomposite catalytic membrane for the removal of H_2S. 1 wt% of silver nanoparticle in PDMS nanocomposite membrane removed more than 60% of H_2S gas from the mixture. The H_2S gas exposed on the Ag-PDMS nanocomposite led to the conversion of Ag into silver sulfide (Ag_2S).

BTEX Removal

Presence of BTEX (benzene, toluene, ethylbenzene and xylene), even in trace levels, reduces the quality of natural gas. Therefore, BTEX needs to be removed from the natural gas streams to meet the quality and safety specifications. In addition, BTEX may also contaminate soil, sediments and groundwater due to accidental oil spill, leakage of gasoline and other petroleum fuels from underground storage tanks and pipelines as well as improper oil-related waste disposal practices, thus, leading to harmful effects on human health [106-109]. Hence, appropriate and efficient extraction methods are required to remove BTEX from natural gas streams and other oil-based aqueous contaminations. Normally, solid phase micro extraction (SPME) methods are used for the removal of BTEX. Recently, many studies have focused on SPME development and preparation of new type of sorbents with remarkable chemical and mechanical stability for specified analytes [110-112]. In this category, polymer based nanocomposites have also been developed as sorbent for the removal of BTEX. Foroutani *et al.* [113] prepared polyaniline-magnetite ($PA@Fe_3O_4$) hollow nanocomposite and used as adsorbent for the headspace solid-phase micro-extraction (HS-SPME) method for removing BTEX from water. It was concluded that the $PA@Fe_3O_4$ hollow nanocomposites were suitable for HS-SPME method for BTEX removal. Gupta and Kulkarni [114] developed polydimethylsiloxane-gold nanocomposite foam and studied the removal of organic contaminates especially BTEX from oil spilled water. The nanocomposite foam acted as an effective adsorbent and exhibited high swelling property against BTEX. Peerakiatkhajorn *et al.* [115] used a photocatalytic approach for the removal of gaseous BTEX. The authors prepared Ag/TiO_2 thin film coated on polyvinyl chloride (PVC) sheet and studied the BTEX degradation under visible light.

0.1% of Ag/TiO$_2$ coated PVC sheet was observed to exhibit higher BTEX re-moval with maximum degradation efficiency of xylene (89%), followed by ethylbenzene (86%), toluene (83%) and benzene (79%). Organic-inorganic hy-brid nanocomposite membranes have also been used for the separation of BTEX from the refinery process water. Zha *et al.* [116] fabricated polyvinyli-dene fluoride (PVDF)/hydrophobic silica sol (Si-R) hybrid hollow fiber mem-branes for the removal of dissolved organics including BTEX from the produced water. The hybrid hollow shell membrane morphology resulted in an increase in collecting velocity and enhanced roughness of both inner and outer surfaces of the membrane. The authors observed that the hybrid hollow fiber membrane exhibited a high regeneration ability with 500 ppm benzene as feed solution.

2.3.2 Oily Wastewater Treatment

Oil and gas industries produce large amounts of unavoidable oily wastewater during refining processes. The oily wastewater can lead to harmful health ef-fects if disposed off without treatment, thus, posing big challenge for the dis-posal without affecting the environment [117]. The oily wastewater also affects the agricultural land and can form oil sludge in the soil. Due to the high viscosity of the oil sludge, it covers the soil particles permanently and affects the water absorption property of the soil, thus, leading to infertility of farming lands. Hence, oily wastewater treatment is essential to avoid the water and soil pollu-tion [118]. The quality of purified water from the oily wastewater mainly de-pends on the crude oil separation process, oil content in water during the sep-aration and primary treatment process [119]. Many conventional methods such as dissolved air flotation (DAF), adsorption, biological treatment, sedimenta-tion in a centrifugal field and hydrocyclones have been reported for the oily wastewater treatment [120-124]. However, the conventional processes have some limitations including low efficiency, high operational costs, and corrosion as well as recontamination problems [119]. In addition, these conventional methods are not effective in removing smaller oil droplets and emulsions from the oily wastewater [125]. Based on these limitations, an alternative based on membrane technology has found more predominant use for the oily wastewater treatment. In fact, polymer based anti-fouling membranes are one of the most important category of materials for treating the oily wastewater. As membrane fouling poses great challenge for effective performance, various methods have been adopted to reduce the fouling of polymeric membranes, such as addition of hydrophilic polymers like polyvinyl acetate (PVAc), cellu-lose acetate (CA) and polyacrylate, etc. By improving the hydrophilicity of membranes, the water molecules form a hydration layer on the membrane surface, which interferes with the adhesion of fouling molecules on the membrane surface [126-128]. Similar to other areas, organic-inorganic hybrid nanocomposite membranes have received more attention recently for water treatment applications due to excellent thermal, mechanical and anti-fouling properties [129-131]. The incorporation of hydrophilic nanomaterials improves the permeability and anti-fouling performance with better surface properties. In refinery processes, polymer nanocomposite membranes have been reported to be used in the primary and secondary water treatment units starting from microfiltration (MF), ultrafiltration (UF), nanofiltration (NF) and reverse osmosis (RO) [132]. Gohari *et al.* [133] prepared ultrafiltration nanocomposite membrane using PSf with inorganic hydrous aluminum oxide (HAO) nanoparticles and studied the separation process from oil/water emulsion. The nanocomposite membrane exhibited almost complete elimination of oil molecules from the emulsion with flux recovery ratio of around 67% after a simple water washing process. This was due to the presence of HAO in the PSf membrane which improved water permeability and also anti-fouling property. Zhao *et al.* [134] fabricated antifouling hybrid membranes using PVDF-g-PTA (TA: [2-(methacryloyloxy)ethyl]trimethylammonium chloride) and TiO_2 nanoparticles. The nanoparticles distributed homogenously into the polymer matrix and led to high strength, underwater superoleophobicity and surface heterogeneity of the nanocomposite membrane. The presence of TiO_2 micro-domains and fluoro-bearing PVDF micro-domains on the membrane surface provided membranes with tunable fouling-resistant and fouling release mechanisms which effectively prevented the oil foulants. The hybrid membrane exhibited near zero flux decline and near complete flux recovery during the oily water separation process. Zeng *et al.* [135] developed ultrafiltration nanocomposite membrane using PVDF and amine functionalized halloysite nanotubes (APTES-HNT). The incorporation of APTES-HNT into the PVDF matrix resulted in an improved pure water flux and decreased the contact angle. The nanocomposite membrane exhibited improved oil removal capacity with oil rejection ratio of more than 90%. Venkatesh *et al.* [136] prepared $PANI/TiO_2$ nanofiber doped PVDF nanocomposite membrane and studied the oil-water emulsion separation. The hydrophilic nature of the $PANI/TiO_2$ nanofiber increased the pure water flux and anti-fouling properties of the nanocomposite membrane. 1 wt% of $PANI/TiO_2$ doped PVDF nanocomposite resulted in the maximum oil rejection of 99% at a 5 bar of operating pressure.

GO has been used as functional filler in the polymer nanocomposite membranes, because it has oxygen-containing functional groups on its basal planes and edges which provides excellent hydrophilicity, strong chemical activity, and uniformity of dispersion in water. The hydrophilic nature of GO results in an improved pure water flux and anti-fouling properties of the polymeric membrane [137-139]. Wang *et al.* [140] prepared GO blended PVDF ultrafiltration membranes. The addition of 0.2 wt% of GO improved the mechanical strength, permeation properties and anti-fouling performance of PVDF membrane resulting in an increased permeability of 94% with a slight change of retention. Comparing oil/water emulsion with surfactant stabilized oil-water emulsion, the latter is quite challenging to be removed due to tiny size of surfactant molecules which lower the removal rate and surfactant adsorption leading to membrane fouling and flux decline [141,142]. Qin *et al.* [143] fabricated a hierarchical membrane using polyether sulfone matrix with GO through phase inversion method. The surface of the nanocomposite membrane was further functionalized with polyvinyl alcohol (PVA). By treating the surfactant stabilized oil/water emulsion on to the nanocomposite membrane resulted in four times higher water flux, significantly higher rejections of both oil (~99.9%) and surfactant (as high as 93.5%) and two-thirds lower fouling ratio (Figure 2.7).

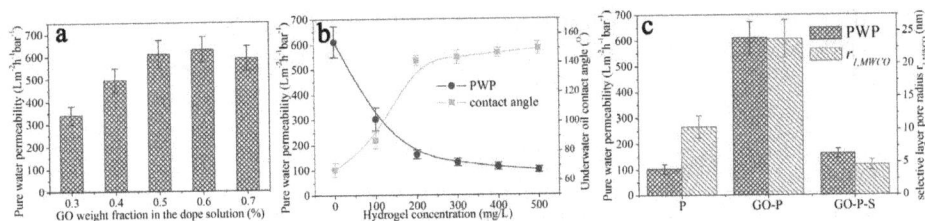

Figure 2.7 Pure water permeability and neutral solute selectivity of the composite membrane; (a) effect of GO weight fraction in nanocomposite dope solution on pure water permeability (PWP) of GO-P (no surface functionalization) membrane; (b) effect of hydrogel concentration on PWP and underwater oil contact angle of GO-P-S (surface functionalized) membrane; (c) membrane pure water permeability (PWP) and selective layer pore radius ($r_{1, MWCO}$, which is calculated from molecular weight cutoff (MWCO)). Reproduced from reference 143 with permission from Nature Publishing.

2.4 Conclusion

In this chapter, the application of polymer nanocomposites in oil and gas industry has been summarized. It also discusses the significance of different nano-reinforcements on the properties and performance of polymer nanocomposites in both upstream and downstream processes. For upstream processes, role of polymer nanocomposites in enhanced oil recovery, advanced drilling muds, inhibition of internal and external pipeline corrosion as well as oil spill cleanup has been reviewed. In the case of downstream processes, the significant contribution of polymer nanocomposites in natural gas sweetening, separation and purification technology as well as oily wastewater treatment has been focused. It can be concluded that applications of polymer nanocomposites have evolved in diverse areas related to oil and gas industry and have contributed significantly to alleviate the difficulties faced by oil and gas industries by providing advanced materials with properties much superior than conventionally used materials.

Abbreviations

Ag	silver
Ag_2S	silver sulfide
Al_2O_3	alumina
APTES-HNT	amine functionalized halloysite nanotubes
BTEX	benzene, toluene, ethylbenzene, xylene
CA	cellulose acetate
CH_4	methane
CNT	carbon nanotubes
CO_2	carbon dioxide
CPN	conducting polymer nanocomposites
CPMR	catalytic polymeric membrane reactors
CRS	cold rolled steel
CuO	copper oxide
DAF	dissolved air flotation
DGEBA	diglycidyl ether of bisphenol A
DMMPF	durable and modified magnetic polystyrene
EOR	enhanced oil recovery
Fe_3O_4	iron oxide
GO	graphene oxide
GPU	gas permeance units
HAO	hydrous aluminium oxide
H_2O	water
HPEI	hyperbranched polyethylenimine
H_2S	hydrogen sulfide

HS-SPME	headspace solid-phase micro-extraction
MF	micro filtration
MgO	magnesium oxide
MOF	metal organic framework
MPa	megapascal
MPG	magnetic polymer graphene
N_2	nitrogen
NaCl	sodium chloride
NF	nanofiltration
NWBM	nano fluid-enhanced water based muds
O_2	oxygen
OBM	oil based muds
PAA	poly(acrylic acid)
PA@Fe_3O_4	polyaniline-magnetite
PAGCs	polyaniline/graphene composites
PAM	polyacrylamide
PANI	polyaniline
PDMS	polydimethylsiloxane
PEI	poly(ethylene imine)
PEG	polyethylene glycol
PIM	polymer of intrinsic micro-porosity
PoPd	poly(o-phenelediamine)
POSS	polyhedral oligomeric silsesquioxane
ppm	parts per million
PPyCB	polypyrrole-carbon block
PS	polystyrene
PSf	polysulfone
PTA	poly([2-(methacryloyloxy)ethyl]trimethylammonium chloride)
PU	polyurethane
PVA	polyvinyl alcohol
PVAc	polyvinyl acetate
PVB	polyvinylbutyral
PVC	polyvinyl chloride
PVDF	polyvinylidene fluoride
RO	reverse osmosis
SiO_2	silica
Si-R	hydrophobic silica sol
SMBA@CNT	polyzwitterion coated CNT

SO₂	sulfur dioxide

SO₂   sulfur dioxide
SPME   solid phase micro extraction
SS   stainless steel
St-DVB   styrene-divinyl benzene
TEOS   tetraethoxysilane
TiO₂   titanium dioxide
TMC   trimesyl chloride
UF   ultrafiltration
WBM   water based muds
ZnO   zinc oxide
ZrO₂   zirconia

References

1. Vengatesan, M. R., Singh, S., Pillai, V. V., and Mittal, V. (2016) Crystallization, mechanical, and fracture behavior of mullite fiber-reinforced polypropylene nanocomposites. *Journal of Applied Polymer Science*, doi:10.1002/app.43725.
2. Kanimozhi, K., Prabunathan, P., Selvaraj, V., and Alagar, M. (2014) Vinyl silane-functionalized rice husk ash-reinforced unsaturated polyester nanocomposites. *RSC Advances*, **4**, 18157-18163.
3. Vengatesan, M. R., Devaraju, S., and Alagar, M. (2011) Studies on thermal, mechanical and morphological properties of organoclay filled azomethine modified epoxy nanocomposites. *High Performance Polymers*, **23**, 3-10.
4. Kumaran, R., Alagar, M., Dinesh Kumar, S., Subramanian, V., and Dinakaran, K. (2015) Ag induced electromagnetic interference shielding of Ag-graphite/PVDF flexible nanocomposites thin films. *Applied Physics Letters*, **107**, 113107.
5. Tjong, S. C. (2012) Polymer nanocomposites for sensor applications. In: *Polymer Composites with Carbonaceous Nanofillers, Properties and Applications*, Wiley-VCH Verlag GmbH & Co. KGaA, Germany, doi:10.1002/9783527648726.ch9.
6. Fink, J. K. (2014) High Performance Polymers, Elsevier, USA.
7. R. C. Advincula, "Polymer materials and nanotechnology for oil and gas", 2014, High Performance Polymers for Oil & Gas. Advincula R. C. (2014) Polymer Materials and Nanotechnology for Oil and Gas. *High Performance Polymers for Oil and Gas Conference*, 2014, Scotland.
8. Cheraghian, G., and Hendraningrat, L. (2014) A review on applications of nanotechnology in the enhanced oil recovery part B: effects of nanoparticles on flooding. *International Nano Letters*, doi:10.1007/s40089-014-0114-7.
9. Wang, T.-L., Hwang, W.-S., and Yeh, M.-H. (2007) Preparation, properties, and anti-corrosion application of poly(methyl methacrylate)/montmorillonite nanocomposites coating on brass via solution polymerization. *Journal of Applied Polymer Science*, **104**, 4135-4143.

10. Mohamadpour, S., Pourabbas, B., and Fabbri, P. (2011) Anti-scratch and adhesion properties of photo-curable polymer/clay nanocomposite coatings based on methacrylate monomers. *Scientia Iranica F*, **18**, 765-771.
11. Li, P., He, X., Huang, T.-C., White, K. L., Zhang, X., Liang, H., Nishimurae, R., and Sue, H.-J. (2015) Highly effective anti-corrosion epoxy spray coatings containing self-assembled clay in smectic order. *Journal of Materials Chemistry A*, **3**, 2669-2676.
12. Waché, R., Klopffer, M.-H., and Gonzalez, S. (2015) Characterization of polymer layered silicate nanocomposites by rheology and permeability methods: Impact of the interface quality. *Oil & Gas Science and Technology*, 70, 267–277.
13. Hu, Z., and Chen, G. (2014) Aqueous dispersions of layered double hydroxide/polyacrylamide nanocomposites: preparation and rheology. *Journal of Materials Chemistry A*, **2**, 13593-13601.
14. Cui, Y., Kundalwal, S. I., and Kumar, S. (2016) Gas barrier performance of graphene/polymer nanocomposites. *Carbon*, **98**, 313-333.
15. Hobson, G. D. and Tiratsoo, E. N. (1975) Introduction to Petroleum Geology, Scientific Press, UK.
16. Maghzi, A., Kharrat, R., Mohebbi, A., and Ghazanfari, M. H. (2014) The impact of silica nanoparticles on the performance of polymer solution in presence of salts in polymer flooding for heavy oil recovery. *Fuel*, **123**, 123-132.
17. Caulfield, M. J., Hao, X., Qiao, G. G., and Solomon, D. H. (2003) Degradation on polyacrylamides. Part I. Linear polyacrylamide. *Polymer*, **44**, 1331-1337.
18. Al-bassama, S. I., and Al-jarraha, M. M. F. (1990) New water-soluble polymer for enhanced oil recovery. *Polymer-Plastics Technology and Engineering*, **29**, 407-415.
19. Kurenkov, K. F., Hartan, H. G., and Lobanov. F. I. (2002) Degradation of polyacrylamide and its derivatives in aqueous solutions. *Russian Journal of Applied Chemistry*, **75**, 1039-1050.
20. Caulfield, M. J., Hao, X., Qiao, G. G., and Solomon, D. H. (2003) Degradation on polyacrylamides: Part II. Polyacrylamide gels. *Polymer*, **44**, 3817-3826.
21. Cheraghian, G., and Hendraningrat, L. (2016) A review on applications of nanotechnology in the enhanced oil recovery part B: effects of nanoparticles on flooding. *International Nano Letters*, **6**, 1-10.
22. Cheraghian, G. (2015) Thermal resistance and application of nanoclay on polymer flooding in heavy oil recovery. *Petroleum Science and Technology*, **33**, 1580-1586.
23. Cheraghian, G., and Khalilinezhad, S. S. (2015) Effect of nanoclay on heavy oil recovery during polymer flooding. *Petroleum Science and Technology*, **33**, 999-1007.
24. Cheraghian, G., Nezhad, S. S. K., Kamari, M., Hemmati, M., Masihi, M., and Bazgir, S. (2015) Effect of nanoclay on improved rheology properties of polyacrylamide solutions used in enhanced oil recovery. *Journal of Petroleum Exploration and Production Technology*, **5**, 189-196.
25. Ye, Z., Qin, X., Lai, N., Peng, Q., Li, X., and Li, C. (2013) Synthesis and performance of an acrylamide copolymer containing nano-SiO_2 as enhanced oil recovery chemical. *Journal of Chemistry*, Article ID 437309, 10 pages.
26. Maurya, N. K., and Mandal, A. (2016) Studies on behavior of suspension of silica nanoparticle in aqueous polyacrylamide solution for application in enhanced oil recovery. *Petroleum Science and Technology*, **34**, 429-436.

27. Rodewald, P. G. (1976) Oil Recovery by Water Flooding Employing Graphite Oxide for Mobility Control, US Patent 3998270.
28. Nguyen, B. D., Ngo, T. K., Bui, T. H., Pham, D. K., Dinh, X. L., and Nguyen, P. T. (2015) The impact of graphene oxide particles on viscosity stabilization for diluted polymer solutions using in enhanced oil recovery at HTHP offshore reservoirs. *Advances in Natural Sciences: Nanoscience and Nanotechnology*, **6**, 015012.
29. Zuniga, C. A., Goods, J. B., Cox, J. R. and Swager, T. M. (2016) Long-term high-temperature stability of functionalized graphene oxide nanoplatelets in Arab-D and API brine. *ACS Applied Materials and Interfaces*, **8**, 1780-1785.
30. Nguyen, T. P., Le, U. T. P., Ngo, K. T., Pham, K. D., and Dinh, L. X. (2016) Synthesis of polymer-coated magnetic nanoparticles from red mud waste for enhanced oil recovery in offshore reservoirs. *Journal of Electronic Materials*, **45**, 3801-3808.
31. Caenn, R., and Chillingar, G. V. (1996) Drilling fluids: State of the art. *Journal of Petroleum Science and Engineering*, **14**, 221-230.
32. Chillingarian, C. V., and Vorbuter, P. (1981) Drilling and Drilling Fluids, Elsevier Scientific Publishing, The Netherlands.
33. Darley, H. C. H., and Gary, G. G. (1998) Composition and Properties of Drilling and Completion Fluids (fifth ed.), Gulf Publishing Co, USA.
34. Brazzel, R. L. (2009) Multi-component Drilling Fluid Additive and Drilling Fluid System Incorporating the Additive, US Patent US7635667 B1.
35. Melbouci, M., and Sau, A. C. (2008) Water-based Drilling Fluids, US Patent 7384892.
36. Pal, S. (2011) Drilling Fluids Engineering (first ed.), Pal Skalle & Ventus Publishing, ApS. Online: http://editorbar.com/upload/ReBooks/2012-9/8c37651495fe6e1cc70b48abe1ff5e05.pdf (assessed 19.01.2017).
37. Jain, R., Mahto, V., and Sharma, V. P. (2015) Evaluation of polyacrylamide-grafted-polyethylene glycol/silica nanocomposite as potential additive in water based drilling mud for reactive shale formation. *Journal of Natural Gas Science and Engineering*, **26**, 526-537.
38. Mao, H., Qiu, Z., Shen, Z., and Huang, W. (2015) Hydrophobic associated polymer based silica nanoparticles composite with core–shell structure as a filtrate reducer for drilling fluid at ultra-high temperature. *Journal of Petroleum Science and Engineering*, **129**, 1-14.
39. Alizadeh, S., Sabbaghi, S., and Soleymani, M. (2015) Synthesis of alumina/polyacrylamide nanocomposite and its influence on viscosity of drilling fluid. *International Journal of Nano Dimension*, **6**, 271-276.
40. Sadeghalvaad, M., and Sabbaghi, S. (2015) The effect of the TiO_2/polyacrylamide nanocomposite on water-based drilling fluid properties. *Powder Technology*, **272**, 113-119.
41. Mao, H., Qiu, Z., Shen, Z., Huang, W., Zhong, H., and Dai, W. (2015) Novel hydrophobic associated polymer based nano-silica composite with core–shell structure for intelligent drilling fluid under ultra-high temperature and ultra-high pressure. *Progress in Natural Science: Materials International*, **25**, 90-93.
42. Madkour, T. M., Fadl, S., Dardir, M. M., and Mekewi, M. A. (2015) High performance nature of biodegradable polymeric nanocomposites for oil-well drilling fluids. *Egyptian Journal of Petroleum*, doi:10.1016/j.ejpe.2015.09.004.

43. William, J. K. M., Ponmani, S., Samuel, R., Nagarajan, R., and Sangwai, J. S. (2014) Effect of CuO and ZnO nanofluids in xanthan gum on thermal, electrical and high pressure rheology of water-based drilling fluids. *Journal of Petroleum Science and Engineering*, **117**, 15-27.
44. Liu, X., and Wang, X. Y. (2014) The research of oil and gas pipeline corrosion and protection technology. *Advances in Petroleum Exploration and Development*, **7**, 102-105.
45. Amadi, S. A., and Ukpaka, C. P. (2014) Performance evaluation of anti-corrosion coating in an oil industry. *Agriculture and Soil Sciences*, **1**, 070-081.
46. Zhou, S., and Wu, L. (2009) Development of nanotechnology-based organic coatings. *Composite Interfaces*, **16**, 281-292.
47. Wold, C. R., and Soucek, M. D. (2000) Viscoelastic and thermal properties of linseed oil-based ceramer coatings. *Macromolecular Chemistry and Physics*, **201**, 382-392.
48. Ni, H., Simonsick Jr, W. J., Skaj, A. D., Williams, J. P., and Soucek, M. D. (2000) Polyurea/polysiloxane ceramer coatings. *Progress in Organic Coatings*, **38**, 97-110.
49. Voevodin, N. N., Grebasch, N. T., Soto, W. S., Arnold, F. E., and Donley, M. S. (2001) Potentio-dynamic evaluation of sol–gel coatings with inorganic inhibitors. *Surface and Coatings Technology*, **140**, 24-28.
50. Hihara, L. H., and Kusada, K. (2011) Corrosion of Bare and Coated Al 5052 – H3 and Al 6061 – T6 in Seawater. Online: http://hinmrec.hnei.hawaii.edu/wp-content/ uploads/ 2010/01/Corrosion-of-Aluminum-Alloys-in-Seawater-_-Progress-Report.pdf (assessed 23.01.2017).
51. Keyoonwong, W., Guo, Y., Kubouchi, M., Aoki, S., and Sakai, T. (2012) Corrosion behavior of three nanoclay dispersion methods of epoxy/organoclay nanocompo-sites. *International Journal of Corrosion*, article ID 924283, 10 pages.
52. Valencaa, D. P., Alves, K. G. B., Melo, C. P., and Bouchonneau, N. (2015) Study of the efficiency of polypyrrole/ZnO nanocomposites as additives in anticorrosion coatings. *Materials Research*, **18**, 273-278.
53. Chang, K. C., Huang, H. H., Lai, M. C., Hung, C. B., Chand, B., Yeh, J. M., and Yu, Y. H. (2009) Comparative electrochemical studies at different operational temperatures for the effect of nanoclay platelets on the anticorrosion efficiency of organo-soluble polyimide/clay nanocomposite coatings. *Journal of Nanoscience and Nanotechnology*, **9**, 3125-3133.
54. Chen, F. and Liu, P. (2011) Conducting polyaniline nanoparticles and their dispersion for waterborne corrosion protection coatings. *ACS Applied Materials and Interfaces*, **3**, 2694-2702.
55. Zaarei, D., Sarabi, A. A., Sharif, F., Kassiriha, S. M., and Gudarzi, M. M. (2010) Preparation and evaluation of epoxy-clay nanocomposite coatings for corrosion protection. *International Journal of Nanoscience and Nanotechnology*, **7**, 126-136.
56. Yeh, J. M., Huang, H. Y., Chen, C. L., Su, W. F., and Yu, Y. H. (2006) Siloxane-modified epoxy resin–clay nanocomposite coatings with advanced anticorrosive properties prepared by a solution dispersion approach. *Surface and Coatings Technology*, **200**, 2753-2763.
57. Yu, Y. H., Yeh, J.-M., Liou, S.-J., and Chang, Y.-P. (2004) Effect of nanoparticles on the anticorrosion and mechanical properties of epoxy coating. *Acta Materialia*, **52**, 475-486.

58. Koch, G. H., Brongers, M. P. H., Thompson, N. G., Virmani, Y. P., and Payer, J. H. (2002) Corrosion Cost and Prevention Strategies in the United States, FHWA-RD-01-156, Office of Infrastructure Research and Development, Federal Highway Administration.

59. Alonso, M. C. I., Valdes, S. S., Vargas, E. R., Tavizon, S. F., Garcia, J. R., Perez, A. S. L., Ramos de Valle, L. F., Fernandez, O. S. R., Martinez, A. B. E., Colunga, J. G. M., and Cabrera-Alvarez, E. N. (2015) Preparation and characterization of polyethylene/clay/silver nanocomposites using functionalized polyethylenes as an adhesion promoter. *Journal of Adhesion Science and Technology*, **29**, 1911-1923.

60. Manjumeena, R., Venkatesan, R., Duraibabu, D., Sudha, J., Rajendran, N., and Kalaichelvan, P. T. (2016) Green nanosilver as reinforcing eco-friendly additive to epoxy coating for augmented anticorrosive and antimicrobial behavior. *Silicon*, **8**, 277-298.

61. Lam, C. K., and Lau, K. T. (2006) Localized elastic modulus distribution of nanoclay/epoxy composites by using nanoindentation. *Composite Structures*, **75**, 553-558.

62. Shi, G., Zhang, M. Q., Rong, M. Z., Wetzel, B., and Friedrich, K. (2003) Friction and wear of low nanometer Si_3N_4 filled epoxy composites. *Wear*, **254**, 784-796.

63. Hartwig, A., Sebald, M., Putz, D., and Aberle, L. (2005) Preparation, characterization and properties of nanocomposites based on epoxy resins – an overview. *Macromolecular Symposia*, **221**, 127-136.

64. Dietsche, F., Thomann, Y., Thomann, R., and Mulhaupt, R. (2000) Translucent acrylic nanocomposites containing anisotropic laminated nanoparticles derived from intercalated layered silicates. *Journal of Applied Polymer Science*, **75**, 396-405.

65. Shi, X., Nguyen, T. A., Suo, Z., Liu, Y., and Avci, R. (2009) Effect of nanoparticles on the anticorrosion and mechanical properties of epoxy coating. *Surface and Coatings Technology*, **204**, 237-245.

66. Bakhshandeh, E., Jannesari, A., Ranjbar, Z., Sobhani, S., and Saeb, M. R. (2014) Anti-corrosion hybrid coatings based on epoxy–silicanano-composites: Toward relationship between the morphologyand EIS data. *Progress in Organic Coatings*, **77**, 1169-1183.

67. Ramezanzadeh, B., Attar, M. M., and Farzam, M. (2011) A study on the anticorrosion performance of the epoxy–polyamide nanocomposites containing ZnO nanoparticles. *Progress in Organic Coatings*, **72**, 410-422.

68. Zhang, C., He, Y., Xu, Z., Shi, H., Di, H., Pan, Y., and Xu, W. (2016) Fabrication of $Fe_3O_4@SiO_2$ nanocomposites to enhance anticorrosion performance of epoxy coatings. *Polymers for Advanced Technologies*, **27**, 740-747.

69. Popoola, A., Olorunniwo, O. E., and Ige, O. O. (2014) Corrosion Resistance Through the Application of Anti-Corrosion Coatings. In: *Developments in Corrosion Protection*, Aliofkhazraei, M. (ed.), InTech, Croatia.

70. Ganash, A. (2014) Anticorrosive properties of poly (o-phenylenediamine)/ZnO nanocomposites coated stainless steel. *Journal of Nanomaterials*, article ID 540276, 8 pages.

71. Chang, C-H., Huang, T.-C., Peng, C.-W., Yeh, T.-C., Lu, H.-I., Hung, W.-I., Weng, C.-J., Yang, T.I., and Yeh, J.-M. (2012) Novel anticorrosion coatings prepared from polyaniline/graphene composites. *Carbon*, **50**, 5044-5051.

72. Yu, Y.-H., Lin, Y.-Y., Lin, C.-H., Chan, C.-C., and Huang, Y.-C. (2014) High-performance polystyrene/graphene-based nanocomposites with excellent anti-corrosion properties. *Polymer Chemistry*, **5**, 535-550.
73. Li, Y., Yang, Z., Qiu, H., Dai, Y., Zheng, Q., Li, J., and Yang, J. (2014) Self-aligned graphene as anticorrosive barrier in waterborne polyurethane composite coatings. *Journal of Materials Chemistry A*, **2**, 14139-14145.
74. Luckachan, G. E., and Mittal, V. (2015) Anti-corrosion behavior of layer by layer coatings of cross-linked chitosan and poly(vinyl butyral) on carbon steel. *Cellulose*, **22**(5), 3275-3290.
75. Fan, F., Zhou, C., Wang, X., and Szpunar, J. (2015) Layer-by-layer assembly of a self-healing anticorrosion coating on magnesium alloys. *ACS Applied Materials and Interfaces*, **7**, 27271-27278.
76. Niratiwongkorn, T., Luckachan, G. E., and Mittal, V. (2016) Self-healing protective coatings of polyvinyl butyral/polypyrrole-carbon black composite on carbon steel. *RSC Advances*, **6**, 43237-43249.
77. Ornitz, B. E., and Champ, M. A. (2002) *Oil Spills First Principles: Prevention and Best Response*, Elsevier Science Ltd., UK.
78. Choi, H. M., and Cloud, R. M. (1992) Natural sorbents in oil spill cleanup. *Environmental Science and Technology*, **26**, 772-776.
79. Mishra, P., and Balasubramanian, K. (2014) Nanostructured microporous polymer composite imprinted with superhydrophobic camphor soot, for emphatic oil–water separation. *RSC Advances*, **4**, 53291-53296.
80. Zhang, C., Li, P., and Cao, B. (2015) Electrospun microfibrous membranes based on PIM-1/POSS with high oil wettability for separation of oil–water mixtures and cleanup of oil soluble contaminants. *Industrial & Engineering Chemistry Research*, **54**, 8772-8781.
81. Nikkhah, A. A., Zilouei, H., Asadinezhad, A., and Keshavarz, A. (2015) Removal of oil from water using polyurethane foam modified with nanoclay. *Chemical Engineering Journal*, **262**, 278-285.
82. Maphutha, S., Moothi, K., Meyyappan, M., and Iyuke, S. E. (2013) A carbon nanotube-infused polysulfone membrane with polyvinyl alcohol layer for treating oil-containing waste water. *Scientific Reports*, **3**, 1509.
83. Liu, Y., Ma, J., Wu, T., Wang, X., Huang, G., Liu, Y., Qiu, H., Li, Y., Wang, W., and Gao, J. (2013) Cost-effective reduced graphene oxide-coated polyurethane sponge as a highly efficient and reusable oil-absorbent. *ACS Applied Materials and Interfaces*, **5**, 10018-10026.
84. Tran, D. N. H., Kabiri, S., Sim, T. R., and Losic, D. (2015) Selective adsorption of oil–water mixtures using polydimethylsiloxane (PDMS)–graphene sponges. *Environmental Science: Water Research & Technology*, **1**, 298-305.
85. Saber, O., Mohamed, N. H., and Arafat, S. A. (2015) Conversion of iron oxide nanosheets to advanced magnetic nanocomposites for oil spill removal. *RSC Advances*, **5**, 72863-72871.
86. Zhang, N., Jiang, W., Wang, T., Gu, J., Zhong, S., Zhou, S., Xie, T., and Fu, J. (2015) Facile preparation of magnetic poly(styrene-divinylbenzene) foam and its application as an oil absorbent. *Industrial & Engineering Chemistry Research*, **54**, 11033-11039.

87. Gomes de Souza, F., Marins, J. A., Rodrigues, C. H. M., and Pinto, J. C. (2010) A magnetic composite for cleaning of oil spills on water. *Macromolecular Materials and Engineering*, **295**, 942-948.
88. Yu, L., Hao, G., Zhou, S., and Jiang, W. (2016) Durable and modified foam for cleanup of oil contamination and separation of oil–water mixtures. *RSC Advances*, **6**, 24773-24779.
89. Liu, C., Yang, J., Tang, Y., Yin, L., Tang, H., and Li, C. (2015) Versatile fabrication of the magnetic polymer-based graphene foam and applications for oil–water separation. *Colloids and Surfaces A: Physicochemical and Engineering Aspects*, **468**, 10-16.
90. Scholes, C. A., Stevens, G. W., and Kentish, S. E. (2012) Membrane gas separation applications in natural gas processing. *Fuel*, **96**, 15-28.
91. Deng, L., Kim, T.-J., Sandru, M., and Hagg, M.-B. (2009) PVA/PVAm Blend FSC Membrane for Natural Gas Sweetening. *Proceedings of the 1st Annual Gas Processing Symposium*, Qatar.
92. Bernardo, P., Drioli, E., and Golemme, G. (2009) Membrane gas separation: A review/state of the art. *Industrial & Engineering Chemistry Research*, **48**, 4638-4663.
93. Ogoshi, T., Itoh, H., Kim, K. M., and Chujo, Y. (2002) Synthesis of organic–inorganic polymer hybrids having interpenetrating polymer network structure by formation of ruthenium–bipyridyl complex. *Macromolecules*, **35**, 334-338.
94. Novak, B. M. (1993) Hybrid nanocomposite materials—between inorganic glasses and organic polymers. *Advanced Materials*, **5**, 422-433.
95. Rafiq, S., Man, Z., Ahmad, F., and Maitra, S. (2010) Silica-polymer nanocomposite membranes for gas separation - a review, Part 1. *International Ceramic Review*, **59**, 341-349.
96. Rafiq, S., Man, Z., Maulud, A., Muhammad, N., and Maitra, S. (2012) Separation of CO_2 from CH_4 using polysulfone/polyimide silica nanocomposite membranes. *Separation and Purification Technology*, **90**, 162-172.
97. Xin, Q., Gao, Y., Wu, X., Li, C., Liu, T., Shi, Y., Li, Y., Jiang, Z., Wu, H., and Cao, X. (2015) Incorporating one-dimensional aminated titania nanotubes into sulfonated poly(ether ether ketone) membrane to construct CO_2-facilitated transport pathways for enhanced CO_2 separation. *Journal of Membrane Science*, **488**, 13-29.
98. Rahman, M. M., Filizn, V., Shishatskiy, S., Abetz, C., Neumann, S., Bolmer, S., Khan, M. M., and Abetz, V. (2013) PEBAX® with PEG functionalized POSS as nanocomposite membranes for CO_2 separation. *Journal of Membrane Science*, **437**, 286-297.
99. Liu, Y., Peng, D., He, G., Wang, S., Li, Y., Wu, H., and Jiang, Z. (2014) Enhanced CO_2 permeability of membranes by incorporating polyzwitterion@CNT composite particles into polyimide matrix. *ACS Applied Materials and Interfaces*, **6**, 13051-13060.
100. Kim, H. W., Yoon, H. W., Yoon, S. M., Yoo, B. M., Ahn, B. K., Cho, Y. H., Shin, H. J., Yang, H., Paik, U., Kwon, S., Choi, J. Y., and Park, H. B. (2013) Selective gas transport through few-layered graphene and graphene oxide membranes. *Science*, **342**, 91-95.

101. Li, H., Song, Z., Zhang, X., Huang, Y., Li, S., Mao, Y., Ploehn, H. J., Bao, Y., and Yu, M. (2013) Ultrathin, molecular-sieving graphene oxide membranes for selective hydrogen separation. *Science*, **342**, 95-98.

102. Dong, G., Zhang, Y., Hou, J., Shen, J., and Chen, V. (2016) Graphene oxide nanosheets based novel facilitated transport membranes for efficient CO_2 capture. *Industrial & Engineering Chemistry Research*, **55**, 5403-5414.

103. Wang, T., Zhao, L., Shen, J., Wu, L., and Bruggen, B. V. (2015) Enhanced performance of polyurethane hybrid membranes for CO_2 separation by incorporating graphene oxide: The relationship between membrane performance and morphology of graphene oxide. *Environmental Science and Technology*, **49**, 8004-8011.

104. Ozdemir, S. S., Buonomenna, M. G., and Drioli, E. (2006) Catalytic polymeric membranes: preparation and application. *Applied Catalysis A*, **307**, 167-183.

105. Nour, M., Berean, K., Chrimes, A., Zoolfakar, A. S., Latham, K., McSweeney, C., Field, M. R., Sriram, S., Kalantar-zadeh, K., and Zhen Ou, J. (2014) Silver nanoparticle/PDMS nanocomposite catalytic membranes for H_2S gas removal. *Journal of Membrane Science*, **470**, 346-355.

106. Wang, B. L., Takigawa, T., Takeuchi, A., Yamasaki, Y., Kataoka, H., Wang, D. H., and Ogino, K. (2007) Unmetabolized VOCs in urine as biomarkers of low level exposure in indoor environments. *Journal of Occupational Health*, **49**, 104-110.

107. Suna, S., Jitsunari, F., Asakawa, F., Hirao, T., Mannami, T., and Suzue, T. (2005) A method for on-site analysis of urinary benzene by means of a portable gas-chromatograph. *Journal of Occupational Health*, **47**, 74-77.

108. Brcic, I., and Skender, L. (2003) Determination of benzene, toluene, ethylbenzene, and xylenes in urine by purge and trap gas chromatography. *Journal of Separation Science*, **26**, 1225-1229.

109. Holcomb, L. C., and Seabrook, B. S. (1995) Indoor concentrations of volatile organic compounds: Implications for comfort, health and regulation. *Indoor Environment*, **4**, 7-26.

110. Abolghasemi, M. M., Hassani, S., Rafiee, E., and Yousefi, V. (2015) Nanoscale-supported heteropoly acid as a new fiber coating for solid-phase microextraction coupled with gas chromatography–mass spectrometry. *Journal of Chromatography A*, **1381**, 48-53.

111. Rahimi, A., Hashemi, P., Badiei, A., Arab, P., and Ghiasvand, A. R. (2011) CMK-3 nanoporous carbon as a new fiber coating for solid-phase microextraction coupled to gas chromatography–mass spectrometry. *Analytica Chimica Acta*, **695**, 58-62.

112. Hashemi, P., Badiei, A., Shamizadeh, M., Ziarani, G. M., and Ghiasvand, A. R. (2012) Preparation of a new solid-phase microextraction fiber by coating silylated nanoporous silica on a copper wire. *Journal of Chinese Chemical Society*, **59**, 727-732.

113. Foroutani, R., Yousefi, V., and Kangari, S. (2015) Synthesis of polyaniline-magnetite hollow nanocomposite as a novel fiber coating for the headspace solid-phase microextraction of benzene, toluene, ethylbenzene and xylenes from water samples. *Analytical Methods*, **7**, 5318-5324.

114. Gupta, R., and Kulkarni, G. U. (2011) Removal of organic compounds from water by using a gold nanoparticle–poly(dimethylsiloxane) nanocomposite foam. *ChemSusChem*, **4**, 737-743.

115. Peerakiatkhajorn, P., Chawengkijwanich, C., Onreabroy, W., and Chiarakorn, S. (2012) Novel photocatalytic Ag/TiO$_2$ thin film on polyvinyl chloride for gaseous BTEX treatment. *Materials Science Forum*, **712**, 133-145.
116. Zha, S., Zhang, G., Dawson, N., Yu, J., Liu, N., and Lee, R. (2016) Study of PVDF/Si-R hybrid hollow fiber membranes for removal of dissolved organics from produced water by membrane adsorption. *Separation and Purification Technology*, **163**, 290-299.
117. Yu, L., Han, M., and He, F. (2013) A review of treating oily wastewater. *Arabian Journal of Chemistry*, doi:10.1016/j.arabjc.2013.07.020.
118. Padaki, M., Surya Murali, R., Abdullah, M. S., Misdan, N., Moslehyani, A., Kassim, M. A., Hilal, N., and Ismail, A. F. (2015) Membrane technology enhancement in oil–water separation: A review. *Desalination*, **357**, 197-207.
119. Rezvanpour, A., Roostaazad, R., Hesampour, M., Nystrom, M., and Ghotbi, C. (2009) Effective factors in the treatment of kerosene–water emulsion by using UF membranes. *Journal of Hazardous Materials*, **16**, 1216-1224.
120. Bensadok, K., Belkacem, M., and Nazzal, G. (2007) Treatment of cutting oil/water emulsion by coupling coagulation and dissolved air flotation. *Desalination*, **206**, 440-448.
121. Ayotamuno, M. J., Kogbara, R. B., Ogaji, S. O. T., and Pobert, S. D. (2006) Petroleum contaminated ground-water: remediation using activated carbon. *Applied Energy*, **83**, 1258-1264.
122. Zhao, X., Wang, Y., Ye, Z., Borthwick, A. G. L., Ni, J. (2006) Oil field wastewater treatment in biological aerated filter by immobilized microorganisms. *Process Biochemistry*, **41**, 1475-1483.
123. Cambiella, A., Bentio, J. M., Pazos, C., and Coca, J. (2006) Centrifugal separation efficiency in the treatment of waste emulsified oils. *Chemical Engineering Research and Design*, **84**, 69-76.
124. Hashmi, K. A., Hamza, H. A., and Wilson, J. C. (2004) CANMET hydrocyclone: an emerging alternative for the treatment of oily waste streams. *Mineral Engineering*, **17**, 643-649.
125. Yan, L., Hong, S., Li, M. L., and Li, Y. S. (2009) Application of the Al$_2$O$_3$–PVDF nanocomposite tubular ultrafiltration (UF) membrane for oily wastewater treatment and its antifouling research. *Separation and Purification Technology*, **66**, 347-352.
126. Shih, Y. J., and Chang, Y. (2010) Tunable blood compatibility of polysulfobetaine from controllable molecular-weight dependence of zwitterionic nonfouling nature in aqueous solution. *Langmuir*, **26**, 17286-17294.
127. Peng, J., Su, Y., Shi, Q., Chen, W., and Jiang, Z. (2011) Protein fouling resistant membrane prepared by amphiphilic pegylated polyethersulfone. *Bioresource Technology*, **102**, 2289-2295.
128. An, Q.-F., Sun, W.-D., Zhao, Q., Ji, Y.-L., and Gao, C.-J. (2013) Study on a novel nanofiltration membrane prepared by interfacial polymerization with zwitterionic amine monomers. *Journal of Membrane Science*, **431**, 171-179.
129. Shao, L., Wang, Z. X., Zhang, Y. L., Jiang, Z. X., and Liu, Y. Y. (2014) A facile strategy to enhance PVDF ultrafiltration membrane performance via self-polymerized

polydopamine followed by hydrolysis of ammonium fluotitanate. *Journal of Membrane Science*, **461**, 10-21.

130. Yan, L., Li, Y. S., and Xiang, C. B. (2005) Preparation of poly (vinylidene fluoride)(pvdf) ultrafiltration membrane modified by nano-sized alumina (Al_2O_3) and its antifouling research. *Polymer*, **46**, 7701-7706.

131. Wang, Z., Yu, H., Xia, J., Zhang, F., Li, F., Xia, Y. and Li, Y. (2012) Novel GO-blended PVDF ultrafiltration membranes. *Desalination*, **299**, 50-54.

132. Munirasu, S., Haija, M. A., and Banat F. (2016) Use of membrane technology for oil field and refinery produced water treatment - A review. *Process Safety and Environmental Protection*, **100**, 183-202.

133. Gohari, R. J., Korminouri, F., Lau, W. J., Ismail, A. F., Matsuura, T., Chowdhury, M. N. K., Halakoo, E., and Jamshidi Gohari, M. S. (2015) A novel super-hydrophilic PSf/HAO nanocomposite ultrafiltration membrane for efficient separation of oil/water emulsion. *Separation and Purification Technology*, **150**, 13-20.

134. Zhao, X., Su, Y., Cao, J., Li, Y., Zhang, R., Liu, Y., and Jiang, Z. (2015) Fabrication of antifouling polymer–inorganic hybrid membranes through the synergy of biomimetic mineralization and nonsolvent induced phase separation. *Journal of Materials Chemistry A*, **3**, 7287-7295.

135. Zeng, G., He, Y., Zhan, Y., Zhang, L., Shi, H., and Yu, Z. (2016) Preparation of a novel poly(vinylidene fluoride) ultrafiltration membrane by incorporation of 3-amino propyl triethoxysilane grafted halloysite nanotubes for oil/water separation. *Industrial and Engineering Chemistry Research*, **55**, 1760-1767.

136. Venkatesh, K., Arthanareeswaran, G., and Chandra Bose, A. (2016) PVDF mixed matrix nano-filtration membranes integrated with 1D-PANI/TiO2 NFs for oil–water emulsion separation. *RSC Advances*, **6**, 18899-18908.

137. Lee, J., Chae, H. R., Won, Y. J., Lee, K., Lee, C. H., Lee, H. H., Kim, I. C., and Lee, J. M. (2013) Graphene oxide nanoplatelets composite membrane with hydrophilic and antifouling properties for wastewater treatment. *Journal of Membrane Science*, **448**, 223-230.

138. Liu, L., Zhao, F., Liu, J., and Yang, F. (2013) Preparation of highly conductive cathodic membrane with graphene (oxide)/PPy and the membrane antifouling property in filtrating yeast suspensions in EMBR. *Journal of Membrane Science*, **437**, 99-107.

139. Wu, H., Tang, B., and Wu, P. (2014) Development of novel SiO_2-GO nanohybrid/polysulfone membrane with enhanced performance. *Journal of Membrane Science*, **451**, 94-102.

140. Wang, Z., Yu, H., Xia, J., Zhang, F., Li, F., Xia, Y., and Li, Y. (2012) Novel GO-blended PVDF ultrafiltration membranes. *Desalination*, **299**, 50-54.

141. Boussu, K., Kindts, C., Vandecasteele, C., and Van der Bruggen, B. (2007) Surfactant fouling of nanofiltration membranes: Measurements and mechanisms. *ChemPhysChem*, **8**, 1836-1845.

142. Kaya, Y., Barlas, H., and Arayici, S. (2011) Evaluation of fouling mechanisms in the nanofiltration of solutions with high anionic and nonionic surfactant contents using a resistance-in-series model. *Journal of Membrane Science*, **367**, 45-54.

143. Qin, D., Liu, Z., Bai, H., Sun, D. D., and Song, X. (2016) A new nano-engineered hierarchical membrane for concurrent removal of surfactant and oil from oil-in water nanoemulsion. *Scientific Reports*, **6**, 24365.

3

Associative Polymer Applications in Chemical Injection for Enhanced Oil Recovery

3.1 Introduction

In the recent years, enhanced oil recovery (EOR) has become a prime field of research especially due to rapid increase in the energy demand. EOR is defined as implementation of various techniques to increase the amount of crude extracted from an oil reservoir. For this purpose, EOR is also called as improved oil recovery or tertiary oil recovery. This oil recovery technique may increase the amount of oil recovery from the reservoir by up to 40%. There are three primary techniques used for EOR applications, which are categorized as gas injection, thermal injection and chemical injection.

3.1.1 Gas Injection

This technique is the most commonly used technique for EOR operations. In the North America, gas injection amounts to more than 60% of EOR production. Gas injection usually involves injection of gases such as nitrogen, carbon dioxide (CO_2) or natural gas [1]. Since this technique uses miscible gases to flood the reservoir, the technique is also called as miscible flooding. The reservoir pressure is maintained due to the miscible displacement. Since the interfacial tension between oil and water is reduced in the reservoir, this technique in turn improves oil displacement, which results in improved oil recovery. This technique allows for total displacement efficiency since the interface between the two interacting fluids is removed [2]. CO_2 is the most commonly used gas for miscible displacement since it is cost effective and also has the ability to increase the viscosity of oil. Oil recovery by miscible displacement primarily depends on the phase behavior of crude oil and gas. The phase behavior characteristics of these fluids are significantly affected by the many factors such as crude oil composition, temperature and pressure of the reservoir, among others [2].

Hariharan Krithivasan* and Vikas Mittal**, The Petroleum Institute (part of Khalifa University of Science and Technology), Abu Dhabi, UAE
*Current address: University of Waterloo, Waterloo, Canada; **Current address: Bletchington, Wellington County, Australia

3.1.2 Thermal Injection

Thermal injection technique amounts for 40% of EOR operations in the North America. This technique involves introduction of heat into the reservoir, which in turn enhances the oil recovery from the reservoir. The most common techniques used are steam flooding and cyclic steam injection. The introduction of steam results in reduced oil viscosity and/or vaporization of part of the crude oil, thus, decreasing the mobility ratio. In addition to this, the increased heat also increases the permeability of the oil due to reduction in the surface tension. In the year 2011, solar power was used for thermal injection in Oman and California, USA. This method involves the same technique as mentioned above except solar power was used to produce steam.

3.1.3 Chemical Injection

Chemical injection is another EOR production method that uses injection of polymers to improve the effectiveness of water flooding. This technique is used to increase the viscosity of water during the water flooding process. The water/oil mobility ratio increases during the chemical injection process, which in turn improves the vertical and areal sweep efficiency of the reservoir. The surface tension between the oil and water can be further reduced by introducing surfactants with the polymers. This improves the microscopic sweep efficiency process by reduction of residual oil saturation. In order to improve the stability of the surfactant polymer formation, co-surfactants, activity boosters and co-solvents are also usually added.

3.2 Scope of Polymers in EOR

The polymer injection or polymer flooding technique was not commonly used compared to the gas and thermal injection techniques until the past decade. This technique involves injection of an alkaline solution, surfactant along with co-surfactants and solvents, and polymers. This method is usually called as alkaline surfactant polymer (ASP) technique [3]. The polymer flooding has proven to be very effective in improving the mobility ratio and, hence, resulting in effective recovery of unswept oil. Many case studies for polymer flooding in the reservoirs can be studied in the literature [4]. Polymer flooding technique has also proven to be cost effective compared to the gas and thermal injection. The efficiency of the polymer injection process falls in the range of 0.75 to 1.75 lb of polymer used for per barrel of oil produced [4].

3.2.1 Working Principle

Hydrogel polymers or associative polymer hydrogels play a very significant role in the EOR process. Firstly, this procedure involves injection of polymer (polymer flooding) for years until 33-50 percent of the pore volume of the reservoir is achieved. Secondly, the oil and the polymer slug is driven towards the production well by flooding the reservoir with water. As the water is injected, it naturally seeks the region of highest permeability. This, in turn, results in the reduced pressure inside the production well. The mobility ratio of the reservoir is one of the primary parameters to be analyzed. This parameter can be used for various screening processes. The *mobility ratio (M)* is a parameter that helps in the understanding of the effects of relative permeability (k) and viscosity (μ) of oil (o) and water (w) in the reservoir on a fractional flow (f). This is represented by Dake's law [5] as

$$f_o = \frac{1}{1+M} = \frac{1}{1 + \frac{\mu_o k_w}{\mu_w k_o}}$$

From this, it can be seen that if the viscosity of the injected water is lower than that of the oil, then the sweep efficiency will be reduced. Polymer injection solves this issue by increasing the viscosity of water. The increased viscosity of water will reduce relative permeability of injected water in the reservoir. This will increase the fractional flow, hence, resulting in increased oil recovery. The displacement of oil in the reservoir by injected water is efficient when the mobility ratio is one or slightly less than one. On the other hand, if the mobility ratio is greater than one, the injected water is likely to leave behind unswept oil, thus, reducing the overall efficiency of the process [6]. When the mobility ratio in the reservoir crosses one, the polymer is injected to increase the viscosity of the water. This will result in reducing the permeability of water through rock pores resulting in increased efficiency [7]. This process is very efficient in reservoirs that have high mobile oil saturation.

3.2.2 Commonly Used Polymers for EOR

The most common characteristics that are expected from a polymer used for EOR are cost effectiveness, high injectivity, low mechanical degradation, effectiveness when mixed with the reservoir brine solutions, resistance to temperature over a long period of time, resistance to microbial degradation, low retention in porous rocks, sensitivity to oxygen (O_2), hydrogen sulfide (H_2S) and other reservoir and oilfield chemicals [8]. The two major polymers used for EOR production are partially hydrolyzed polyacrylamide (HPAM) and xanthan gum. The chain extension and physical entanglement of the solvated chains in these polymers are used for the viscosity enhancement mechanism.

HPAM results in higher viscosity because of its ability to cause chain expansion as a result of repulsion in the ionic groups. The molecular weight of HPAM is directly proportional to the viscosity of the solution. If the molecular weight of the HPAM increases, it results in an increase in the viscosity of the solution [9]. On the other hand, the higher molecular weight HPAM tends to degrade irreversibly during higher shear rates as the backbone of the polymer breaks down. Due to this issue, the higher molecular weight HPAM tends to degrade in pumps and near the bore well regions [10,11]. The EOR process, however, requires high molecular weight HPAM to achieve high viscosity at low concentration of polymers. In addition to degradation, the HPAM also results in rapid reduction of viscosity of the solution in saline environment [12]. In the saline environment, the HPAM's polymer chains contract due to the shielding of ionic groups in HPAM. There is also a possibility of polymer precipitation in the presence of divalent ions [13], thus, resulting in loss of performance.

Xanthan gum is a rigid polysaccharide used for EOR production. Unlike the aforementioned HPAM, xanthan gum is not degraded due to shear and is insensitive to saline environment and divalent ions [14]. On the other hand, the cost of xanthan gum is high as compared to HPAM. Xanthan gum is also susceptible to biodegradation inside the reservoirs. In addition, there is also a potential of microbial berries from the manufacturing process. Due to its properties, this polysaccharide is used exclusively in drilling fluids [15]. However, the solutions of both xanthan gum and HPAM show reduced viscosity at higher temperatures.

3.3 Associating Polymers

The very first associating polymers were developed to study the behavior of proteins [16]. Later, associating polymers were developed for polymer coatings applications [17]. The development of hydrophobically associating polymers was not undertaken until two decades later. The first hydrophobically associating polymers were prepared by partial esterification of maleic anhydride/styrene copolymers with non-ionic ethoxylated alcohol surfactants [18]. The use of these polymers was limited though, due to the factors like thermal degradation and alkaline hydrolysis. The other issues faced by long chain polyethers were the decomposition in the presence of O_2 and alkaline environment. To address this issue, further works led to the development of maleic anhydride/styrene/vinyl benzyl polyglycol ether copolymers [19,20]. Later to meet the demands of enhanced oil recovery, in the 1980's, acrylamide based hydrophobically associating polymers were developed extensively, leading to the generation of advanced category of EOR polymers. In a recent study, development of associating self-assembly polymer (SAP) systems were derived from HPAM and xanthan gum (XG) using an EOR surfactant and β-cyclodextrin [21]. The polymers, named as SAP-HPAM and SAP-XG, were analyzed for EOR operation. Figure 3.1 demonstrates differential pressure along

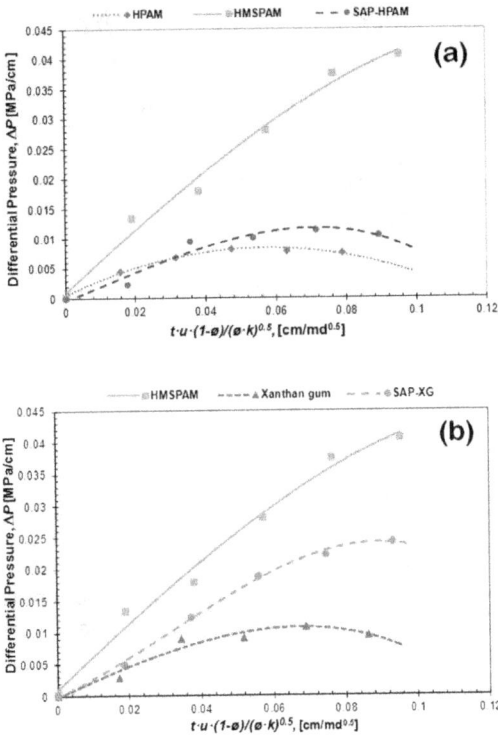

Figure 3.1 Differential pressure along the sandpack during heavy oil displacement by polymer flooding for (a) HPAM, SAP-HPAM, and HMSPAM and (b) xanthan gum, SAP-XG, and HMSPAM. Reproduced from Reference 21 with permission from American Chemical Society.

the sandpack during heavy oil displacement by polymer flooding for these polymers in comparison with HPAM and commercial hydrophobically modified polyacrylamide (HMSPAM). It was observed that SAP-HPAM exhibited effective propagation of these polymers through the sandpack system, which was better in performance than SAP-XG. In another study, novel hydrophobically associating acrylamide polymer incorporating cucurbit[7]uril ester (ACE) and hexadecyl dimethyl allyl ammonium chloride ($C_{16}DMAAC$) was reported for EOR purposes, as shown in Figure 3.2.

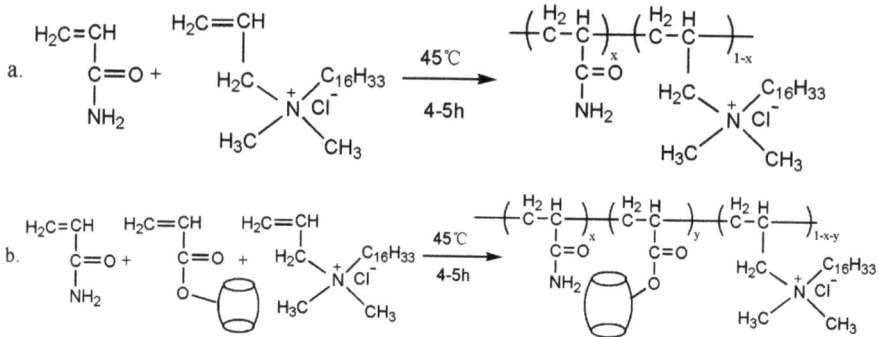

Figure 3.2 Synthesis of (a) AM/C_{16}DMAAC, and (b) AM/ACE/C_{16}DMAAC associating copolymers. Reproduced from Reference 22 with permission from American Chemical Society.

3.3.1 Synthesis

The preparation of associating polymers can be broadly categorized into two general techniques. In the first method, the associating polymer can be obtained by co-polymerization of water-soluble and hydrophobic monomers. In the second method, the polymers are modified after polymerization process to introduce the hydrophobic or hydrophilic groups. The first method where the associating polymers are obtained by co-polymerization using micellar polymerization in free-radical mode is the most commonly used method. In this technique, sodium dodecyl sulphate (SDS) is used as a surfactant in the aqueous solution to solubilize hydrophobic monomer. This results in micelles of SDS containing hydrophobic monomer. Acrylamide is commonly used as the water-soluble monomer because of its ability to produce water-soluble copolymers that are effective at polymer concentrations below 1% mass [23-

25]. Associating acrylamide polymers are, thus, synthesized by the copolymer- ization of acrylamide with a hydrophobic monomer and other monomers like acrylic acid. Alternatively, in order to incorporate carboxylic acid groups in the associating polyacrylamide base, hydrolysis can also be achieved after the polymerization process. However, to achieve better compositional heteroge- neity, copolymerization technique is predominantly used. The rheological properties of the resulting polymers depend largely on the compositional het- erogeneity. Figure 3.3 shows an example of polymer nanoparticles generated

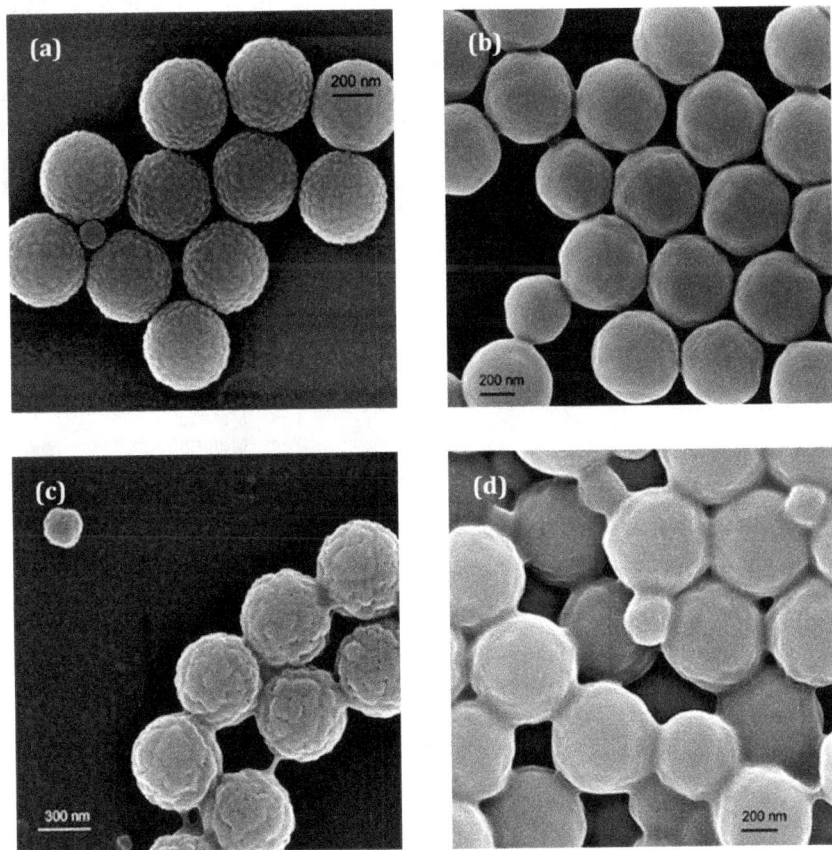

Figure 3.3 Microscope images of the particles formed by emulsion copolymerization of styrene and N-isopropylacrylamide. Reproduced from Reference 26 with permission from Wiley.

by copolymerization of styrene with poly(n-isopropylacrylamide) [26]. Figure 3.4 also shows the kinetics of swelling and de-swelling of the various copolymer materials generated [26]. Similarly, Figure 3.5 also shows the schematics of growth of poly(N-isopropylacrylamide) from the surface of polystyrene particles using controlled polymerization [27]. The different monomers used for the free radical polymerization of acrylamide based associating polymers are also listed in the Table 3.1.

As mentioned above, associating polymers are also prepared by modification of the polymer after polymerization process. The major drawback in this process is the difficulty in mixing and achieving reaction homogeneity during reactions that involve viscous polymer solutions. For instance, the associating polymers prepared by modification of styrene/maleic anhydride copolymers have been reported [43]. The water solubility is achieved by fully sulfonating the copolymer. Then, the hydrophobic group can be incorporated by a reaction with alkyl amine.

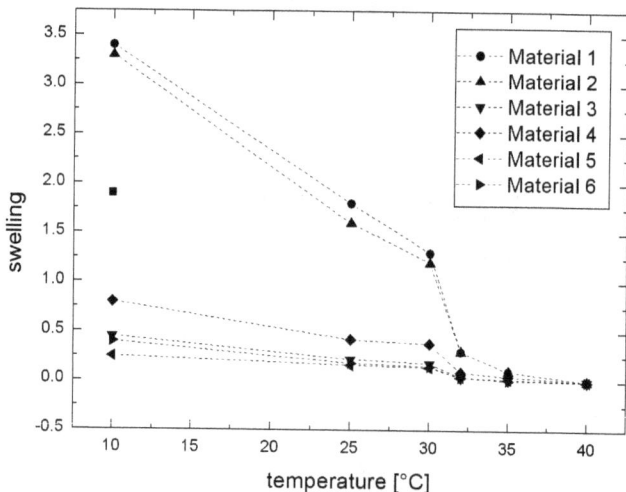

Figure 3.4 Swelling de-swelling kinetics of emulsion copolymerization of styrene and N-isopropylacrylamide. Reproduced from Reference 26 with permission from Wiley.

The efficiency of the process of incorporating hydrophobic monomer into the polymer has been reported in a large number of studies related to medical drug delivery research in the past two decades [44-48]. These research applications can also be extended for application in the EOR process. From these studies, it can be seen that the reaction rate for the acrylamide polymerization in the aqueous solution is very similar to that of the muscular solution. A positive increase in the rate of incorporation of the hydrophobic group in the polymer can be seen due to the solubilization of the hydrophobic group monomer

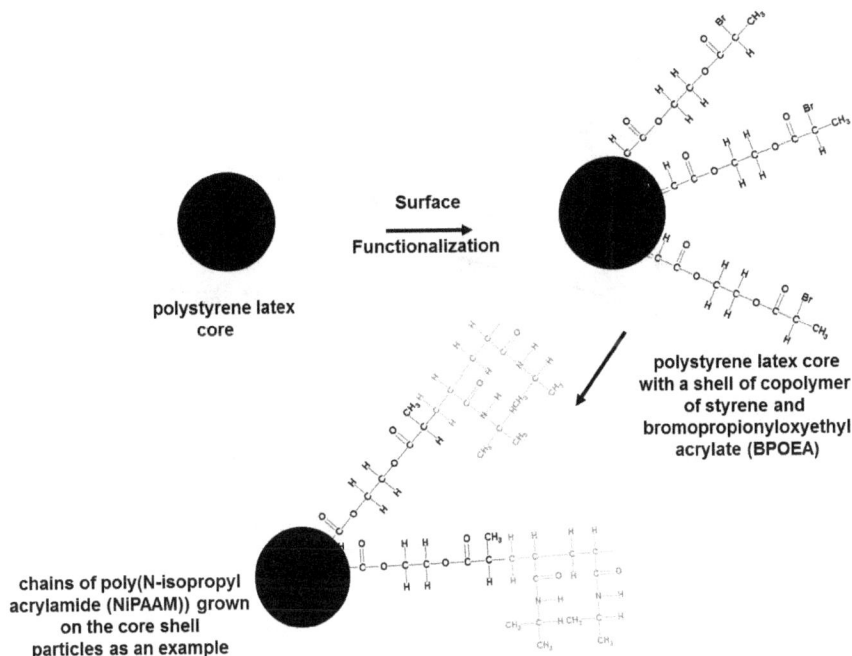

Figure 3.5 Representation of BPOEA functionalization of particles and subsequent growth of poly(N-isopropylacrylamide) brushes by controlled polymerization. Reproduced from Reference 27 with permission from Elsevier.

within the micelles. The rate of incorporation is directly proportional to the number of hydrophobic monomers per micelle. This also means that the hydrophobic monomer can be depleted before complete polymerization, if the value of the ratio of hydrophobic monomer to micelle is greater than one. Figure 3.6 demonstrates an example of the synthesis of copolymer of hydrophobic l-lactide with hydrophilic ethylene glycol (PLA/PEG; (b) PLA/PEG/PLA; (c) a multiblock copolymer of L-lactide and ethylene oxide) [44].

Table 3.1 Monomers for free radical polymerization

Monomer	Reference(s)
Vinyl sulfonate	[28]
2-Acrylamido-2-methyl-1-propanesulfonic acid (AMPS)	[29]
N-vinyl pyrolidinone	[30-32]
N-alkyl acrylates and methacrylates	[33,34]
Acrylamides	[23,26,27,35,36]
Styrene	[26,27,37-39]
Anionic monomers	[40,41]
Cationic monomers	[42]

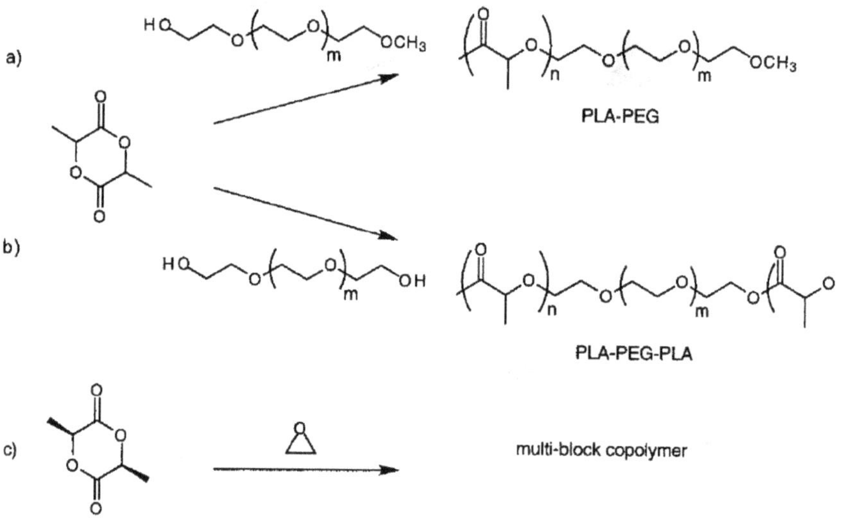

Figure 3.6 Synthesis of copolymers: (a) PLA/PEG; (b) PLA/PEG/PLA; (c) a multiblock copolymer of L-lactide and ethylene oxide. Reproduced from Reference 44 with permission from American Chemical Society.

The final stage of synthesis is the polymer purification. The polymers have to be purified after the micellar polymerization process. The most basic method used is the dilution of the polymer solution and using it without further modification [49-51]. Another common technique used is the precipitation of polymer in a non-solvent like methanol. Subsequently, the polymer is filtered and dried under vacuum [52,53]. In some cases, the polymer is dialyzed with distilled water for up to three days. Afterwards, the low molecular weight impurities are removed by dilution.

3.4 Characterization of Polymer Interactions

3.4.1 Hydrophobic Interactions

The most common technique used for the characterization of hydrophobic interactions is by probing the hydrophobic groups with fluorescent probes. The most successful probe reported in a large number of research studies is pyrene [54-61]. The advantages of pyrene are low solubility in water and a characteristic fluorescence spectrum. The use of pyrene results in an increase in the fluorescence and a change in the emission spectrum, as it is solubilized by the micelles or the associating polymers [60,61]. An excimer is formed when there are two or more pyrene molecules present in a hydrophobic region. The excimer has a different fluorescence emission spectrum compared to the pyrene's emission spectrum. Based on this difference in the spectra, it has been calculated that a minimum of ten hydrophones are required to form a hydrophobic region in an acrylamide/dodecyl methacrylate copolymer. From these measurements, it is also seen that for the aggregation to occur, the concentration should be lower than the critical aggregation concentration. Above the critical aggregation concentration, no more aggregation can be seen. In addition to this, it is observed that the polymer behavior in a solution is dominated by the intermolecular hydrophobic associations.

The characterization of hydrophobic interactions leads to classification of the concentration regions for the associating polymers into three parts. Firstly, at low concentration, no more intermolecular association can be seen. Secondly, at intermediate concentration, sporadic association can be seen, but there is no evidence of networking. Finally, at the high concentration region, the viscosity is increased due to extensive intramolecular associations and networking.

3.4.2 Stability under Reservoir Conditions

The hydrolytic stability of the polymers is another important characteristic to be analyzed. This characterization can be used to analyze the stability of the polymers at reservoir conditions. The hydrolytic stability of some associating polymers were examined by Bock *et al.* [62] and Hickner *et al.* [63]. Polymer solutions at 2000 ppm in brine solution were heated at 93 °C for 100 days. On incorporating 2-acrylamido-2-methyl-1-propanesulfonic acid (AMPS) in the polymer, it was observed that the degree of hydrolysis was much lower. For example, a polymer containing polyacrylamide (PAM) group, 78% of amide groups were hydrolyzed as opposed to 36% of amide groups in polymers with AMPS. N-vinyl-2-pyrrolidinone (NVP) exhibited even lower degree of hydrolysis as only 5% of amide groups were hydrolyzed for polymers with NVP under the same conditions. The rate of hydrolysis could further be reduced by the associating groups in the polymer compared to an acrylamide/NVP copolymer.

3.5 Rheology of Associating Polymers

Several factors contribute to the rheological properties of the associating polymers. The major factors are the average molecular weight, hydrophobe distribution, degree of hydrolysis, type of hydrophobe used and degree of incorporation of the hydrophobe [64,65]. An increase in the hydrophobe content reduces the solubility of the associating polymers. In addition, an increase in the molecular weight of the associating polymer increases the chain length, due to which the chain length of the hydrophobe also increases. The increase in the chain length of the hydrophobe reduces the amount of hydrophobe required to make the polymer insoluble. In addition to this, the solubility of the associating polymer is further reduced when fluorocarbon containing hydrophobic groups are used. To overcome this issue, ionic character is introduced in the backbone of the polymer chain, which, in turn, increases the solubility of the polymer. Along with solubility, the rheological properties of the associating polymers can also be affected due to the introduction of ionic groups.

The flow behavior of the associating polymers is also affected due to the introduction of hydrophobic groups [66]. The change in the flow characteristics depends mainly on the intermolecular and intramolecular associations. In this work by English *et al.* [66], rheological properties of hydrophobically modified alkali-swellable emulsion (HASE) polymer were studied in alkaline media containing nonionic ethoxylated surfactants (NP6 and NP10). As shown in Figure 3.7, NP10 due to its miscibility formed mixed junction domains with the polymer hydrophobes, whereas in the case of NP6, the system segregated into polymer-rich and surfactant-rich phases. The work by Schulz *et al.* [67] established the relation between viscosity and polymer concentration. On the plot between reduced viscosity against associating and non-associating polymer concentration, it was observed that the associating polymer enhanced the viscosity after a critical concentration point. If the hydrophobic groups were introduced below this critical concentration point, there was a slight decrease in the viscosity due to intramolecular associations.

It is clear that the effect of hydrophobic group depends directly on the polymer concentration. The rheological properties of the associating polymers are examined under two different conditions, (i) the dilute state, where the concentration of the polymer is below the critical concentration, and (ii) the semi-dilute state, where the concentration of the polymer is above the critical concentration.

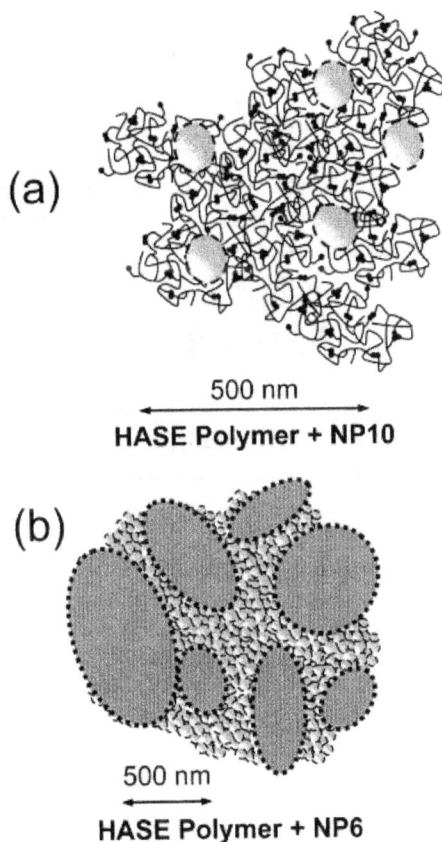

Figure 3.7 Schematic illustrations of (a) HASE polymer with NP10 and (b) HASE polymer with NP6. Reproduced from Reference 66 with permission from American Chemical Society.

3.5.1 Viscosity in Dilute State

The intrinsic viscosity k and Huggin's constant η parameters can be used to study the rheological properties of the associating polymers. These parameters are used to determine the degree of hydrophobic interactions and the molecular weight of the polymer [68].

For the dilute state, according to the Flory-Huggins equation, the reduced viscosity is a linear function of the polymer concentration [69]

$$\frac{\eta - \eta_0}{\eta_0 c} = k[\eta]^2 c$$

where c is the polymer concentration and η_0 is the solvent viscosity. The viscosity measurements of polymer solutions with low polymer concentration can be used to measure the values of η and k. At a constant molecular weight, the intrinsic viscosity decreases and the Huggins constant η increases as the hydrophobic content increases. The intrinsic viscosity is related to the average molecular weight of the polymer through the equation [70] as

$$[\eta] = K[M_w]^a$$

The increase in the Huggins constant and the decrease in the intrinsic viscosity due to hydrophobic content is mainly due to the polymer chain contraction because of intramolecular association. Hence, the Huggins constant can be used a measurement for hydrophobic interactions in the associating polymers. A value of η greater than 0.8 indicates that there is intramolecular associations in the polymer solution.

Bock *et al.* [71] and Qian *et al.* [72] studied the effect of introducing ionic character to the associating polymers. In two samples of associating polymers with same molecular weight and hydrophobe content, one sample was introduced with ionic character on its backbone. Due to the introduction of the ionic character, an electrostatic repulsion between the negative charges of the carboxylate group was observed. This resulted in the increase of the hydrodynamic volume in the polymer chain. With the increase in the ionic content on the backbone, the intrinsic viscosity of the polymer solution increased and the Huggins constant decreased. The polymer solvent interactions were improved due to the electrostatic repulsion. This resulted in a lower Huggins number, hence, it was clear that the ionic character and the hydrophobic group had opposite effects on the intrinsic viscosity and the Huggins constant.

3.5.2 Viscosity in Semi-Dilute State

As the polymer concentration is above the critical concentration, the polymer-solvent mixture reaches a semi-dilute state. The rheological characteristics of the polymer solution in this state are different than that of the dilute state, as mentioned above. In this state, the increase in the polymer concentration results in intermolecular association and, hence, enhances the viscosity of the polymer solution. For example, if the hydrophobe content is increased from 0.8 mol% to 1 mol%, the viscosity of the solution increases significantly. In addition, at a given concentration of the hydrophobe, increasing the molecular weight of the associating polymer results in increased viscosity of the solution [73,71].

Based on the studies by Bock *et al.* [71] and Grattoni *et al.* [74], the increase in the hydrophobe content resulted in the increase in the viscosity of the solution due to the intermolecular interactions. In addition, it was also observed that the hydrophobes containing phenyl group enhanced the viscosity significantly, especially at high hydrophobe concentrations.

The apparent viscosity of the solution stays constant at concentrations above 3000 ppm. Further, the viscosity increases with the increase in shear, which is called shear thickening. This increase in viscosity stops at a particular point and starts to reduce after a certain period of shear exposure, which is called shear thinning. This phenomenon occurs due to shifting of the intermolecular and intramolecular association with the shear rate. In addition, it can also been seen that the interchain associations are favored by temperature and, hence, the viscosity increases with increase in temperature. However, at very high temperatures, the viscosity decreases due to the degradation of the associating polymer. Figure 3.8 shows the shear thickening behavior of hydrophobically associating acrylamide polymers as a function of hydrophobe level [25].

3.5.3 Chemical Interaction

The analysis of chemical interactions with salts is very critical for the associating polymers because of the conditions in the reservoir. In the presence of a saline environment, the viscosity of associating polymer can increase, this effect is more significant at high polymer concentrations. This phenomenon is similar to that of the micelle formation in the presence of ionic surfactants. In saline environment, the critical micelle concentration (CMC) reduces due to enhanced aggregation. Thus, the enhanced association describes the effect on viscosity of the associating polymer in a saline environment [22,75]. The sensitivity of HPAM to saline environment is considered as a disadvantage at high hydrophobe concentrations.

As discussed in the previous sections, the surfactant concentration also affects the viscosity of the associating polymer systems. Conditions where the surfactant concentration is closer to the CMC, a peak in the viscosity of the associating polymer systems can be observed [76]. Below CMC, the polymer cross-linking efficiency increases due to the presence of more than one type of hydrophobe in the micelles. The cross-linking between polymers, in turn, increases the viscosity of the system. On the other hand, when the surfactant concentration is above the CMC, the number of micelles increases, affecting the cross-linking, thus, resulting in reduced viscosity [77].

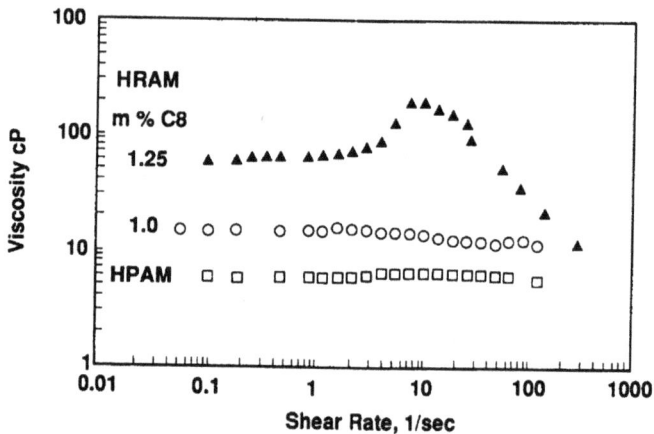

Figure 3.8 Shear thickening behavior of hydrophobically associating acrylamide polymers as a function of hydrophobe level. Reproduced from Reference 25 with permission from American Chemical Society.

3.5.4 Flow of Associating Polymers in Porous Rocks

The study of the flow of the associating polymers in porous media is another factor to be considered for the use of polymers in EOR. The resistance of the associating polymers should be low for the EOR applications. The resistance factor is calculated by Bock *et al.* [62], which is described by the following equation:

$$R = \frac{\Delta P_{polymer}}{\Delta P_{brine}}$$

where ΔP refers to the pressure drop across the core for the respective solution and R represents the resistance factor. The resistance factor can increase if the hydrophobe levels are too high. This is because of the polymer adsorption through the pores, thus, leading to plugging. In order to deal with this issue, modification is performed on the polymer to reduce the level of adsorption. The resistance factor for HPAM peaks at injection velocities at 10 ft/day and drops at higher velocities due to degradation caused by high shear rates [62].

3.6 Conclusion

The primary and secondary oil recovery methods result in 20 to 40% oil recovery from the reservoirs, while adapting to tertiary recovery methods (or EOR) will result in 40 to 60% recovery of reservoir's oil. Though CO_2 injection technique has been used widely across North America and Russia, the applications of associating polymers have been overlooked for the past decade. This review paper summarizes the synthesis, characterization and rheology of associating polymers and their scope in EOR. In addition, this report adds to the previous review on the subject [78] by reviewing the recent developments on the application of associative polymers in EOR.

Micellar polymerization is required to produce hydrophobic associating polymers that are soluble in water. The synthesized associating polymers are then characterized using fluorescence techniques by using pyrene as fluorescent probes. The rheological properties of the polymers depend on various factors like molecular weight, hydrophobe distribution, temperature, etc. An optimum has to be achieved among the molecular weight, ionic characterization, polymer concentration and hydrophobe content.

Polyacrylamide based polymers are used commonly for the synthesis of associating polymers that are used for EOR. Even though other techniques like gas injection and thermal injection are cheaper than the polymer injection technique, the sweep efficiency of the polymer injection technique is significantly higher than other techniques.

This review paper serves as a precursor to the research in the field of synthesis and characterization of associative polymers for achieving enhanced oil recovery.

References

1. Walsh, M. P., and Lake, L. W. (2003) *A Generalized Approach to Primary Hydrocarbon Recovery*, Volume 4, Elsevier Science, The Netherlands.
2. Gozalpour, F., Ren, S., and Tohidi, B. (2005) CO_2 EOR and storage in oil reservoir. *Oil and Gas Science and Technology*, **60**(3), 537-546.
3. Tabary, R., and Bazin, B. (2007) Advances in Chemical Flooding. In: IFP-OAPEC Joint Seminar Improved Oil recovery (IOR) Techniques and Their Role in Boosting the Recovery Factor, France.
4. Pope, G. A. (2007) Overview of Chemical EOR. Online: http://www.uwyo.edu/eori/_files/misc_download/overview%20of%20chemical%20eor.pdf (assessed 31.01.2017).
5. Dake, L. P. (1983) *Fundamentals of Reservoir Engineering*, Volume 8, 1st Edition, Elsevier Science Publishers, The Netherlands.
6. Sastry, N., Dave, P., and Valand, M. (1999) Dilute solution behaviour of polyacrylamides in aqueous media. *European Polymer Journal*, **35**(3), 517-525.
7. Littman, W. (1998) *Polymer Flooding*, Developments in Petroleum Science, Elsevier Science Publishers, The Netherlands.
8. Abidin, A., Puspasari, T., and Nugroho, W. (2012) Polymers for enhanced oil recovery   technology. *Procedia Chemistry*, **4**, 11-16.
9. Chauveteau, G., and Sorbie, K. (1991) Mobility control by polymers. In: *Basic Concepts in Oil Recovery Processes*, Elsevier Science Publishers, USA.
10. Liu, D.-X., Zhao, X.-T., and Qiu, G.-M. (2008) Review on the degradation of polyacrylamide used for flooding. *Polymer Materials Science and Engineering*, **9**, 8.
11. Li, J., Yang, X., Yu, X., Xu, L., Kang, W., Yan, W., Gao, H., Liu, Z., and Guo, Y. (2009) Rare earth oxide-doped titania nanocomposites with enhanced photocatalytic activity towards the degradation of partially hydrolysis polyacrylamide. *Applied Surface Science*, **255**(6), 3731-3738.
12. Li, K., Baohui, W., Xuejia, Z., Qin, R., Wei, J., and Xue, Y. (2008) Adsorption of polyacrylamide on soil. *Environmental Pollution and Control*, **1**, 9.
13. Zaitoun, A., and Potie, B. (1983) Limiting Conditions for the Use of Hydrolyzed Polyacrylamides in Brines Containing Divalent Ions. *SPE Oilfield and Geothermal Chemistry Symposium*, Society of Petroleum Engineers, USA, doi:10.2118/11785-MS.
14. Lopez, M., Vargas-Garcia, M., Suarez-Estrella, F., and Moreno, J. (2004) Properties of xanthan obtained from agricultural wastes acid hydrolysates. *Journal of Food Engineering*, **63**(1), 111-115.
15. Garcia-Ochoa, F., Santos, V., Casas, J., and Gomez, E. (2000) Xanthan gum: production,   recovery, and properties. *Biotechnology Advances*, **18**(7), 549-579.
16. Strauss, U. P., and Jackson, E. G. (1951) Polysoaps. I. viscosity and solubilization studies on an n-dodecyl bromide addition compound of poly-2-vinylpyridine. *Journal of Polymer Science*, **6**(5), 649-659.
17. Dubin, P., and Strauss, U. P. (1967) Hydrophobic hypercoiling in copolymers of maleic acid and alkyl vinyl ethers. *The Journal of Physical Chemistry*, **71**(8), 2757-2759.

18. Beihoffer, T. W., Lundberg, D. J., and Glass, J. E. (1989) Influence of polarity in water- soluble polymer synthesis. In: *Polymers in Aqueous Media*, American Chemical Society, USA, pp. 151-164.

19. Corson, F. P., and Evani, S. (1977) Aqueous Thickening Agents Derived from Vinyl Benzyl Ether Polymers, US Patent 4,008,202.

20. Corson, F. P., and Evani, S. (1977) Vinyl Benzyl Ethers and Nonionic Water Soluble Thickening Agents Prepared Therefrom, US Patent 4,029,873.

21. Wei, B., Romero-Zeron, L., and Rodrigue, D. (2014) Evaluation of two new self-assembly polymeric systems for enhanced heavy oil recovery. *Industrial and Engineering Chemistry Research*, **53** (43), pp 16600-16611.

22. Zou, C., Gu, T., Xiao, P., Ge, T., Wang, M., and Wang, K. (2014) Experimental study of cucurbit[7]uril derivatives modified acrylamide polymer for enhanced oil recovery. *Industrial and Engineering Chemistry Research*, **53**(18), 7570-7578.

23. Emery, O., Lalot, T., Brigodiot, M., and Marechal, E. (1997) Free-radical polymerization of acrylamide by horseradish peroxidase-mediated initiation. *Journal of Polymer Science, Part A: Polymer Chemistry*, **35**(15), 3331-3333.

24. Odian, G. G., and Odian, G. (2004) *Principles of Polymerization*, Wiley-Interscience, USA.

25. Bock, J., Siano, D. B., Valint, P. L., and Pace, S. J. (1989) Structure and properties of hydrophobically associating polymers. In: *Polymers in Aqueous Media: Performance through Association*, Glass, J. E. (ed.), Advances in Chemistry Series 223; American Chemical Society, USA, Chapter 22.

26. Mittal, V., Matsko, N. B., Butte, A., and Morbidelli, M. (2008) Swelling deswelling behavior of PS-PNIPAAM copolymer particles and PNIPAAM brushes grafted from polystyrene particles & monoliths. *Macromolecular Materials and Engineering*, **293**, 491-502.

27. Mittal, V., Matsko, N. B., Butte, A., and Morbidelli, M. (2008) Functionalized polystyrene latex particles as substrates for ATRP: Surface and colloidal characterization. *Polymer*, **48**, 2806-2817.

28. Hawker, C. J. (1997) "Living" free radical polymerization: A unique technique for the preparation of controlled macromolecular architectures. *Accounts of Chemical Research*, **30**(9), 373-382.

29. Greesh, N., Hartmann, P. C., Cloete, V., and Sanderson, R. D. (2008) Impact of the clay organic modifier on the morphology of polymer–clay nanocomposites prepared by in situ free-radical polymerization in emulsion. *Journal of Polymer Science, Part A: Polymer Chemistry*, **46**(11), 3619-3628.

30. Hawker, C. J., Elce, E., Dao, J., Volksen, W., Russell, T. P., and Barclay, G. G. (1996) Well-defined random copolymers by a "living" free-radical polymerization process. *Macromolecules*, **29**(7), 2686-2688.

31. Wan, D., Satoh, K., Kamigaito, M., and Okamoto, Y. (2005) Xanthate-mediated radical polymerization of n-vinylpyrrolidone in fluoroalcohols for simultaneous control of molecular weight and tacticity. *Macromolecules*, **38**(25), 10397-10405.

32. Ray, B., Kotani, M., and Yamago, S. (2006) Highly controlled synthesis of poly (n-vinylpyrrolidone) and its block copolymers by organostibine-mediated living radical polymerization. *Macromolecules*, **39**(16), 5259-5265.

33. Haddleton, D. M., Kukulj, D., Duncalf, D. J., Heming, A. M., and Shooter, A. J. (1998) Low-temperature living "radical" polymerization (atom transfer polymerization) of methyl methacrylate mediated by copper (i) n-alkyl-2-pyridylmethanimine complexes. *Macromolecules*, **31**(16), 5201-5205.

34. Strehmel, V., Laschewsky, A., Wetzel, H., and Gornitz, E. (2006) Free radical polymerization of n-butyl methacrylate in ionic liquids. *Macromolecules*, **39**(3), 923-930.

35. Ganachaud, F., Monteiro, M. J., Gilbert, R. G., Dourges, M.-A., Thang, S. H., and Rizzardo, E. (2000) Molecular weight characterization of poly (n-isopropylacrylamide) prepared by living free-radical polymerization. *Macromolecules*, **33**(18), 6738-6745.

36. Xiao, D., and Wirth, "M. J. (2002) Kinetics of surface-initiated atom transfer radical poly- merization of acrylamide on silica. *Macromolecules*, **35**(8), 2919-2925.

37. Buback, M., Kowollik, C., Kurz, C., and Wahl, A. (2000) Termination kinetics of styrene free-radical polymerization studied by time-resolved pulsed laser experiments. *Macromolecular Chemistry and Physics*, **201**(4), 464-469.

38. Zhang, H., Hong, K., and Mays, J. W. (2002) Synthesis of block copolymers of styrene and methyl methacrylate by conventional free radical polymerization in room temperature ionic liquids. *Macromolecules*, **35**(15), 5738-5741.

39. Jakubowski, W., Min, K., and Matyjaszewski, K. (2006) Activators regenerated by electron transfer for atom transfer radical polymerization of styrene. *Macromolecules*, **39**(1), 39-45.

40. Nakano, T., Takewaki, K., Yade, T., and Okamoto, Y. (2001) Dibenzofulvene, a 1, 1-diphenylethylene analogue, gives a π-stacked polymer by anionic, free-radical, and cationic catalysts. *Journal of the American Chemical Society*, **123**(37), 9182-9183.

41. Sumerlin, B. S., Donovan, M. S., Mitsukami, Y., Lowe, A. B., and McCormick, C. L. (2001) Water-soluble polymers. 84. Controlled polymerization in aqueous media of anionic acrylamido monomers via raft. *Macromolecules*, **34**(19), 6561-6564.

42. Peiffer, D. (1990) Hydrophobically associating polymers and their interactions with rod- like micelles. *Polymer*, **31**(12), 2353-2360.

43. Cook, R. L., King Jr, H., and Peiffer, D. G. (1992) Pressure-induced crossover from good to poor solvent behavior for polyethylene oxide in water. *Physical Review Letters*, **69**, 3072.

44. Uhrich, K. E., Cannizzaro, S. M., Langer, R. S., and Shakesheff, K. M. (1999) Polymeric systems for controlled drug release. *Chemical Reviews*, **99**(11), 3181- 3198.

45. Svec, F., Peters, E. C., Sykora, D., and Frechet, J. M. (2000) Design of the monolithic polymers used in capillary electrochromatography columns. *Journal of Chromatography A*, **887**, 3-29.

46. Rudzinski, W., Dave, A., Vaishnav, U., Kumbar, S., Kulkarni, A., and Aminabhavi, T. (2002) Hydrogels as controlled release devices in agriculture. *Designed Monomers and Polymers*, **5**(1), 39-65.

47. Kickelbick, G. (2003) Concepts for the incorporation of inorganic building blocks into organic polymers on a nanoscale. *Progress in Polymer Science*, **28**(1), 83-114.

48. Fevola, M. J., Hester, R. D., and McCormick, C. L. (2003) Molecular weight control of

polyacrylamide with sodium formate as a chain-transfer agent: Characterization via size exclusion chromatography/multi-angle laser light scattering and determination of chain-transfer constant. *Journal of Polymer Science Part A: Polymer Chemistry*, **41**(4), 560-568.
49. Evani, S., and van Phung, K. (1988) Amphiphilic Monomer and Hydrophobe Associative Composition Containing a Polymer of a Water-soluble Monomer and Said Amphiphilic Monomer, US Patent 4,728,696.
50. Evani, S., and van Phung, K. (1990) Hydrophobe Associative Composition Containing a Polymer of a Water-soluble Monomer and an Amphiphilic Monomer, US Patent 4,921,902.
51. Geise, G. M., Lee, H.-S., Miller, D. J., Freeman, B. D., McGrath, J. E., and Paul, D. R. (2010) Water purification by membranes: the role of polymer science. *Journal of Polymer Science, Part B: Polymer Physics*, **48**(15), 1685-1718.
52. Edwards, J. H., Feast, W. J., and Bott, D. C. (1984) New routes to conjugated polymers: 1. a two-step route to polyacetylene. *Polymer*, **25**(3), 395-398.
53. Li, H., Liu, Y., Zhang, Z., Liao, H., Nie, L., and Yao, S. (2005) Separation and purification of chlorogenic acid by molecularly imprinted polymer monolithic stationary phase. *Journal of Chromatography A*, **1098**(1), 66-74.
54. Dualeh, A. J., and Steiner, C. A. (1990) Hydrophobic microphase formation in surfactant solutions containing an amphiphilic graft copolymer. *Macromolecules*, **23**(1), 251-255.
55. Koussathana, M., Lianos, P., and Staikos, G. (1997) Investigation of hydrophobic inter- actions in dilute aqueous solutions of hydrogen-bonding interpolymer complexes by steady-state and time-resolved fluorescence measurements. *Macromolecules*, **30**(25), 7798-7802.
56. Jorand, F., Boue-Bigne, F., Block, J., and Urbain, V. (1998) Hydrophobic/hydrophilic properties of activated sludge exopolymeric substances. *Water Science & Technology*, **37**, 307-315.
57. Yessine M.-A., and Leroux J.-C. (2004) Membrane-destabilizing polyanions: interac- tion with lipid bilayers and endosomal escape of biomacromolecules. *Advanced Drug Delivery Reviews*, **56**(7), 999-1021.
58. Liu, S. Q., Tong, Y., and Yang, Y.-Y. (2005) Incorporation and in-vitro release of doxorubicin in thermally sensitive micelles made from poly isopropylacrylamide n,n-dimethylacrylamide)-poly (d,l-lactide-glycolide) with varying compositions. *Biomaterials*, **26**(24), 5064-5074.
59. Lo, C.-L., Lin, K.-M., and Hsiue, G.-H. (2005) Preparation and characterization of intelligent core-shell nanoparticles based on poly (d,l-lactide)-g-poly (n-isopropyl acrylamide-co-methacrylic acid). *Journal of Controlled Release*, **104**(3), 477-488.
60. Varadaraj, R., Bock, J., Brons, N., and Pace, S. (1993) Probing hydrophobic microdomains of hydrophobically associating acrylamide-n-alkylacrylamide copolymers in solution using a solvatochromic absorption dye probe. *The Journal of Physical Chemistry*, **97**, 12991-12994.
61. Varadaraj, R., Branham, K. D., McCormick, C. L., and Bock, J. (1994) Analysis of hydrophobically associating copolymers utilizing spectroscopic probes and labels.

In: *Macro- molecular Complexes in Chemistry and Biology*, Dubin, P., Bock, J., Davis, R., Schulz, D. N., Thies, C. (eds.), Springer-Verlag, Germany, pp. 15-31.
62. Bock, J., Pace, S. J., and Valint, P. L. (1987) Enhanced Oil Recovery with Hydrophobically Associating Polymers Containing Sulfonate Functionality, US Patent 4,702,319.
63. Hickner, M. A., Ghassemi, H., Kim, Y. S., Einsla, B. R., and McGrath, J. E. (2004) Alternative polymer systems for proton exchange membranes (pems). *Chemical Reviews*, **104**(10), 4587-4612.
64. Annable, T., Buscall, R., Ettelaie, R., and Whittlestone, D. (1993) The rheology of solutions of associating polymers: Comparison of experimental behavior with transient network theory. *Journal of Rheology*, **37**(4), 695-726.
65. Aubry, T., and Moan, M. (1994) Rheological behavior of a hydrophobically associating water soluble polymer. *Journal of Rheology*, **38**(6), 1681-1692.
66. English, R. J., Laurer, J. H., Spontak, R. J., and Khan, S. A. (2002) Hydrophobically modified associative polymer solutions: Rheology and microstructure in the presence of nonionic surfactants. *Industrial and Engineering Chemistry Research*, **41**(25), 6425-6435.
67. Schulz, D., and Bock, J. (1991) Synthesis and fluid properties of associating polymer systems. *Journal of Macromolecular Science - Chemistry*, **28**(11-12), 1235-1243.
68. Ma, X., and Pawlik, M. (2007) Intrinsic viscosities and Huggins constants of guar gum in alkali metal chloride solutions. *Carbohydrate Polymers*, **70**(1), 15-24.
69. Sakai, T. (1968) Huggins constant k for flexible chain polymers. *Journal of Polymer Science, Part A2: Polymer Physics*, **6**(8), 1535-1549.
70. Bock, J., Valint. Jr., P. L., Pace, S. J., Siano, D. B., Schulz, D. N., and Turner, S. R. (1988) Hydrophobically associating polymers. In: *Water-soluble Polymers for Petroleum Recovery*, Stahl, G. A., Schulz, D. N. (eds.), Springer, USA, pp. 147-160.
71. Bock, J., Siano, D., Valint Jr, P., and Pace, S. (1989) Structure and properties of hydrophobically associating polymers. *Polymers in Aqueous Media: Performance through Association, Advances in Chemistry Series*, **223**, 411-424.
72. Bai, Y., Qian, J., Sun, H., and An, Q. (2006) Dilute solution behavior of partly hydrolyzed poly(vinyl acetate) in selective solvent mixtures and the pervaporation performance of their membranes in benzene/cyclohexane separation. *Journal of Membrane Science*, **279**, 418-23.
73. Rossi, S., Ferrari, F., Bonferoni, M. C., and Caramella, C. (2001) Characterization of chitosan hydrochloride-mucin rheological interaction: influence of polymer concentration and polymer: mucin weight ratio. *European Journal of Pharmaceutical Sciences*, **12**(4), 479-485.
74. Grattoni, C. A., Al-Sharji, H. H., Yang, C., Muggeridge, A. H., and Zimmerman, R. W. (2001) Rheology and permeability of crosslinked polyacrylamide gel. *Journal of Colloid and Interface Science*, **240**(2), 601-607.
75. Chiu, Y.-L., Chen, S.-C., Su, C.-J., Hsiao, C.-W., Chen, Y.-M., Chen, H.-L., and Sung, H.-W. (2009) Ph-triggered injectable hydrogels prepared from aqueous n-palmitoyl chitosan in vitro characteristics and in vivo biocompatibility. *Biomaterials*, **30**(28), 4877-4888.

76. Glass, J., Lundberg, D., Zeying, M., Karunasena, A., and Brown, R. (1990)
Viscoelasticity and High Shear Rate Viscosity in Associative Thickener Formulations. *Proceedings of the Water-Borne and Higher-Solids Symposium*, USA, pp. 102-120.

77. Piculell, L., Guillemet, F., Thuresson, K., Shubin, V. and Ericsson, O. (1996) Binding of surfactants to hydrophobically modified polymers. *Advances in Colloid and Interface Science*, **63**, 1-21.

78. Taylor K. C., and Nasr-El-Din, H. A. (1998) Water-soluble hydrophobically associating polymers for improved oil recovery: A literature review. *Journal of Petroleum Science and Engineering*, **19**(3), 265-280.

4

Xanthan Gum as Advanced Polymer System for Enhanced Oil Recovery

4.1 Introduction

Gum is the terminology used for describing the hydro-colloidal gels, which have water affinity and show binding characteristics with water as well as with other organic/inorganic materials. Chemically, the gums are polysaccharides or carbohydrate polymers, and exist in entire life forms. The first microbial polysaccharide used commercially was dextran, which was developed in early 1940 [1]. Xanthan, a natural polysaccharide, is the second microbial polysaccharide utilized for the commercial purposes. Xanthan gum or polysaccharide B-1459 is an extracellular high molecular hetero-polysaccharide, and is prepared by the xanthomonas campestris NRRL B-1459 bacterium. It was initially developed at the Northern Regional Research Center of United States Agriculture Department in the year 1963 [2]. Substantial production of xanthan gum began in 1964 and numerous developments in the manufacturing processes of xanthan gum have been made since then. Nowadays, xanthan gum is regarded as the most critical microbial polysaccharide commercially. It has been widely studied due of its characteristics that permit it to supplement several other natural as well as synthetic water soluble gums. Due to exclusive rheological properties of xanthan gum, it is employed as stabilizer for emulsions as well as suspensions, and acts as a rheological control agent in the aqueous systems [3]. Xanthan gum solutions are very pseudo-plastic and display excellent suspending properties [4]. A beneficial characteristic of xanthan gum is its capability to generate highly viscous aqueous solutions as well as dispersions, which can retain their viscosity in the occupancy of various salts. The xanthan gum solutions are non-Newtonian in nature and its apparent viscosity varies remarkably when the diversified shear stresses are enforced. The viscosity of xanthan gums decreases at greater shear [5]. The safety as well as toxicological properties of this material for the pharmaceutical and food applications have also been widely studied. As a result, the commercialization of xanthan gum as stabilizer and thickener has continuously developed

Haleema Saleem and Vikas Mittal, The Petroleum Institute (part of Khalifa University of Science and Technology), Abu Dhabi, UAE*
Current address: Bletchington, Wellington County, Australia
© 2018 Central West Publishing, Australia

at an annual rate of 5–10% due to its superior physico-chemical characteristics, when compared to several other commercial polysaccharides.

In the petroleum industry, biopolymer xanthan gum is used for several applications such as oil drilling, pipeline cleaning, fracturing as well as workover and completion. Xanthan gum has good compatibility with salt and good resistance against thermal degradation, thus, leading to its application in the drilling fluids [6]. High viscosity of xanthan gum at low shear helps to generate drilling fluids with low concentration of suspended solids. Xanthan gum can retain this characteristic even at extreme conditions like high concentration alkali, acid and salt in the solution as well as high temperature. This characteristic is very important in the application of xanthan gum in the offshore drilling operations or other severe conditions. The pseudoplasticity of the xanthan gum solutions exhibits lower viscosity at the drill bit, where higher shear rate occurs, whereas high viscosity is generated in the annulus region, where the shear is depressed. Hence, the biopolymer helps in multi-purpose mode by allowing rapid penetration at the bit along with suspending cuttings in the annulus region.

For each barrel of the oil produced, on an average, almost two barrels remain in the ground. Hence, enhanced oil recovery (EOR) has become a substantial application of xanthan gum in the recent years. The fundamental principle practiced is to enhance the separation of water and oil, thus, increasing the oil recovery. Nevertheless, the quality of xanthan gum is an important consideration, as many impurities present might enhance the complication during the oil refining process. During the utilization of xanthan gum in the tertiary oil recovery process using micellar-polymer flooding, the polymer thickened brine solution is employed to drive the surfactant slug through the porous reservoir rock for mobilizing the residual oil. Polymer restricts the by-passing of water through the surfactant as well as guarantees good area sweeping [7]. Xanthan gum is compatible with most of the surfactants as well as several additional injection fluid additives, which are employed in the tertiary oil recovery formulations [8]. For both the aforementioned applications, the objective of the polymer is to lower the motility of the injected water by viscosity enhancement. The thermal resistance property of xanthan gum makes it an ideal mobile control agent as well as a reliable displacing agent. Other different specialty usages for xanthan gum include rust removal, wet slag, welding rods and cleaning waste from the gas pipelines.

In this review chapter, application of the biopolymer xanthan gum in the oil and gas industries is focused. The structure, properties and production method of xanthan gum are also briefly outlined. An up-to-date review of the application of xanthan gum in the EOR process, oil drilling operation (as dispersant and tackifier) and in fracturing fluids is carried out. In addition to this, applications of xanthan gum for the chemical absorption of carbon dioxide and corrosion inhibition in the mild steel are also mentioned.

4.2 Structure

Xanthan gum has a complex molecular structure, with the primary structure comprising of replicated pentasaccharide units which are made up of two units of glucose, two units of mannose and one unit of glucuronic acid, present in the molar ratio 2.8:2.0:2.0. The main chain contains β-D-glucose units, which are connected at the positions 1 and 4. The primary chain chemical structure is identical to the cellulose structure. The tri-saccharide side chains consist of D-glucuronic acid units between the two D-mannose units, which are connected at the O-3 position of intermittent glucose residue present in the primary chain [9,10]. Almost 50% of the terminal D-mannose consists of pyruvic acid, which are connected by the keto group to the 4 and 6 positions. The D-mannose units are connected to the primary chain containing an acetyl group at O-6 position. The existence of pyruvic as well as acetic acids generates an anionic polysaccharide type [11].

The trisaccharide units are jointly aligned with the polymeric backbone, as a result, a stiff chain is generated which might exist as a triple, double or single helix [12,13]. The chain interacts with other polymer chains, thereby leading to the formation of a complex. Molecular weight distribution range of xanthan gum is 2×10^6 - 20×10^6 Da. This depends on the organization between the chains to form diverse distribution of individual chains. Molecular weight of xanthan is greatly influenced by the changes in the fermentation conditions utilized in the production process. Xanthan solution generated by the dissolution process at moderate temperatures exhibits extreme viscosity. Thus, the overall viscosity is greatly influenced by the dissolution temperature, molecular conformations as well as the emergence of ordered structures [14]. The structure of xanthan gum is also depicted in Figure 4.1 [15].

The preparation of xanthan gum is considered to be identical to that of the preparation of exopolysaccharide by other gram-negative bacteria. The preparation process can be split into three sections (1) uptake of simple sugars, subsequently converting to nucleotidal derivatives, (2) aggregation of pentasaccharide units that are connected to iso-pentyl pyrophosphate carrier, and (3) the polymerization of pentasaccharide repeating units, followed by secretion [16].

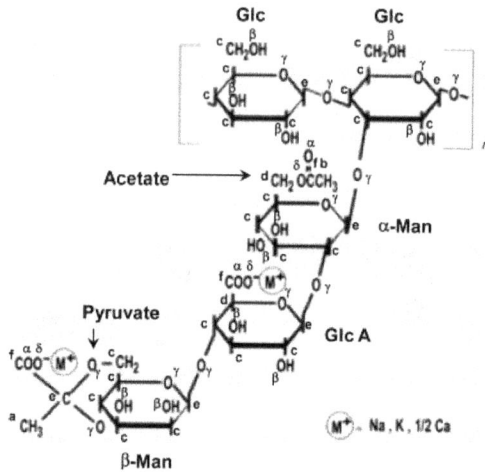

Figure 4.1 Structure of xanthan gum. Reproduced from Reference 15 with permission from American Chemical Society.

4.3 Production

Most of the commercial xanthan gum production processes employ glucose or invert sugars, and the majority of the industries prefer the batch production processes rather than the continuous processes [17]. Different alternate substrates like hydrolyzed rice, acid whey, sucrose, barley, coconut juice, sugar cane molasses, sugar cane etc., are available, however, glucose is still regarded as the best in terms of supply, product quality and product yield [18,19]. It was confirmed that the production as well as properties of the biopolymer xanthan gum is greatly influenced by culture medium [20], bacterial strain [21], pH [22], temperature [23], agitation rate [24] and time of fermentation [25]. For the generation of xanthan gum, the microbial strain is initially conserved for attaining long term storage by utilizing established methods for maintaining the appropriate characteristics [14]. Minute quantity of preserved culture is developed by growth on the surface of solid or in liquid media for attaining the inoculum for large bioreactors. The microorganism growth and the generation of xanthan gum is affected by various factors like the mode of operation (continuous or batch), type of bioreactor employed, medium composition as well as culture conditions (concentration of dissolved oxygen, pH and temperature). At the end of the fermentation process, the broth consists of bacterial cells, xanthan and several other chemicals. For the recovery of xanthan, cells are normally separated first, by either centrifugation or filtration [26]. The purification might include precipitation employing water-miscible non-solvents (acetone, ethanol, isopropanol etc.), incorporation of several salts as well as pH adjustments [26]. Followed by the precipitation process, the product is dewatered and dried.

4.4 Properties and Surface Modifications

Xanthan is a white to cream colored free-flowing powder and tasteless biopolymer. The biopolymer is soluble in cold as well as hot water, however, the material is insoluble in most of the organic solvents. The industrial significance of this material is dependent on its capability to govern the rheology of water based systems. Xanthan is utilized as a thickener in various industries, stabilizing the emulsions and suspensions in textile and paper mill industries [27] and in enhanced oil recovery processes. Xanthan is also commonly used in the food industry due to its higher viscosity at lower concentrations, solubility in cold or hot water, good stability in acid systems and very less change in viscosity with temperature variation. In addition to this, the material also displays good suspending properties due to its higher yield value and capacity to contribute great freeze-thaw stability [28,29]. The solution of xanthan has good stability over a wide range of temperatures (till 90 °C), salt concentrations (till 150 g/l NaCl) and pH (2-11) [30]. The rheological properties of xanthan solutions change with the nature of polymer, i.e., average molecular weight, and pyruvate as well as acetate contents [31,32]. In the molecule, the levels of acetal and pyruvic acid substitutions may change with operational conditions [33], bacterial development [34], and type of *Xanthomonas sp.* utilized [35]. Xanthan gum with a greater level of pyruvilation and acetylation degree enhances the viscosity of its water solutions due to intermolecular associations [36]. Nevertheless, pyruvate free xanthan is considered to be more appropriate for the generation of mobility control solutions, which are utilized in the enhanced oil recovery processes [37]. It was proved that by the incorporation of salt (monovalent or divalent), the xanthan gum chains were subjected to a co-operative conformational transition from a disarranged conformation to an arranged and highly rigid structure [38,39].

At different pH values, xanthan gum exists in an ordered conformation, where the acetal links are secured from the chemical attack. Normally, xanthan gum of 1000 ppm concentration range are utilized for the EOR application [40]. At these concentrations, xanthan gum does not degrade by the application of typical shear forces existing in the petroleum reservoirs. Further, this material is comparatively insensitive to the greater electrolyte concentrations and does not precipitate in the presence of divalent ions [41]. Dissolving xanthan gum in solution is a two-stage slow process. During the first stage, a gel is generated when solvent molecules begin to dissolve and connect with the polymer chains. Second stage greatly depends on the type of mixing. A good dispersion is necessary for preventing minute pore plugging during the field usage [42]. Xanthan solution properties are influenced by the measurement and dissolution temperature, and additional non-xanthan polymers.

A drawback of xanthan gum is its sensitivity to the bacterial degradation. It was established that the salt tolerant anaerobic as well as aerobic microorganisms can cause the degradation of xanthan gum chains, thereby, leading to a loss in the solution viscosity [43]. Biocides are employed for suppressing the growth of microorganisms that degrade xanthan gum. In majority of the cases, formaldehyde is used as an efficient biocide. Nevertheless, the application of biocides for protecting xanthan gum makes the lesser environmental impact of xanthan gum polymer somewhat controversial.

Roy *et al.* [44] reported modified amphiphilic xanthan having hydrophobic octyl moieties onto the carboxylic acid functional groups using a grafting reaction. In the course of the reaction, the reagents (octylamine, carbodiimide (EDAC) and N-hydroxysuccinimide (NHS)) were incorporated in the same stoichiometry with reference to the carboxylic functional groups. The derivative xanthan gum followed similar organization as the precursor, however, the intermolecular hydrophobic interactions greatly affected the transient network lifetime. The rheological as well as structural characteristics of these derivatives were also analyzed employing the rheological and fluorescence analyses. It was observed that an enhancement in the fluorescence intensity occurred for both xanthan samples as the concentration of polymer increased. This event was, however, more distinct for the modified xanthan as compared to the precursor. Figure 4.2 also demonstrates the rheological performance of the precursor and modified xanthan gum sample.

In other study describing the physical modification of xanthan gum, Wang *et al.* [45] synthesized aerogel composites based on sodium montmorillonite clay (Na+-MMT) and xanthan gum employing an environmental benign freeze-drying technique. For enhancing the characteristics of XG/clay aerogels, a bio-based polysaccharide was utilized. Fourier transform infrared (FTIR) spectroscopy confirmed the molecular interactions between biopolymer and clay in the aerogel composite. The thermo-gravimetric analysis (TGA) confirmed that the clay enhanced the thermal stability of the aerogels. The bio

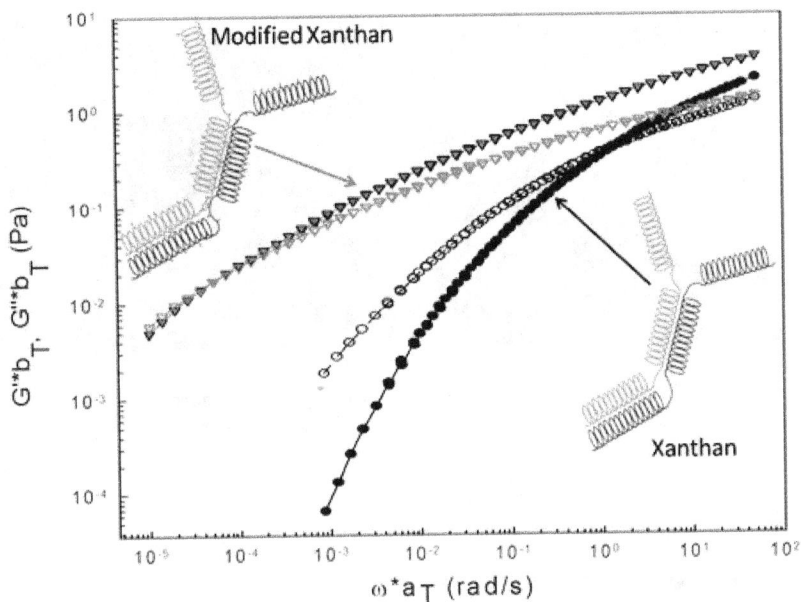

Figure 4.2 Rheological properties of the two xanthan gum samples as a function of angular viscosity. Reproduced from reference 44 with permission from American Chemical Society.

based aerogels exhibited identical degradation pattern with two major stages: bound water loss as well as aerogel matrix degradation. It was further concluded that the incorporated glycerol was removed between the temperatures 180 and 250 °C, which overlapped with the onset degradation temperatures of xanthan at approximately 235 °C and 240 °C, respectively. Figure 4.3 also shows the scanning electron micrographs of the various aerogels at high magnification [45].

Jampala *et al.* [15] also investigated the structure and rheological properties of xanthan gum modified in a cold plasma environment. The modification was carried out using dichlorosilane (DS)-plasma conditions and subsequently *in-situ* amination by ethylenediamine (ED). Network strength of the polymer was observed to increase after plasma treatment. Optimization of plasma treatment time and polymer concentration was performed to obtain stable gels of xanthan gum. Figure 4.4 shows the high-resolution electron spectroscopy for chemical analysis (ESCA) spectra of samples treated with DS-plasma for 15 min, followed by *in-situ* ED-grafted.

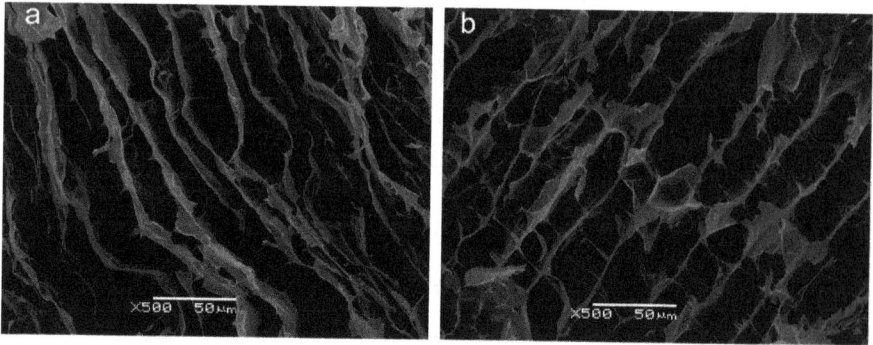

Figure 4.3 SEM micrographs of two different xanthan gum based aerogels. Reproduced from Reference 45 with permission from American Chemical Society.

4.5 Applications

4.5.1 Enhanced Oil Recovery Process

Almost 60-70% of the original oil in place (OOIP) is generally not generated by the conventional methods, which depends on hydrocarbon as well as reservoir drive. The secondary recovery is generally water flooding which recovers approximately 20-50% of the OOIP. The EOR normally involves the recovery beyond the secondary stage. The EOR methods achieved particular attention due to depleting resources of crude oil worldwide. The different EOR processes are chemical and gas flooding, combustion, steam and electric heating. The chemical flooding methods are recognized by the particular chemical, which is injected. Chemical EOR processes are one of the predominantly used methods for the petroleum production because of feasible additional production costs and easy application. The chemical EOR techniques employ interfacial tension minimization and/or the mobility ratio improvement for mobilizing the oil [46]. These EOR processes are contingent on the polymer flooding (PF) or the alkali-surfactant-polymer (ASP) flooding [47]. The existence of the surfactant decreases the interfacial tension between the hydrocarbon and aqueous phases. The PF aqueous solutions have been regarded as an interesting alternative to the conventional water flooding due to its ability for the recovery of extra oil from the reservoirs [48,49]. The application of aqueous solutions in the EOR stage is very useful for keeping the rate of crude oil production at the necessary economical levels, after the conventional methods have

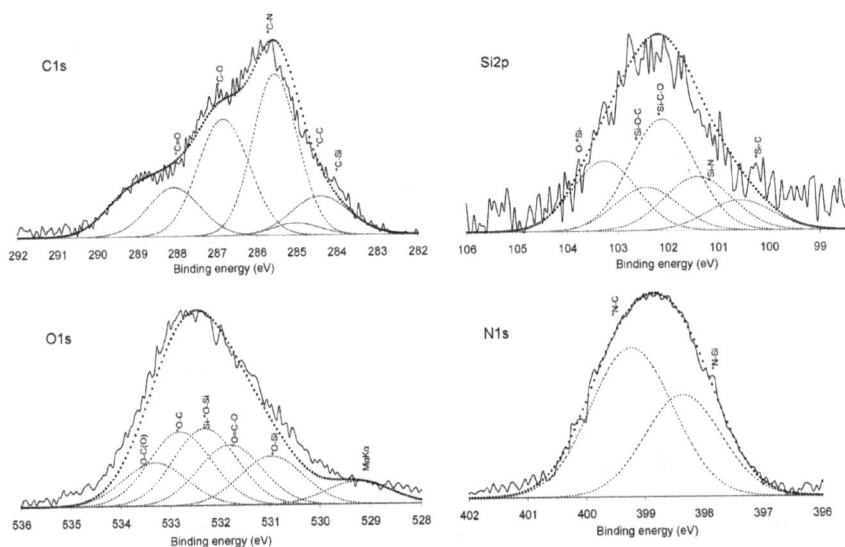

Figure 4.4 High-resolution ESCA spectra of 15-min DS-plasma-treated and subsequently in situ ED-grafted. Reproduced from Reference 15 with permission from American Chemical Society.

been disabled. In the PF process, when the water soluble polymers are incorporated into the injection water, it can enhance the water phase viscosity along with improvement in the mobility ratio. It can also increase the swept volume, thereby, increasing the efficiency of oil recovery [49, 50]. The custom made EOR injection fluids generally consist of water soluble polymers like polyacrylamides. These polymers lower permeability contrast and enhance the aqueous phase viscosity. The operational hazard related with PF involves polymer degradation, loss of polymer to porous media and loss of injectivity.

In EOR applications, the most commonly employed synthetic water soluble polymer is partially hydrolyzed polyacrylamide (HPAM). Though the polymer has been successfully used in the recovery of tertiary oil [49,50], the EOR polymers based on polyacrylamide have some disadvantages. Due to the presence of recurrent electric charges on the backbone, the HPAM collaborates with the charged species existing in the solution. HPAM is also very sensitive towards the existence of surfactants and hydrocarbons in the reservoir. From an environmental outlook, the application of synthetic polymer in EOR is troublesome. In most cases, after the injection, HPAM will remain in the reser-

voir. Due to the environmental regulations, the industrial processes are required to minimize the spillage, and the use of non-biodegradable polymer
can lead to bio-accumulation during accidental environmental release. Hence,
the biopolymer based EOR technology will be more promising in the future,
especially if the existing limitations are taken care of. During the chemical EOR
processes, other environmentally beneficial materials like several polysaccharides and nano-cellulose have the potential to fully or partially substitute
HPAM in the coming years. The important physical properties of biopolymer
materials are lesser sensitivity towards the shear degradation and greater viscosifying efficiency at lesser additive concentration. Xanthan gum is regarded
as one of important biopolymers which can be practiced commercially in the
petroleum reservoirs [51]. For the effective replacement of EOR polymers like
HPAM, the biopolymer materials should have properties such as greater salinity tolerance, higher shear and mechanical stability, good adsorption and rheological properties, and excellent viscosifying efficacy as well as thermal tolerance. For instance, despite the fact that xanthan gum is high-priced and
more susceptible to biodegradation, it provides excellent viscosity generating
capacity and very good stability under the mechanical shear and high salinity
conditions, when compared to HPAM [52]. However, in the presence of oxidizing agents, xanthan gum is exposed to the oxidative degradation process. This
type of degradation has a vital role in the viscosity loss of the xanthan gum solutions [53]. In addition to this, various bonds present in the xanthan gum
molecule, like the acetyl groups [54], pyruvate-ketals and glycosidic linkages
[55], are exposed to cleavage by the hydrolysis process, which causes remarkable viscosity loss. Jang *et al.* studied the promising application of xanthan
gum and HPAM as the PF agents for the purpose of heavy oil recovery under a
range of saline conditions [56]. The rheological analysis was performed to analyze the variation in shear viscosity, when the polymer was subjected to a
range of reservoir conditions. As the xanthan gum concentration was increased, the shear viscosity was also observed to increase. For the polymer
flooding in EOR, the enhanced shear viscosity of xanthan gum can raise the
sweep efficiency as well as the residual resistance factor [57]. The results indicated that the Newtonian viscosity area observed at lower shear rates were
not seen at the higher concentrations. On the other hand, remarkable shear
thinning property was observed, which was due to disentanglement and
alignment of the xanthan gum chains along the direction of flow. Overall, the
rheological test analysis revealed that the HPAM solution was more sensitive
to higher temperature and salinity than the xanthan gum solution. The xanthan gum structure was reported to have helix as well as random coil confor-

mations, which greatly depends on the temperature of dissolution [16]. A flood test was also carried out with xanthan gum solution at 3 wt% salinity, and it was observed that the recovery of oil was 7.2% greater than HPAM solution. The solution of xanthan gum at 3 wt% and 10 wt% salinity exhibited no remarkable difference in the oil recovery. As the main objective of employing the polymers in EOR operation is to enhance the solution viscosity and lower the rock permeability, and both of these changes help to enhance the sweep efficiency by lowering the mobility of displacing fluid [58]. Thus, the results obtained in this analysis confirmed the superiority of xanthan gum over HPAM for polymer flooding applications for enhanced oil recovery, especially for the applicability of xanthan gum at high salinity reservoir conditions. Thus, the use of biopolymers like xanthan gum for EOR applications has the potential to provide superior performance in comparison with the synthetic polymers.

The mixing of xanthan gum with surfactants was also studied by Taugbol *et al.* [59] and the combination was proved to be very useful for EOR. The authors demonstrated that by using xanthan gum along with the surfactant alkyl propoxy-ethoxy sulfate ($C_{12-15}(PO)_4$-$(EO)_2$-OSO_3-Na^+), greater than 50% of residual oil could be recovered. However, it was noted that the recovery of residual oil by utilizing only a surfactant solution was lesser. In addition to this, it was also observed that the consolidation of xanthan gum and dodecyl-*o*-xylene sulfonate surfactant recovered even lesser residual oil when compared to the case where only surfactant solution was utilized [60]. The authors suggested the explanation for this phenomenon as the development of large micellar aggregates that exhibit an adverse effect on the flow behavior of surfactant through the porous media. In a study by Wei *et al.* [61], the chemical formulation as well as the rheological properties of a self-assembling biopolymer (SAP) system were studied. The aim was to ascertain the suitability of the system as a substitute in the polymer flooding process for oil recovery, specifically under extreme reservoir conditions of salinity, temperature and hardness. The viscoelastic SAP system was generated by the non-bonding associations of xanthan gum side branches with the EOR surfactant b-cyclodextrin inclusion complex in aqueous solutions. It has been reported that the water soluble polymers can link with the ionic surfactants through the non-bonding interactions, which can enhance the rheological and physico-chemical properties of the basic polymer [62,63]. Thus, the improved viscoelastic system displays higher thermal and mechanical stability, along with increased hardness and resistance to brine salinity because of the interlocking effect. From the sand pack flood tests performed at the reservoir conditions, it was observed that this SAP system generated remarkably greater resistance factors than xanthan gum during the flow in porous media. This clearly indicated its efficiency in governing the motility of the displacing fluid. Further, the SAP system displayed good capability as the *in-situ* permeability modifier, thus, confirming the superior performance.

Sveistrup *et al.* [64] examined the polysaccharides starch, xanthan gum and scleroglucan, and confirmed these materials to be environmentally friendly EOR agents. In this study, the viscosifying properties of these polysaccharide based biopolymers were examined, especially viscometric and interfacial tension assessments. The polysaccharides exhibited robust shear tolerance, supportive viscosifying performance, electrolyte tolerance as well as temperate interactions with the surfactants. These polysaccharides were also observed to exhibit encouraging viscosifying accomplishment in semi-dilute regime, due to the rigid helical structure of the chains, which are characterized by the extensive persistence lengths. Further, both xanthan and scleroglucan displayed sufficient tolerance towards the electrolyte, particularly at moderate and greater shear rates, where the biopolymers remained in shear aligned states characterized by the scattered inter-chain entanglements. The xanthan gum, which is a polyelectrolyte, has side chains which are connected with backbone under deserted electrostatic conditions, which results in marginal reduction in the hydrodynamic volumes and viscosities during the complete range of measured shear rates. At low and moderate shear rates, the polyelectrolyte backbone recantation occurs because of the existence of electrolyte. However, at the higher rates of shear, the shear alignment synthetically expands the polyelectrolyte chain, contrasting the coil recantation effects.

The biopolymer-surfactant interactions are generally much more complicated, compared to the biopolymer-electrolyte interactions. For xanthan gum, there exists only minute effect of extra anionic surfactant at greater salt content, when compared to the surfactant-free brine. The viscosity of xanthan gum is clearly dependent on temperature, with a double helical configuration at lesser temperatures to a disarranged coil structure at greater temperatures [65]. The xanthan gum viscosities are observed to be reduced in the presence of surfactant. The noticed viscosity variations are attributable to the polymer-surfactant interactions that affect the biopolymer's secondary helical structure. In the case of xanthan gum, unbound as well as bound anionic surfactants result in electrostatic repulsive interplay on the polyelectrolyte backbone, which results in the polymer recantation and lower inter-chain entanglements. Non-electrostatic interactions are also observed between the surfactants and xanthan gum. At moderate and higher shear rates, viscosity has been observed to be reduced, as the non-electrostatic surfactant-polymer interactions operate mainly on the xanthan gum's inherent backbone, oppositely affecting the side-chain association property. These non-electrostatic interactions are capable to promote a structural transition to a weaker as well as less arranged conformation, from a rigid ordered double helix. The interfacial tension analysis has also been observed to confirm the binding of anionic surfactant to the backbone of xanthan gum. Overall, xanthan gum has proved to an excellent environmental friendly EOR agent, which is appropriate for oil recovery particularly under extreme reservoir conditions of temperature, hardness and salinity. Consequently, xanthan gum continues to draw research interest for enhanced utilization in the EOR processes.

4.5.2 Dispersant for Drilling

The demand for oil and gas has increased manifold in the past several decades because of industrialization, increased population and enhanced living standards. However, numerous producing oilfields have turned into mature fields, which causes a loss in the production. The development of new oilfields is also being undertaken, even in challenging reservoir environments, like deep offshore regions or high pressure-high temperature (HPHT) reservoir conditions. Thus, it has become very challenging to design the drilling fluids. In accordance with the American Petroleum Institute (API), the drilling fluid is described as a circulating fluid which is employed in the rotary drilling for performing any or all of the different functions necessary during the drilling operations. Several kinds of drilling fluids like oil based drilling mud (OBM), water based drilling mud (WBM) and their derivatives are being utilized, along with appropriate clay as well as polymer additives. The important functions of the drilling fluids include (1) regulation of the downhole formation pressures, (2) preventing damage of producing formation, (3) cooling and lubricating drill bit, (4) overcoming the formation fluid pressure, and (5) removal of cuttings, developed by drill bit from bore hole [66]. For this, an effective drilling fluid should possess certain characteristics like desired rheological behavior, temperature stability, prevention of fluid loss, stability under different pressure and operating conditions [67].

The classification of drilling muds as either WBMs or OBMs, depends on the nature of continuous phase of the mud [42]. OBMs normally employ hydrocarbon oil as the major liquid component along with alternate materials like clays or colloidal asphalts, which are incorporated to contribute the required viscosity together with polymers, emulsifiers and additives such as weighting agents. WBMs generally consists of viscosifiers, weighting agents, fluid loss control agents, emulsifiers, lubricants, salts, pH control agents and corrosion inhibitors. For the rheology control function, various polymers like xanthan gum and partially hydrolyzed polyacrylamide (PHPA) are generally employed. Sea water mud is a type of WBM which is designed for the offshore drilling application, where make-up water is collected from ocean. In the salt water muds, dissolved sodium chloride (NaCl) acts as the main component. Some of the most commonly used type of sea water mud additives are bentonite clay, lignite, lignosulfonate and caustic soda. Biopolymer xanthan gum can be utilized instead of bentonite. In order to eliminate the problems related to the reduction of viscosity in polymer-based aqueous fluids, formates like sodium formate and potassium formate are generally incorporated in the fluids for increasing their thermal stability. Nevertheless, the technology of utilizing formates leads to high price. The stability of wellbore treatment can be preserved at temperatures till 135-160°C (275-325 °F), by adding different polysaccharides to the fluids.

Nano-fluids demonstrate potential usage in upstream oil and gas industry for increasing the performance of different processes like exploration, drilling as well as completion, production and EOR operations [68, 69]. Some of the current research studies addressed the application of nanoparticles for solving drilling related issues such as lost circulation, pipe sticking, torque and drag [70]. The advancement of appropriate nano-fluids for the WBM needs an acceptable dispersion medium for the nanoparticles, which should have good compatibility with the oilfield chemicals. Biopolymers like xanthan gum and guar gum are well known for their usage in the oil and gas industries as a drilling fluid additive, fluid loss controlling agent, rheological modifier and emulsion stabilizer. Xanthan gum generates exclusive rheological properties like high low shear viscosity as well as strong shear thinning behavior, due to the existence of a complicated network in the solution originated from the stiff polymer chains.

In a recent study, the ability of xanthan gum and guar gum for generating the stabilized iron-based nano-fluids was confirmed [71]. The study aimed to amend the guar gum solutions with small quantity of xanthan gum in order to remarkably enhance the biopolymer capability for stabilizing highly concentrated iron nano- and micro-particle suspensions. The cooperative effect between xanthan gum and guar gum developed a viscoelastic gel which had the ability to maintain 20 g/L iron particles suspension for over 24 h. The xanthan gum/guar gum viscoelastic gel was characterized by a marked shear thinning property. This behavior, along with lower biopolymer concentration, regulated lower viscosity values at greater shear rates, thereby, expediting the injection in porous media. In addition to this, the thermo-sensitivity of soft elastic polymeric network stimulates greater stability and prolonged storage times at low temperatures and sudden decline in the viscosity at greater temperatures. This property can be utilized for improving the flowability and delivery of suspensions to the target, as well as to efficiently tune and control the release of iron particles. In another recent study by Ponmani *et al.* [72], zinc oxide (ZnO) and copper oxide (CuO) nano-fluids were generated, using the two-step method in a 0.4 wt% xanthan gum aqueous solution as a base. The CuO and ZnO nano-fluids were characterized for dispersion and stability, utilizing the dynamic light scattering and scanning electron microscopy. It was observed that the stability of ZnO nano-fluid was lower as compared to the CuO nano-fluid in the xanthan gum aqueous solution. Xanthan gum formed a jelly structure over the nanoparticles, keeping the particles suspended in solution. The study also confirmed that the electrical and thermal conductivities of these nano-fluids enhanced as the nanoparticle concentration was increased. The thermal and the electrical conductivities increased by almost 25% and 50%, for the ZnO and CuO nano-fluid, respectively. This signifies the advancements in the nano-fluid based technology with xanthan gum as the primary additive in upstream oil and gas industries.

Karen *et al.* [73] generated nano-fluid enhanced WBM (NWBM) by employing the nano-fluids of CuO and ZnO in the xanthan gum aqueous solution as the base fluid. The concentration of nanoparticles in the nanofluids were 0.1 wt%, 0.3 wt% and 0.5 wt% [72]. The generated nano-fluids were added to WBM as an additive (1% by volume). An enhancement in the concentration of nanoparticles increased the thermal conductivity in both the nano-fluid based drilling fluid. The improvement in the thermal conductivity of ZnO based nano-fluids was observed to be from 12% to 23%. However, in the case of CuO based nano-fluids, greater enhancement in the thermal conductivity was observed, i.e., from 28% to 53%. In the case of ZnO nano-fluid systems with WBM, an improvement in the thermal conductivity from 17% to 34% was seen. For the CuO nano-fluid system with WBM, the thermal conductivity enhancement was from 20% to 38%. This increased thermal conductivity of the fluids was a clear indication of the capability of mud to cool down faster as it proceeds up to surface. Further, high pressure rheological analysis was performed on NWBM to determine the influence of nanofluids on the rheological behavior at different pressures (0.1 and 10 MPa) and temperatures (25 °C, 70 °C, 90 °C and 110 °C). The remarkable performance at high pressures and temperatures revealed high potential of NWBM of oil drilling applications.

4.5.3 Tackifier for Drilling

Both xanthan gum and polyacrylamide gum are very common polyelectrolyte polymers used in various industries [74]. The anionic polyacrylamide (HPAM) is commonly used to treat industrial and municipal wastewater due to its low cost and effective flocculence. Large number of polar groups are present in PAM molecular chain that can adsorb the solid particles which are suspended in the drilling fluid. PAM is also commonly used as an industrial oil displacement agent due to its water solubility, ionic properties and several other functions [75]. Interaction of xanthan gum with other polymers like PAM, the performance can even improve further. For instance, PAM/xanthan gum solution has excellent salt resistant characteristic, and hence can be employed as a better salt resistant thickening agent as well as flocculant. As HPAM encounters xanthan gum, the dipole reaction of the hydrogen groups and hydrophobic groups induces molecular chain association, thus causing the formation of macromolecular structure.

During the drilling of the natural gas hydrate (NGH) core samples, the temperature of the drilling fluid in the bore circulation should be lowered below the freezing point to make sure that the hydrate core is preserved in stable and balanced condition. In a study performed by Zhao *et al.* [76], the influence of K^+, Na^+, Mg^{2+} and Ca^{2+} cations, at different concentrations, on the rheological behavior of PAM containing cationic and anionic PAMs (CPAM and HPAM) as well as xanthan gum solution was studied. The analysis was performed after the macromolecules were generated due to the interaction between the xanthan gum (tackifier) and polyacrylamide (flocculant). Further, the crosslinking reaction mechanism of polyacrylamide/xanthan gum macromolecular system and the cation disruption mechanism of crosslinking process were also examined. It was observed that the CPAM/xanthan gum solution exhibited ehnaced cation resistant property than the HPAM/xanthan gum solution. On comapring the viscosity values of two different types of PAM/xanthan gum solutions in varying proportions (f_{PAM}), at a total concentration of 0.02%, it was noted that the viscosity of the solution increased as the PAM content developed. This was due to the fact that the PAM is a flexible chain molecule, while xanthan gum is a rigid chain molecule. The variation in conformation was not difficult for flexible chain, where the inherent tangle was collapsed through the segmental motion, thereby, reducing the flow resistance.

Thus, xanthan gum, in interaction with other polymers, has a strong potential for use as tackifier for drilling operations. Especially, the useful performance is achieved at low concentrations.

4.5.4 Corrosion Inhibitor for Mild Steel

Mild steel (MS) has a wide variety of industrial applications, hence, the corrosion stability of MS in acidic media is an important concern for the material technologists and corrosion scientists [77]. Metals as well as alloys are exposed to immensely corrosive acids during industrial processes like acid cleaning, acid descaling, oil well acidizing and acid pickling [78]. Such extreme conditions lead to accelerated corrosion rates and material degradation, resulting in large economic losses. Natural polymers which show a high affinity for the metal surfaces and exhibit excellent corrosion inhibition effectiveness have been focus of research due to these properties, in addition to the environmental friendly nature [79]. In addition, grafted or modified polysaccharides have also the capability to function as corrosion inhibitors as these polymeric materials possess numerous adsorption sites, which can adsorb more efficiently to the surfaces, thus, helping to achieve effective corrosion inhibition.

In a study by Biswas *et al.* [80], xanthan gum and xanthan gum-grafted-polyacrylamide (XG-*g*-PAM) were analyzed as corrosion inhibitors for mild steel in 15% HCl medium. Different characterization like potentiodynamic polarization, gravimetric analysis and electrochemical impedance spectroscopic analysis were performed for ascertaining the effectiveness of the aforementioned inhibitors. It was revealed that both of these inhibitors followed the Langmuir adsorption isotherm. It was observed that the inhibition efficiency ($\eta\%$) enhanced with the increase in the concentration of inhibitor. At 0.4 g/L concentration, xanthan gum exhibited 90.2% inhibition efficiency, whereas, the efficiency of XG-g-PAM was 92.7%. Further, there was no remarkable change in $\eta\%$ as the concentration was increased. The superior efficiency of XG-g-PAM was because of the fact that it enclosed the surface of metal more effectively than the xanthan gum, thereby, retarding the dissolution of the mild steel. The SEM analysis indicated the protection of surface of metal by xanthan gum and XG-g-PAM. The microscopy images also revealed that xanthan gum as well as XG-g-PAM formed a barrier film on the surface of metal, which prevented the spread of corrosion. Overall, XG-g-PAM was concluded to have better corrosion inhibition ability than xanthan gum for mild steel in the acidic conditions.

Thus, in addition to the large variety of applications of xanthan gum in oil and gas industry, it along with its derivatives have also strong potential of large scale application for the corrosion inhibition of metallic surfaces under extreme corrosion conditions.

4.5.5 Chemical Absorption of Carbon Dioxide

The gas-liquid mass transfer in the non-Newtonian liquid is a relevant example of absorption of gas in the pseudo-plastic flow of the industrial operations like fluidized bed, fermentation broth and slurry [81,82]. Park *et al.* [83] measured the absorption rate of carbon dioxide (CO_2) in the aqueous solution of xanthan gum in the range 0-0.15 wt%, which consisted of NaOH concentration of 0-2.0 kmol/m^3 in a flat-stirred vessel with an agitation speed of 50 rpm at 0.101 MPa and temperature of 25 °C. Volumetric liquid side mass transfer coefficient of CO_2, which was connected to the viscosity and the elastic behavior of xanthan gum solution, was employed to determine the CO_2 chemical absorption rate (R_A). The aqueous solution of xanthan gum with the elastic property of the non-Newtonian liquid caused increase in R_A, when compared to the Newtonian liquid based on the same viscosity of the solution.

4.5.6 Fracturing Fluids

Fracturing is considered to be a hydraulic process, in which overpressure is practiced for cracking the rock and for enhancing the flow of oil. In the interest of avoiding the collapse of the stone, the proppants are settled in these fractures, which perform as spacers. As a consequence, the viscous properties of the fracturing fluid are generally considered to be the most significant factor. The ideal fracturing fluid is required to be efficient with lesser fluid loss. This can be generally achieved by mixing the high fluid viscosity with the fluid loss additives. The most commonly used types of fracturing fluids are water based fluids and oil based fluids. The linear fracturing fluids are water based fracturing fluids with no chemically crosslinked structures. Xanthan gum is one of the most commonly used types of linear gels employed as the fracturing fluids. Due to the high low-shear viscosity of xanthan, it is very productive to maintain the proppants in the suspension and to transport them as far as possible [84]. Xanthan gum is a vicsosifier, which can be employed as a linear gel system as well as a crosslinked fluid. The main application of xanthan gum in the stimulation has been as a thickener for the hydrochloric acid (HCl). The application of this biopolymer is restricted to the acid concentrations till 15% and temperature of 200 °F (93 °C). A breaker is defined as an additive which facilitates a viscous fracturing fluid to be degraded manageably to a thin fluid which could be generated back out of the fracture. Xanthan can act as an enzyme gel breaker that can cause degradation of the polymer chains by breaking the backbone structure of the thickeners and ultimately of the fluid loss additive. Due to their infinite polymer-degrading activity and inherent specificity, the enzyme breakers have numerous benefits in several other breaker systems. Xanthan gum can also perform as stabilizer for foaming treatments, which are incorporated into the fracturing fluid, to enhance the foam half-life, specifically at the elevated temperature.

4.5.7 Removal of Dyes from Aqueous Solutions

Xantan gum can also be used as an effective adsorbent for the removal of dyes commonly present in aqueous wastewater generated in oi land gas industries. In a recent study, Ghorai *et al.* [85] reported novel nanocomposite consisting of hydrolyzed polyacrylamide (PAM) grafted onto XG along with incorporated nano-silica (h-XG/SiO$_2$). The composite was observed to be a promising adsorbent for the separation of toxic methyl violet (MV) and methylene blue (MB) from the aqueous solution. The preparation process involved the saponification of grafted PAM and the *in-situ* generation of nano-scale SiO$_2$ by the sol-gel process. The kinetics of MV and MB adsorption from the aqueous solution confirmed that the adsorption of dyes took place very quickly, in accordance with Langmuir adsorption isotherm and pseudo-second-order kinetics. The process was also observed to be entropy driven. Figure 4.5 presents the adsorption mechanism of MB and MV using the nanocomposite. The significant adsorption ability of dyes on the nanocomposite (99.4% efficiency of MB removal and 99.1% efficiency of MV removal) was justified on the basis of hydrogen bonding interactions as well as the electrostatic and dipole-dipole interactions between the cationic dye and anionic adsorbent. Due to the superior regeneration ability of the nanocomposite, such bio-based nanocomposites have the potential of large scale usage for the adsorption of toxic dyes from the wastewater in oil and gas industries.

4.6 Summary and Outlook

Xanthan is an outstanding commercially produced industrial gum, obtained by the fermentation process. Compared to carboxymethylcellulose, polyacrylamide, modified starch as well as some plant polysaccharides (guar gum), xanthan gum has fair technical advantage in the oilfield development. Overall, the extensive range of applications for xanthan gum is due to its remarkable properties such as (1) greater viscosity at lower concentrations (600-2000 ppm), (2) non-Newtonian property, (3) mechanical degradation resistance, (4) low sensitivity of viscosity to the variation in salinity, (5) environmental friendly material as it is a biopolymer, and (6) good temperature stability (till 90 °C). Xanthan gum is considered as the one of the few successful biopolymers which can be practiced commercially in the petroleum reservoirs. During the application of xanthan gum in beach, ocean, permafrost drilling and halogen layer, this biopolymer has apparent effect on the blowout prevention. Even lower concentration of solution of xanthan gum can retain the viscosity of water based drilling fluids. The polymer offers good suspension performance, to enable it to restrict the collapse of borehole, prevent the formation blowout and expedite discharging the cutting gravel out of the wells.

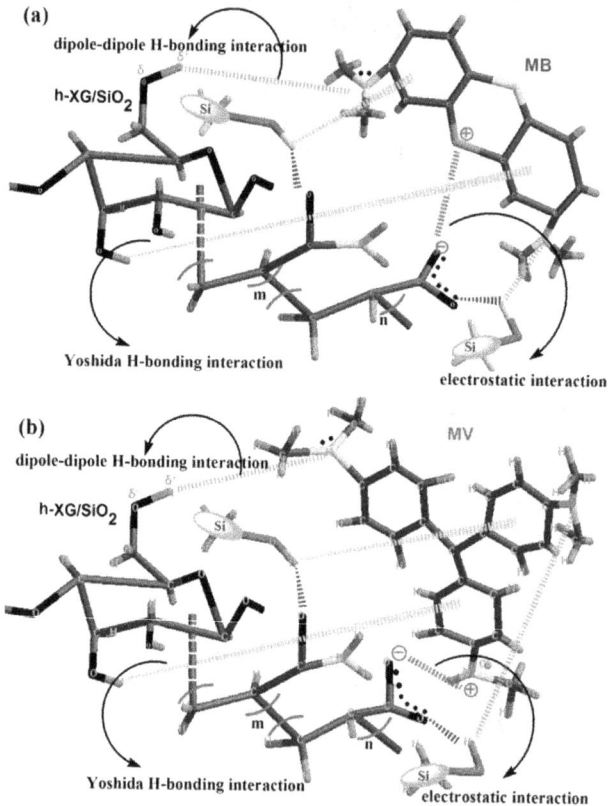

Figure 4.5 Adsorption mechanism of (a) MB and (b) MV using xanthan gum-silica nanocomposites. Reproduced from Reference 85 with permission from American Chemical Society.

The observations made in this review clearly indicate that a large number of studies have been undertaken on establishing the application of xanthan gum in the drilling and EOR applications. Xanthan gum has been proved to be an excellent environmental friendly EOR agent suitable for effective oil recovery, particularly under extreme reservoir conditions of temperature, hardness and salinity. It provides excellent viscosity generating capacity and good stability under mechanical shear and high salinity conditions. Xanthan gum has also been reported to have application in the nano-fluids, which demonstrate potential usage in upstream oil and gas industry for enhancing the performance of different processes like exploration, drilling as well as completion, production and EOR operations. Further, the polymer has been observed to exhibit corrosion inhibiting ability for mild steel in acidic conditions. It is also considered as one of the most commonly used types of linear gels used as fracturing fluids. In summary, there is a significant anticipation that xanthan gum can play a vital role in increasing the production of oil since the utilization of polymer flooding has displayed recovery of more than 20% additional oil from the OOIP. In the coming years, more application of xanthan gum are also anticipated to be developed.

References

1. Sarwat, F., Ul Qader, S. A., Aman, A., and Ahmed, N. (2008) Production and characterization of a unique dextran from an indigenous Leuconostoc mesenteroides CMG713. *International Journal of Biological Sciences*, **4**, 379-386.
2. Margaritis, A., and Zajic, J. E. (1978) Biotechnology review: mixing mass transfer and scale-up of polysaccharide fermentations. *Biotechnology and Bioengineering*, **20**, 939-1001.
3. W. H. Mcneely (1967) Biosynthetic polysaccharides. In: *Microbial Technology*, Peppler, H. J. (ed.), Reinhold Publishing Corp., USA, p. 381.
4. Milas, M., Reed, W. F., and Printz, S. (1996) Conformations and flexibility of native and renatured xanthan in aqueous solutions. *International Journal of Biological Macromolecules*, **18**, 211-221.
5. Katzbauer, B. (1998) Properties and applications of xanthan gum. *Polymer Degradation and Stability*, **59**, 81-84.
6. Rosalam, S., and England, R. (2006) Review of xanthan gum production from unmodified starches by Xanthomonas comprestris sp. *Enzyme and Microbial Technology*, **39**(2), 197-207.
7. Lee, B. H. (2014) *Fundamentals of Food Biotechnology*, 2nd edition, Wiley-Blackwell, USA.
8. Abidin, A. Z., Puspasari, T., and Nugroho, W. A. (2012) Polymers for enhanced oil recovery technology. *Procedia Chemistry*, **4**, 11-16.

9. Melton, L. D., Mindt, L., Rees, D. A., and Sanderson, G. R. (1976) Covalent structure of the polysaccharide from Xanthomonas campestris: evidence from partial hydrolysis studies. *Carbohydrate Research*, **46**, 245-257.

10. Jansson, P. E., Kenne, L., and Lindberg, B. (1975) Structure of the exocellular polysaccharide from Xanthomonas campestris. *Carbohydrate Research*, **45**, 275-282.

11. Sandford, P. A., and Baird, J. (1983) Industrial utilization of polysaccharides. In: *The Polysaccharides*, Aspinall, G. O. (ed.), volume 2, Academia Press, USA, pp. 411-490.

12. Milas, M., and Rinaudo, M. (1979) Conformational investigation on the bacterial polysaccharide xanthan. *Carbohydrate Research*, **76**, 189-196.

13. Morris, E. R. (1977) Molecular origin of xanthan solution properties. In: *Extracellular Microbial Polysaccharides*, ACS Symposium Series, volume 45, USA, pp. 81-89.

14. Garcia-Ochoa, F., Santos, V. E., Casas, J. A., and Gomez, E. (2000) Xanthan gum: production, recovery, and properties. *Biotechnology Advances*, **18**, 549-579.

15. Jampala, S. N., Manolache, S., Gunasekaran, S., and Denes, F. S. (2005) Plasma-enhanced modification of xanthan gum and its effect on rheological properties. *Journal of Agricultural and Food Chemistry*, 53, 3618-3625.

16. Palaniraj, A., and Jayaraman, V. (2011) Production, recovery and applications of xanthan gum by Xanthomonas campestris. *Journal of Food Engineering*, **106**(1), 1-12.

17. Letisse, F., Chevallereau, P., Simon, J. L., and Lindley, N. D. (2001) *Applied Microbiology and Biotechnology*, **55**, 417-422.

18. Rosalam, S., and England, R. (2006) Review of xanthan gum production from unmodified starches by Xanthomonas camprestris sp. *Enzyme and Microbial Technology*, **39**, 197-207.

19. El-Salam, M. H. A., Fadel, M. A., and Murad, H. A. (1994) Bioconversion of sugarcane molasses into xanthan gum. *Journal of Biotechnology*, **33**, 103-106.

20. Amanullah, A. S., Satti, S., and Nienow, A. W. (1998) Enhancing xanthan fermentations by different modes of glucose feeding. *Biotechnology Progress*, **14**, 265-269.

21. Rodriguez, H., and Aguilar, L. (1997) Detection of Xanthomonas campestris mutants with increased xanthan production. *Journal of Industrial Microbiology and Biotechnology*, 18(4), 232-234.

22. Esgalhado, M. E., Roseiro, J. C., and Amaral, C. M. T. (1995) Interactive effects of pH and temperature on cell growth and polymer production by Xanthomonas campestris. *Process Biochemistry*, **30**(7), 667-671.

23. Shu, C. H., and Yang, S. T. (1991) Kinetics and modeling of temperature effects on batch xanthan gum fermentation. *Biotechnology and Bioengineering*, **37**, 567-574.

24. Amanullah, A., Tuttiett, B., and Nienow, A. W. (1998) Agitator speed and dissolved oxygen effects in xanthan fermentations. *Biotechnology and Bioengineering*, **57**(2), 198-210.

25. Cacik, F., Dondo, R. G., and Marques, D. (2001) Optimal control of a batch bioreactor for the production of xanthan gum. *Computers and Chemical Engineering*, **25**, 409-418.

26. Flores Candia, J.-L., and Deckwer, W.-D. (1999) Xanthan gum. In: *Encyclopedia of*

Bioprocess Technology: Fermentation, Biocatalysis, and Bioseparation, Flickinger, M. C., and Drew, S. W. (eds.), volume 5, Wiley, USA, pp. 2695-2711.

27. Kennedy, J. F., and Bradshaw, I. J. (1984) Production, properties and applications of xanthan. *Progress in Industrial Microbiology*, **19**, pp. 319-371.

28. Rinaudo, M., and Milas, M. (1978) Polyelectrolite behaviour of a bacterial polysaccharide from Xanthomonas campestris: comparison with CMC. *Biopolymers*, **17**, pp. 2663-2678.

29. Margaritis, A., and Zajic, J. E. (1978) Biotechnology review: mixing, mass transfer and scale-up of polysaccharide fermentation. *Biotechnology and Bioengineering*, **20**, pp. 939-1001.

30. Lee, B. H. (1996) *Fundamentals of Food Biotechnology*, VCH Publishers Inc., USA.

31. Moraes, I. C. F., Fasolin, L. H., Cunha, R. L., and Menegalli, F. C. (2011) Dynamic and steady-shear rheological properties of xanthan and guar gums dispersed in yellow passion fruit pulp. Brazilian Journal of Chemical Engineering, 28(3), doi: 10.1590/S0104-66322011000300014.

32. Kang, F. S., and Pettit, D. J. (1993) Xanthan, gellan, welan, and rhamsan. In: *Industrial Gums. Polysaccharides and their Derivatives*, Whistler, R. L., and BeMiller, J. N. (eds.), 3rd edition, Academic Press, USA, pp. 341-399.

33. Trilsbach, G. F., Pielken, P., Hamacher, K., and Sahm, H. (1984) Xanthan formation by X. campestris under different culture conditions. *Third European Congress on Biotechnology*, Verlag Chemie. Germany, pp. 65-70.

34. Souw, P., and Demain, A. (1979) Nutritional studies on xanthan production by Xanthomonas campestris NRL B-1459. *Applied Environmental Microbiology*, **37**, 1186-1192.

35. Tait, M. I., Sutherland, I. W., and Sturman, C. (1986) Effect of growth conditions on the production, composition and viscosity of Xanthomonas campestris exopolysaccharide. *Journal of General Microbiology*, **132**, 1483-1492.

36. Tako, M., and Nakamura, S. (1984) Rheological properties of deacetylated xanthan in aqeous media. *Agricultural and Biological Chemistry*, **48**, 2987-2993.

37. Holzwarth, G., and Ogletree, J. (1979) Pyruvate-free xanthan. *Carbohydrate Research*, **76**(1), 277-280.

38. Morris, E. R., Rees, D. A., Young, G., Walkinshaw, M. D., and Darke, A. (1977) Order–disorder transition for a bacterial polysaccharide in solution-role for polysaccharide conformation in recognition between Xanthomonas pathogen and its plant host. *Journal of Molecular Biology*, **110**, 1-16.

39. De Mello Luvielmo, M., Borges, C. D., de Oliveira Toyama, D., Vendruscolo, C. T., Scamparini, A. R. P. (2016) Structure of xanthan gum and cell ultrastructure at different times of alkali stress. *Brazilian Journal of Microbiology*, **47**(1), doi: 10.1016/j.bjm.2015.11.006.

40. Chatterji, J., and Borchardt, J. K. (1981) Application of water soluble polymers in the oil field. *Journal of Petroleum Technology*, **33**(11), 2042-2056.

41. Taylor, K. C., and Nasr-El-Din, H. A. (1998) Water-soluble hydrophobically associating polymers for improved oil recovery: A literature review. *Journal of Petroleum Science and Engineering*, **19**, 265-280.

42. Nashawi, I. S. (1991) Laboratory Investigation of the Effect of Brine Composition on Polymer solution – Part 2 Xanthan gum (XG) Case. Society of Petroleum Engineers.

Online: https://www.onepetro.org/download/general/SPE-23534-MS?id=general%2FSPE-23534-MS (assessed 29th January 2017).

43. Cadmus, M. C., Jackson, L. K., Burton, K. A., Plattner, R. D., and Slodki, M. E. Biodegradation of xanthan gum by bacillus sp. *Applied and Environmental Microbiology*, **44**(1), 5-11.

44. Roy, A., Comesse, S., Grisel, M., Hucher, N., Souguir, Z., and Renou, F. (2014) Hydrophobically modified xanthan: An amphiphilic but not associative polymer. *Biomacromolecules*, **15**, 1160-1170.

45. Wang, I., Schiraldi, D. A., and Sanchez-Soto, M. (2014) Foamlike xanthan gum/clay aerogel composites and tailoring properties by blending with agar. *Industrial and Engineering Chemistry Research*, **53**(18), 7680-7687.

46. Thomas, S. (2008) Enhanced oil recovery - An overview. *Oil and Gas Science and Technology – Rev. IFP*, **63**(1), pp. 9-19.

47. Samanta, A., Ojha, K., and Mandal, A. (2011) Interactions between acidic crude oil and alkali and their effects on enhanced oil recovery, *Energy and Fuels*, **25**(4), 1642-1649.

48. Sorbie, K. S. (1991) *Polymer-Improved Oil Recovery*, Springer, USA.

49. Chelaru, C., Diaconu, I., and Simionescu, I. (1998) Polyacrylamide obtained by plasma-induced polymerization for a possible application in enhanced oil recovery. *Polymer Bulletin*, **40**, 757-764.

50. Wang, W., Liu, Y., and Gu, Y. (2003) Application of a novel polymer system in chemical enhanced oil recovery (EOR). *Colloid and Polymer Science*, **281**, 1046-1054.

51. Standnes, D. C., and Skjevrak, I. (2014) Literature review of implemented polymer field projects. *Journal of Petroleum Science and Engineering*, **122**, 761-775.

52. Guo, X. H., Li, D. W., Tian, J., and Liu, Y. Z. (1999) Pilot Test of Xanthan Gum Flooding in Shengli Oilfield. SPE 57294, *SPE Asia Pacific Improved Oil Recovery Conference*, Malaysia.

53. Harris, M. J., Herp, A., and Pigman, W. (1971) Depolymerization of polysaccharides through the generation of free radicals at a platinum surface: a novel procedure for the controlled production of free-radical oxidations. *Archives of Biochemistry and Biophysics*, **142**(2), 615-622.

54. Ash, S. G., Clarke-Sturman, A. J., Calvert, R. and Nisbet T. M. (1983) Chemical Stability of Biopolymer Solutions. SPE 12085, *58th Annual Technical Conference and Exhibition*, USA.

55. Seright, R. S., and Henrici, B. J. (1990) Xanthan stability at elevated temperatures. *SPE Reservoir Engineering*, **5**(1), 52-60.

56. Jang, H. Y., Zhang, K., Chon, B. H., and Choi, H. J. (2015) Enhanced oil recovery performance and viscosity characteristics of polysaccharide xanthan gum solution. *Journal of Industrial and Engineering Chemistry*, **21**, 741-745.

57. Xu, L., Xu, G., Liu, T., Chen, Y., and Gong, H. (2013) The comparison of rheological properties of aqueous welan gum and xanthan gum solutions. *Carbohydrate Polymers*, **92** (2013), 516-522.

58. Quy, N. M., Ranjith, P. G., Choi, S. K., Giao, P. H., and Jasinge, D. (2009) Analytical assessment of horizontal well efficiency with reference to improved oil recovery of the south-east Dragon oil field southern offshore of Vietnam. *Journal of Petroleum Science and Engineering*, **66**(3-4), 75-82.

59. Taugbol, K., Ly, T. V., and Austad, T. (1995) Chemical flooding of oil-reservoirs. 3. Dissociative surfactant–polymer interaction with a positive effect on oil-recovery. *Colloids and Surfaces A*, **103**, 83-90.
60. Austad, T., and Taugbol, K. (1995) Chemical flooding of oil-reservoirs. 2. Dissociative surfactant–polymer interaction with a negative effect on oil-recovery. *Colloids and Surfaces A*, **103**, 73-81.
61. Wei, B., Romero-Zerón, L., and Rodrigue, D. (2015) Improved viscoelasticity of xanthan gum through self-association with surfactant: β-cyclodextrin inclusion complexes for applications in enhanced oil recovery. *Polymer Engineering and Science*, **55**, 523-532.
62. Ritacco, H., Albouy, P. A., Bhattacharyya, A., and Langevin, D. (2000) Influence of the polymer backbone rigidity on polymer-surfactant complexes at the air/water interface. *Physical Chemistry and Chemical Physics*, **2**, 5243-5251.
63. Thuresson, K., Nystroem, B., Wang, G., and Lindman, B. (1995) Effect of surfactant on structural and thermodynamic properties of aqueous-solutions of hydrophobically-modified ethyl(hydroxyethyl)cellulose. *Langmuir*, **11**(10), 3730-3736.
64. Sveistrup, M., van Mastrigt, F., Norrman, J., Picchioni, F., and Paso, K. (2016) Viability of biopolymers for enhanced oil recovery. *Journal of Dispersion Science and Technology*, **37**(8), 1160-1169.
65. Choppe, E., Puaud, F., Nicolai, T., and Benyahia, L. (2010) Rheology of xanthan solutions as a function of temperature, concentration and ionic strength. *Carbohydrate Polymers*, **82**(4), 1228-1235.
66. Brazzel, R. L. (2009) Multi-component Drilling Fluid Additive, and Drilling Fluid System Incorporating the Additive, US patent 7635667.
67. Melbouci, M., and Sau, A. C. (2008) Water-based Drilling Fluids, US patent 7384892.
68. Kong, X., Ohadi, M. M. (2010) Applications of Micro and Nano Technologies in the Oil and Gas Industry - An Overview of the Recent Progress. *Proceedings of the Abu Dhabi International Petroleum Exhibition Conference*, UAE, pp. 1-4.
69. Friedheim, J., Young, S., Stefano, G., Lee, J., and Guo, Q. (2012) Nanotechnology for Oilfield Applications - Hype or Reality? SPE 157032, *SPE International Oilfield Nanotechnology Conference*, The Netherlands. Online: http://www.nioclibrary.ir/Reports/9140.pdf (assessed 31st January 2017).
70. Amani, M., Al-Jubouri, M., and Shadravan, A. (2012) Comparative study of using oil-based mud versus water based mud in HPHT fields. *Advances in Petroleum Exploration and Development*, **4**(2), 18-27.
71. Xue, D., and Sethi, R. (2012) Viscoelastic gels of guar and xanthan gum mixtures provide long-term stabilization of iron micro- and nanoparticles. *Journal of Nanoparticle Research*, **14**, 1239.
72. Ponmani, S., William, J. K. M., Samuel, R., Nagarajan, R., Sangwai, J. S. (2014) Formation and characterization of thermal and electrical properties of CuO and ZnO nanofuids in xanthan gum. *Colloids and Surfaces A: Physicochemical and Engineering Aspects*, **443**, 37-43.
73. William, J. K. M., Ponmani, S., Samuel, R., Nagarajan, R., Sangwai, J. S. (2014) Effect of CuO and ZnO nanofluids in xanthan gum on thermal, electrical and high pressure rheology of water-based drilling fluids. *Journal of Petroleum Science and Engineering*, **117**, 15-27.

74. Kennedy, J. R. M., Kent, K. E., and Brown, J. R. (2015) Rheology of dispersions of xanthan gum, locust bean gum and mixed biopolymer gel with silicon dioxide nanoparticles. *Materials Science and Engineering C*, **48**, 347-353.

75. Zhang, R., Shi, W., Yu, S., Wang, W., Zhang, Z., Zhang, B., Li, L., and Bao, X. (2015) Influence of salts, anion polyacrylamide and crude oil on nanofiltration membrane fouling during desalination process of polymer flooding produced water. *Desalination*, **373**, 27-37.

76. Zhao, D., Liu, H., Guo, W., Qu, L. and Li, C. (2016) Effect of inorganic cations on the rheological properties of polyacrylamide/xanthan gum solution. *Journal of Natural Gas Science and Engineering*, **31**, 283-292.

77. Qiu, L. G., Wu, Y., Wang, Y. M., and Jiang, X. (2008) Synergistic effect between cationic gemini surfactant and chloride ion for the corrosion inhibition of steel in sulphuric acid. *Corrosion Science*, **50**, 576-582.

78. Abiola, O. K., and James, A. O. (2010) The effects of aloe-vera extract on corrosion and kinetics of corrosion process of zinc in HCl solution. *Corrosion Science*, **52**, 661-664.

79. de Souza, F. S., and Spinelli, A. (2009) Caffeic acid as a green corrosion inhibitor for mild steel. *Corrosion Science*, **51**, 642-649.

80. Biswas, A., Pal, S., and Udayabhanu, G. (2015) Experimental and theoretical studies of xanthan gum and its graft co-polymer as corrosion inhibitor for mild steel in 15% HCl. *Applied Surface Science*, **353**, 173-183.

81. Ranade, V. R., and Ulbrecht, J. J. (1978) Influence of polymer additives on the gas-liquid mass transfer in stirred tanks. *AIChE Journal*, **24**(5), 796-803.

82. Nakanoh, M., and Yoshida, F. (1980) Gas absorption by Newtonian and non-Newtonian liquids in a bubble column. *Industrial and Engineering Chemistry Process Design and Development*, **19**, 190-195.

83. Park, S.-W., Choi, B.-S., and Lee, J.-W. (2008) Chemical absorption of carbon dioxide into aqueous elastic xanthan gum solution containing NaOH. *Journal of Industrial and Engineering Chemistry*, **14**(3), 303-307.

84. Hublik, G. (2012) Xanthan. In: *Polymer Science: A Comprehensive Reference*, volume 10, Matyjaszewski, K., and Moller, M. (eds.), Elsevier, USA, pp. 221-229.

85. Ghorai, S., Sarkar, A., Raoufi, M., Panda, A. B., Schonherr, H., and Pal, S. (2014) Enhanced removal of methylene blue and methyl violet dyes from aqueous solution using a nanocomposite of hydrolyzed polyacrylamide grafted xanthan gum and incorporated nanosilica. *ACS Applied Materials and Interfaces*, **6**(7), 4766-4777.

5

Polymeric Materials for Oil Spill Clean Up and Phenol Removal

5.1 Introduction

Today, large amount of oil-wastewater is generated by the oil and gas industry [1], which is hazardous to the environment [2]. In addition, refined petroleum products and crude oil are accidently spilled into water bodies from the sources such as offshore platforms, oil cargo ships and tankers, which causes serious damage to marine environment [3]. There are many instances of oil spill accidents like Torrey Canyon oil spill in 1967, Mexico oil spill in 2010 (210 million gallons of oil spilled into sea) and Bohai Bay oil spill in 2011. In order to protect the marine ecosystem, oil industries and environment protection authorities are in continuous need of effective methods for successful separation of the water-oil mixture. Oil spills lead to different forms with water, such as stable water-in-oil emulsion, unstable water-in-oil emulsion and free-floating oil [4]. Separation of unstable oil-in-water emulsion or free-floating oil is applicable by the conventional techniques (flotation, gravity separation, coagulation, ultrasonic separation, skimming and centrifugation), however, stable oil-in-water emulsions cannot be separated by applying these techniques [5-7]. High operational cost and low efficiency are the other disadvantages of the conventional techniques. To overcome these challenges, some promising membrane separation based methods have also been developed, such as reverse osmosis [8], ultrafiltration [9,10], microfiltration [11], dehydration of oil emulsion by pervaporation [12], membrane distillation [13] and flocculation followed by microfiltration [14]. Current day techniques employed for the cleaning of oil spills are of three types such as extraction of oil from the surface of water using absorbents [15-19], applying dispersant agents for oil degradation [20,21], and controllable burning of oil spill. Among these techniques, oil removal by absorbents is considered a promising method for allowing effective removal of oil from water. Oil absorbents are basically categorized into three classes, i.e. inorganic mineral product (zeolites [22,23], perlite [24] and silica [25,26]), synthetic polymers [27, 28] and natural organics (rice straw [29], corn cob [30], wool fiber [31], kapok fiber [32] and milkweed fiber [18]). For oil absorption

Naman Arora and Vikas Mittal*, The Petroleum Institute (part of Khalifa University of Science and Technology), Abu Dhabi, UAE
*Current address: Bletchington, Wellington County, Australia

process, absorbents are facilitated with distinct advantages such as simple operation, cost effectiveness, environment friendliness and ability to clean oil spills without any secondary pollution [5,33,34]. An ideal oil absorbent material should also possess low density, high selectivity, high oil absorption ability and excellent reusability [15]. Research studies have also underlined the advantages of special wettability-controlled oil-water mixture separation method over traditional separation methods. Superhydrophobic or superoleophobic surfaces possess a water or oil contact angle (CA) higher than 150° and sliding angle less than 10° [35-38]. These materials have generally been fabricated by either designing with superoleophobic-superhydrophilic surface or developing with superhydrophobic - superoleophilic surface. There are two strategies to fabricate superhydrophobic - superoleophilic materials, either by constructing a rough structure on surface or modifying a surface with low surface energy chemicals [39]. Generally, superoleophobic surface exhibits superhydrophobicity [40-42] and superhydrophilic surface shows superoleophilicity [43,44]. However, it has been observed that hydrophilic and oleophobic properties can be generated simultaneously due to some stimuli-responsive surfaces, based on unfavourable interaction with non-polar liquids (hexadecane) and a favourable interaction with polar liquids (water) [45-47]. Materials with superhydrophobic - superoleophilic or superoleophobic-superhydrophilic surfaces have a significant role in oil-water separation application.

Phenols and its derivatives are one of the major organic pollutants which lead to significant water pollution. Presence of phenolic compounds in water is very concerning due to their high oxygen demand, low biodegradability and high toxicity. For the treatment of industrial wastewater containing phenolic derivatives, many techniques have been employed such as adsorption, catalytic oxidation and separation via membrane [48-52]. Phenolic compounds are introduced into water bodies by effluents of industries like petroleum, plastic, pharmaceutical, coke manufacturing, rubber, textiles and pesticides manufacturing [53-55]. Long-term intake of toxic phenol and its derivatives can cause severe chronic insomnia, loss of appetite, rapid fatigue and headache [56]. Overall, adsorption technology is widely employed for organic and inorganic micro-pollutants removal from industrial wastewater.

In this article, various polymer based sponges/foams (Section 5.2), aerogels (Section 5.3), fibers (Section 5.4), membranes (Section 5.5) and other polymer absorbents (Section 5.6) have been reviewed for the purpose of removal of oil from water. In Section 5.7, polymeric adsorbents for the adsorption of phenol and its derivatives from aqueous and other solutions have also been reviewed in detail.

5.2 Polymer Based Sponges and Foams for Oil Spill Clean Up

A wide variety of sponges and foams have been employed as oil absorbents for the efficacious absorption of oil from water bodies. Wang *et al.* [57] reported the preparation of carbon nanotube (CNT)/polydimethylsiloxane (PDMS) coated polyurethane (PU) sponge by using dip coating method to absorb variety of organic impurities or oils from water. PU sponges as 3D porous materials are commercially available with the capability to absorb water as well as organic solvents. By anchoring CNT/PDMS coatings onto the frame of hydrophilic PU sponge, the wettability of sponge was changed from hydrophilic to superhydrophobic. It was observed that the water contact angle of the PDMS-coated PU sponge was $140\pm3°$ which could not be classified as superhydrophobic material. However, the combination of micro-scale surface roughness of PU sponge and hydrophobic CNT/PDMS nanocomposites provided the hybrid with a water contact angle of $162\pm2°$. The resulting superhydrophobic sponge exhibited an absorption capacity 15 - 25 times its own weight to remove the organic solvents or oils, depending on their viscosities and densities. The surface of sponge revealed its crater like nanostructures which featured Cassie-Baxter surfaces. This kind of structure increased the roughness of surface and further led to a composite interface where the air got trapped within the grooves beneath the liquid and as a result, induced the superhydrophobicity. In order to achieve continuous removal of 20 L of oil (gasoline, n-hexane, or n-hexadecane) from water surface, a small piece (0.42 g) of CNT/PDMS-coated PU sponge (2.0 x 2.0 x 3.0 cm^3) was connected as a filter in a tube in conjunction with a vacuum system (Figure 5.1). The resulting assembly was capable to separate large amount of oil up to 35000 times its own weight from water surface. Jiang *et al.* [58] reported the fabrication of amine functionalized silica nanoparticles coated polyurethane/polytetrafluoroethylene (PTFE) sponge. In this study, commercially available PU sponge was deposited with PTFE. Subsequently, PU/PTFE sponge was immersed into A-SiO$_2$ (amine treated) nanoparticles suspension followed by the immersion in SiO$_2$ nanoparticles suspension. The resulting sponge was the treated with silane (octyltrichlorosilane) by using vapour phase deposition technique, providing a large surface roughness. It resulted in increasing the water contact angle (WCA) and roll-off angle which confirmed it as superhydrophobic material to remove oil or organic solvent from water. Zhang *et al.* [59] reported the synthesis of nanocellulose sponges which are flexible, ultra-light in weight and hydrophobic by using a novel and efficient silylation process in water. The materials had very high porosity (≥99%), which was engineered by freeze drying of the nanofibrillated cellulose (NFC) water

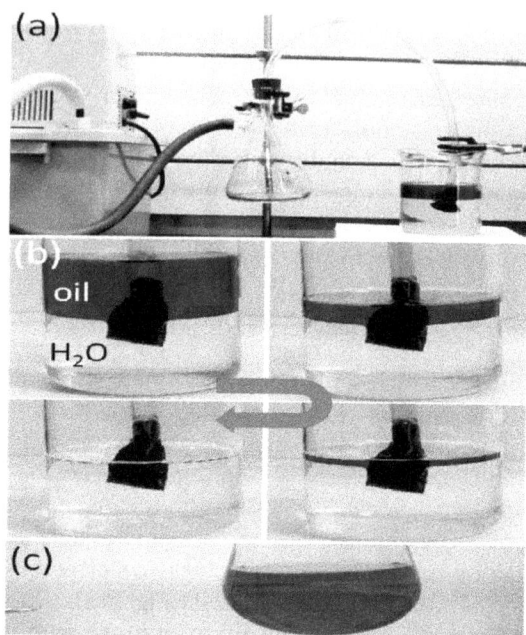

Figure 5.1 Images of (a) continuous oil/water separation system, (b) progress of the continuous absorption and removal of an organic solvent from the water surface, and (c) oil collected in part b. Reproduced from reference 57 with permission from American Chemical Society.

suspension in the presence of methylmethoxysilane (MTMS) sol. The silylated sponges were able to remove dodecane spills from water (Figure 5.2). The authors also investigated the absorption capacity of sponge for various oils and other organic solvents which were found to be 49-102 (g/g) depending on the nature of the liquids. Wang *et al.* [60] reported the microfibrillated cellulose fibers (MCF sponges) modified with MTMS (methylmethoxysilane) via a chemical vapour deposition process to make it a superhydrophobic material. The sponges exhibited excellent recyclability (more than 30 times), elasticity, flexibility and porosity (up to 99.84%). The absorption capacities of sponge for pump oil, motor oil, white oil and silicone oil were determined as 197, 198, 178 and 228 g/g, respectively. The sponge also exhibited high absorption capacities for a variety of organic solvents, ranging from 88 g/g to 163 g/g depending on the viscosity and density of the liquid. Wu *et al.* [61] reported the polyurethane/silica nanoparticles (PU/SiO$_2$ NPs) sponge subsequently modified with

Unmodified

Silylated (18.9 wt% Si)

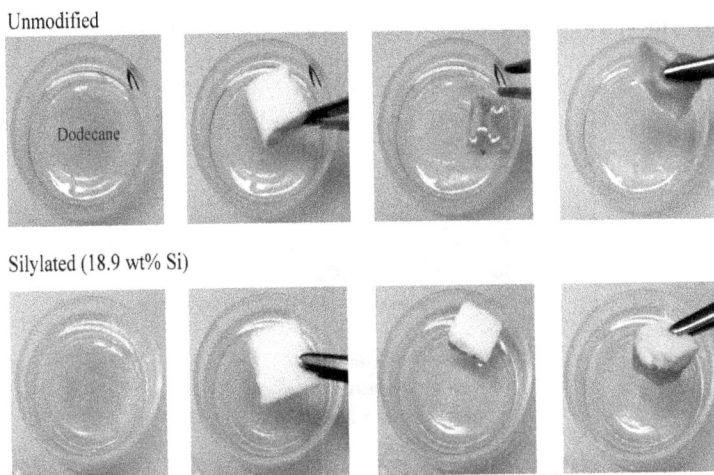

Figure 5.2 Removal of a red colored dodecane spill (0.02 g) from water with the silylated NFC sponge (0.02 g). In comparison, the unmodified material was not selective and lost its original shape. Reproduced from Reference 59 with permission from American Chemical Society.

gasoline as a hydrophobic material. The absorption capacity of the sponge for diesel, peanut oil and motor oil were calculated as 95±3, 108±4, and 103±3 g/g, respectively. The resultant sponge was subjected to 15 absorption-squeeze cycles, which gradually reduced its oil sorption capacity from 103 to 76 g/g. Liu *et al.* [62] reported reduced graphene oxide (rGO) coated polyurethane sponge fabricated by using a facile method. Thermo-gravimetric analysis (TGA) revealed that the amount of rGO coated onto the PU sponge was too less to change the decomposition behaviour of the PU sponge. The sponge exhibited good strength, elasticity, low density (0.0088 g/cm³), porosity (99.29%), excellent recyclability (50 times without deterioration) and high hydrophobicity, the properties which confirmed its suitability as absorbent for oil spill clean up. It was subjected to variety of organic solvents and oils, the absorption capacities were observed to be higher than 80 g/g for all the liquids and the highest value was recorded as 160 g/g for chloroform (Figure 5.3).

Pan *et al.* [63] reported hydrophobic polyvinyl-alcohol formaldehyde (PVF) sponges. The authors modified the PVF sponge by treating it with the mixture of acetonitrile and stearoyl chloride in the presence of pyridine at different temperatures. Hydrophobicity of sponge was enhanced with increasing the reaction temperature, as confirmed by the solubility parameter measurements. The successful anchoring of hydrophobic stearoyl groups onto the PVF frame was confirmed by ATR-IR spectroscopy. The resultant hydrophobic PVF sponges exhibited absorption capacities for oil products from 13.7 g/g to 56.6 g/g, maintaining high porosity (≥94.8%) and good reusability for 35 absorption- squeeze

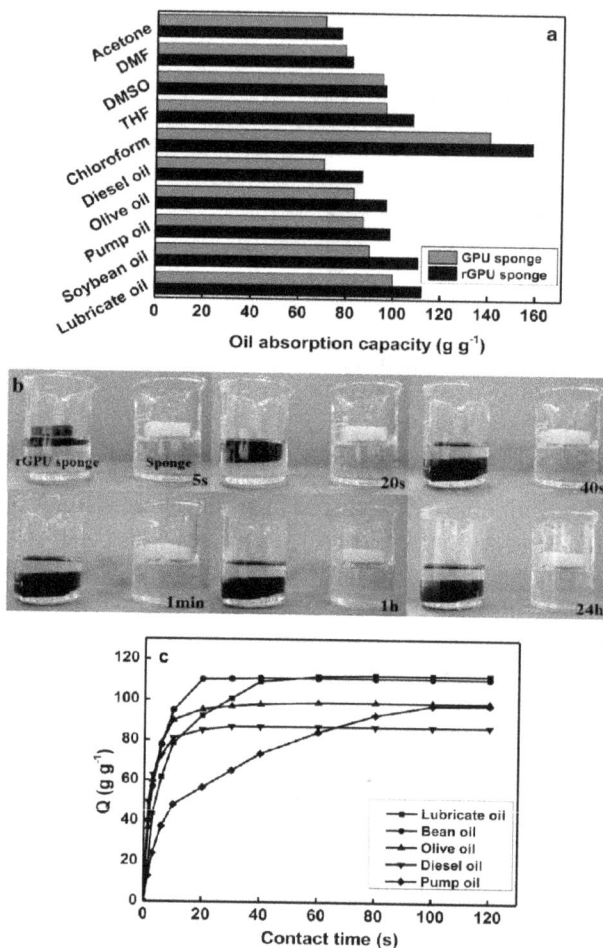

Figure 5.3 (a) Absorption capacity of the GPU and rGPU sponges for various organic liquids, (b) images of the absorption processes for the PU and rGPU sponges, (c) absorption capacities of the rGPU sponge versus contact time with various oils. Reproduced from Reference 62 with permission from American Chemical Society.

cycles. Nguyen *et al.* [64] fabricated graphene based melamine sponge via dip coating and facile method. Anchoring of hydrophobic graphene onto the sponge skeleton occurred due to mechanical flexibility of graphene and strong van der Waals interactions between the graphene and sponge skeleton, which led the water contact angle to change from 0° to 162°. The authors observed that the sponge with graphene loading ≤ 5.1% did not restrict the absorption of water, however, as the loading of graphene was increased to ≥7.3%, water droplets attained the water contact angle of about 160° with quasi-spherical shapes on the surface of the sponge, representing superhydrophobicity. The sponge exhibited superhydrophobic and superoleophilic surfaces because of the combined behaviour of the graphene micro/nano-framework, hydrophobic property of graphene and the micro-porous structure of melamine sponge. The absorption capacities of resultant sponge were up to 165 times its own weight for different liquids, along with high selectivity and excellent recyclability. However, during recycling of such coated sponges, the linkage between graphene and sponge skeleton could be affected which can further reduce the hydrophobic property of the coated sponge. For ensuring the adhesion between graphene nano-sheets and sponge skeleton, the sponge was immersed in dilute solution of PDMS in xylene (0.25 mg/ml), followed by overnight drying in vacuum oven at 120 °C. The formation of PDMS thin film onto the graphene-coated sponge cemented graphene with sponge skeleton without affecting the superhydrophobicity and superoleophilicity of the resultant sponge. Zhu *et al.* [65] reported robust superhydrophobic PU sponge, fabricated by one-step solution immersion method. The polyurethane sponge was poured into hexane solution containing methyltrichlorosilane (0.5 - 3%, v/v) for 30 min followed by air drying with a relative humidity of 30 - 70% to evaporate hexane and further dried for 2 h at 60 °C to obtain a water contact angle in the range of 142-157°. The sponge could be reused for more than 300 cycles to remove oil from aqueous solution and exhibited absorption capacities about 12 - 25 times its own weight for oil and organic solvents.

Turco *et al.* [66] reported embedded magnetic multiwalled carbon nanotubes (MWNTs) into a PDMS porous sponge prepared by polymerizing the PDMS prepolymer in the presence of magnetic MWNTs covered hard template. The authors generated 3% MWNTs - PDMS and 1% MWNTs - PDMS sponges, with water contact angles of 153.4±6.9° and 146.6±4.6°, respectively. The presence of magnetic MWNTs improved the oil absorption capacity, mechanical strength, chemical stability and thermal stability of the resultant sponges. Scanning electron microscopy (SEM) revealed that superoleophilic and superhydrophobic properties of PDMS-MWNTs sponge occurred due to the combination of

MWNTs and methyl groups of PDMS. Basically, presence of methyl groups on the surface of sponge reduced the surface energy of sponge, along with enhancement of this effect by MWNTs also enhanced this effect, thereby effectively increasing the surface roughness of the sponge. The reusability of sponges was checked up to 10 cycles by using simple squeezing method, suggesting that no significant variation of oil absorption capacity and water contact angle after each cycle were noticed. The oil mass absorption ability of the PDMS - MWNT sponges were determined in the range of 8.5 to 20 g/g depending on the surface tension, viscosity and density of oils and organic solvents.

Zhang et al. [67] reported swellable porous PDMS oil absorbent, which had three dimensional interconnected porous structure, good selectivity for oil/water separation, excellent reusability for 20 absorbing-recovering cycles without losing absorption capacity. High absorption capacities of 4 - 34 g/g were determined for various oils and organic solvents depending on their viscosity and density. For crude oil, diesel and gasoline, the absorption capacities were determined as 9, 12, and 22 g/g, respectively. Swellable porous PDMS was prepared by a modified sugar-template method. The method involved the direct curing of a PDMS prepolymer in the solution of p-xylene in the presence of commercially available sugar particles. The authors also observed that on increasing the amount of p-xylene from 0 to 2 times that of PDMS, the absorption capacity of PDMS oil absorbent for petroleum ether also increased from 13.8 to 20.8 g/g. More importantly, a gel state of PDMS occurred after curing due to the presence of p-xylene, which resulted in a swellable structure of PDMS. The porosity of swellable porous PDMS was observed in the range of 50 - 81% depending on the sugar-particle template size. The water contact angle of the sponge was determined as 144°.

Nagappan et al. [68] reported bio-inspired hybrid- PDMS based melamine sponge with high selectivity for oil/water separation, excellent recyclability, water contact angle > 175° and high absorption capacities ranging from 2000 to 3600 w/w%. The sponge was prepared by using polymethylhydoxysiloxane (PMHOS), lotus leaf powder and phenyl substituted silica ormosil (PSiOr). The authors observed that the superhydrophobic property of hybrid occurred due to the presence of PMHOS and PSiOr additionally enhanced the surface roughness, surface area, adhesiveness and water contact angle of hybrid on the substrate. Bio-inspired micro/nano-composites hybrid could, thus, be confirmed as an effective and environmental friendly absorbent for oil spill clean up application. In another study reporting the use of sponges for oil spill clean up, Zhu et al. [69] studied the fabrication of copper-film coated superhydrophobic and superoleophilic PU sponges by using solution-immersion method. It was observed that the as-prepared sponges did not sink beneath the water surface, while the original sponge submerged into the water bath. Silver-mirror like surface appeared when the as-prepared sponges were partially or totally immersed in water by an external force, which exhibited Cassie-Baxter non-wetting behaviour owing to a continuous air layer between the water and superhydrophobic surface (Figure 5.4).

Figure 5.4 Images of a piece of the (a) as-prepared sponge, (b) original sponge, (c) as-prepared sponge placed on a water bath and (d) as-prepared sponge immersed in the water bath by an external force. Reproduced from reference 69 with permission from American Chemical Society.

Li *et al.* [70] reported three-dimensional graphene/polypyrrole (PPy) foams for the clean up of oil spills. Tanobe *et al.* [71] reported the modification of post-consumed polyurethane foams by grafting with polystyrene, thereby generating an oil absorbent for the cleaning of oil spills. Calcagnile *et al.* [72] reported the functionalization of commercially available polyurethane foam with oleic acid-capped colloidal iron oxide nanoparticles, followed by the treatment with PTFE particles to obtain superhydrophobic and superoleophilic material for the absorption of oil from polluted regions.

The details of various polymeric sponges or foams are also listed in Table 5.1.

Table 5.1 Polymeric sponges or foams for oil spill clean up

Authors	Sponge/Foams	Method	Oils/Organic solvents	Absorption capacity (g/g) or (times) or (w/w%)	WCA	OCA	Ref.
Wang et al.	CNT-PDMS coated PU sponge	Dip-coating	n-Hexadecane n-Hexane Gasoline Soybean oil Used motor oil Diesel	[15-25 times its own weight]	162±2°	0° 0° 0° - - -	[57]
Jiang et al.	PU/PTFE/A-SiO$_2$/SiO$_2$ coated PU sponge	Vapour phase deposition	[n-Hexane, Pentane, Heptane, Benzene, Toluene, Silicone oil]	-	~ 165°	-	[58]
Zhang et al.	Silylated nano-fibrillated cellulose sponge	Freeze drying method	[Dodecane, Acetone, Silicone oil, Chloroform, Toluene, Ethanol, Mineral oil, Dichloromethane]	[49-102 g/g]	136°	-	[59]
Wang et al.	MTMS modified micro-fibrillated cellulose sponge	Freeze drying method	White oil Silicone oil Pump oil Motor oil [Cyclohexane, DMF, Acetone, DMSO, Hexane, Ethanol, Toluene, Dichloromethane, Isopropyl alcohol]	178 g/g 228 g/g 197 g/g 198 g/g [88-163 g/g]	154°	-	[60]

Table 5.1 (continued)

Authors	Sponge/Foams	Method	Oils/ Organic solvents	Absorption capacity (g/g) or (times) or (w/w %)	WCA	OCA	Ref.
Wu et al.	SiO$_2$ NPs/polyurethane sponge modified with gasoline	Dip coating method	Motor oil Peanut oil Diesel	103±3 g/g 108±4 g/g 95±3 g/g	126°	-	[61]
Liu et al.	Reduced graphene oxide coated PU sponge	Dip coating method	[Acetone, DMF, DMSO,THF, Chloroform, Diesel, Olive oil, Pump oil, Soybean oil, Lubricate oil]	[80-160 g/g]	127°	0°	[62]
Pan et al.	Hydrophobic PVF sponge	Wet chemical synthesis	[Toluene, n-Hexane, Kerosene, soybean oil, Hydraulic oil, Crude oil]	[13.7-56.6 g/g]	138±1.5°	-	[63]
Nguyen et al.	Graphene based Melamine sponge	Dip coating method	[Chloroform, DMF, Hexane, Acetone, Ethanol, Methanol, Used pump oil, Pump oil, Soybean oil, Motor oil]	[54-165 times its own weight]	162°	-	[64]
Zhu et al.	Methyltrichlorosilane treated polyurethane sponge	one-step solution immersion method	[Decane, Dodecane, Octane, Crude oil, Gasoline, Bean oil, Lubricating oil]	[15-25 times its own weight]	157°	0°	[65]

Table 5.1 (continued)

Authors	Sponge/Foams	Method	Oils/ Organic solvents	Absorption capacity (g/g) or (times) or (w/w %)	WCA	OCA	Ref.
Turco et al.	MWNT-PDMS sponge	Template assisted	[Dichloromethane, Petroleum ether, Hexane, Chloroform, THF, Toluene, Gasoline]	[8.5 to 20 g/g]	153.4±6.9°	0°	[66]
Zhang et al.	Swellable porous PDMS absorbent	modified sugar-template method	Crude oil Diesel Gasoline [Chloroform, Cyclohexane, dichloromethane, Petroleum ether, n-Hexane, Toluene, Acetone]	9 g/g 12 g/g 22 g/g [4-34 g/g]	144°	0°	[67]
Nagappan et al.	Bio-inspired hybrid-PDMS/Melamine sponge	Dip coating method	[Chloroform, Toluene, Benzene, Decane, Diesel oil, Canola oil, Corn oil, Soybean oil]	[2000 to 3600 w/w%]	> 175°	<5°	[68]
Zhu et al.	Copper film-coated PU sponge	Solution-immersion process	[Lubricating oil, Octane, Decane and Dodecane]	[13 times its own weight]	> 170°	-	[69]
Li et al.	3D-graphene/polypyrrole foam	Vacuum freeze drying method	Kerosene Diesel	101.8 g/g 108.76 g/g	-	-	[70]
Calcagnile et al.	PU/NPs/PTFE based foam	Deposition technique	Mixed oil with blue dye	-	>160°	-	[72]

5.3 Polymer Based Aerogels for Oil Spill Clean Up

Aerogels have been employed as effective oil absorbents due to their low density, large porosity and high specific surface area. These lowest-density solids are composed of about 99.98% air by volume, but are still extremely rigid. Aerogels can easily bear many times more weight than their own weight and have high ability for oil spill clean up [73]. Chin *et al.* [74] reported the preparation of cellulose aerogel with magnetic and hydrophobic properties via *in-situ* incorporation of magnetic Fe_3O_4 nanoparticles (NPs) in it and further followed by TiO_2 thin layer coating on the surface of the aerogels. After the CO_2 supercritical point drying of cellulose gel, the magnetic cellulose aerogel was obtained. TEM revealed that the mean particle size of Fe_3O_4 NPs ranged between 7 and 10 nm and were encapsulated between the cellulose fibers. SEM analysis also exhibited that the magnetic cellulose aerogel was comprised of nanofiber networks and the coating of TiO_2 led to nanofibers' surfaces rougher and thicker. Hydrophobicity and oleophilicity of the aerogel was attributed to the thin layer of TiO_2 which was coated onto the cellulose aerogel. The oil absorption capacity of the magnetic cellulose aerogel for paraffin oil was up to 28 times its own weight and it could be easily removed by the usage of external magnet during the absorption process. Korhonen *et al.* [17] reported the nanocellulose aerogels possessing high porosity, selectivity for oil-water separation and good recyclability for up to 10 cycles without deterioration of absorption capacity and weight. The authors functionalized the cellulose nanofibrils of aerogel with uniform nanoscopic coating of TiO_2 NPs (7 nm), which was responsible for the hydrophobic and oleophilic properties of nanocellulose aerogel. Absorption capacities of the aerogel varied from 20-40 g/g, depending on the density of the different liquids. Different oils and organic solvents were examined for absorption capacity of the nanocellulose aerogel such as hexadecane, paraffin oil, petroleum benzene, mineral oil, octane, etc.

Nguyen *et al.* [73] reported highly porous aerogel prepared from the cellulose fibers of paper waste. The authors functionalized the cellulose aerogel with methyltrimethoxysilane (MTMS) to develop its hydrophobic and oleophilic properties. MTMS-coated cellulose aerogel exhibited water contact angles of 143° and 145°, as shown in Figure 5.5. In this study, Ruby (RB), Te Giac Trang (TGT) and Rang Dong (RD) crude oils were used to examine the oil absorption capacity of the aerogel at 25 °C. The absorption capacity of RB, TGT and RD were determined as 18.4, 18.5 and 20.5 g/g, respectively. The absorption capacity of the absorbent is controlled by the hydrophobic interaction between the oils and absorbents, oil viscosity, van der Waals forces, pore morphology and capillary effect [16, 75-79]. This is probably a reason that RB and TGT had almost similar oil absorption capacity because of their comparable viscosity values (0.0090 and 0.0088, respectively) and the absorption capacity of RD was observed to be higher because of lower viscosity value (0.0062) of RD crude oil. The authors also examined the effect of temperature on oil absorption capacity of MTMS-coated cellulose aerogel. As the temperature was increased from 10 to 25 °C, the crude oil (RB) absorption capacity also increased from 13.9 to 18.4 g/g, achieving the highest oil sorption value (24.4 g/g) at 40 °C and subsequently decreased to 19.9 g/g at 60 °C. Sai *et al.* [80] also reported surface modified bacterial cellulose aerogel, which exhibited high porosity (~99.6%), high surface area (≥169.1 m^2/g) and low density (≤6.77 mg/cm^3). The water contact angle and absorption capacities for various oils of the aerogel were observed to be 146.5° and ~185 g/g, respectively. Si *et al.* [81] reported superhydrophobic-

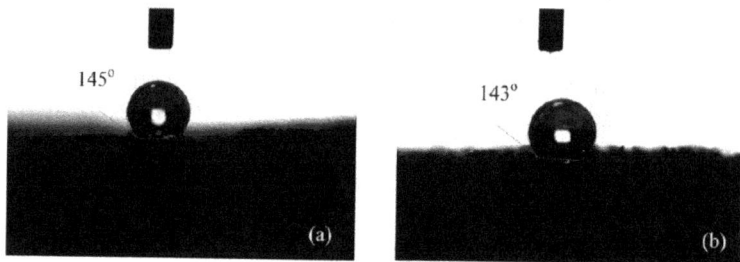

Figure 5.5 (a) Water contact angle on the external surface of the coated aerogel and (b) on the cut surface of the coated aerogel. Reproduced from Reference 73 with permission from American Chemical Society.

superoleophilic polyacrylonitrile (PAN)/SiO$_2$ nanofiber based aerogels for the separation of the water-in-oil emulsions. SEM revealed that the fiber aerogels had open-cell geometry with a major cellular pore size of 10-50 μm. Due to the numerous minor cellular pores (1-5 μm sizes), the major cells were observed to be highly interconnected. With increasing content of SiO$_2$ (from 0.01 to 2 wt%), the average pore size of the aerogels got reduced from 18.9 to 4.2 μm and in-contrast, the specific surface area increased from 2.66 to 76.54 m^2/g. After introducing the SiO$_2$ nanoparticles into the aerogels, the surface roughness, wettability and WCA (162°) were increased. Fibrous aerogels also exhibited anti-fouling property for long usage, excellent reusability and super elasticity. Jiang *et al.* [82] also reported the ultra-porous (99.5 to 99.9%) and ultra-light (1.7 to 8.1 mg/cm^3) triethoxyl(octyl)silane)-coated aerogels, prepared by using cellulose nanofibrils (defibrillated from rice straw cellulose). Zheng *et al.* [83]

prepared the polyvinyl alcohol (PVA)-cellulose nanofibril (CNF) based aerogels which were treated with methyltrichlorosilane by using thermal chemical vapour deposition, rendering the water contact angle of ($150.3\pm1.2°$). The authors observed that methyltrichlorosilane-coated PVA/CNF aerogels were highly porous (>98%) and had ultra-low densities (< 15 Kg/m^3). The absorption capacities ranging from 44 to 96 times its own weight for various oils and organic solvents such as diesel oil, gasoline, crude oil, corn oil, pump oil, hexane, chloroform and toluene were observed.

In summary, large number of polymer based aerogels have been developed for oil spill cleaning purposes. Various polymer based aerogels for cleaning of oil spills are also listed in Table 5.2.

5.4 Polymer Based Fibers for Oil Spill Clean Up

Synthetic fibers have potential application in the removal process of oil spills from water. Zhu *et al.* [19] employed electrospinning process to fabricate cost-effective polyvinyl chloride (PVC)/polystyrene (PS) fibers with low density (2-3 mg/cm^3) and high porosity of ~99.7%. SEM analysis revealed that the sorbent was made up of fibers having diameter in the range of 1.5 to 3 μm which enhanced the surface area and oil adsorption on the surface of the fibers. Additionally, large number of interconnected voids (size ranging between 20-70 μm) were observed among fibers (Figure 5.6). These voids had the tendency to store large amounts of oil, thus, avoiding the escape of oil from PVC/PS sorbent at the time of draining. The authors calculated the absorption capacities of PVC/PS sorbent for ethylene glycol, diesel, peanut oil and motor oil as 81 g/g, 38 g/g, 119 g/g and 146 g/g, respectively. The voids among the fibers were considered as crucial for high absorption capacity of the sorbent. In another study reporting the use of fibers as an absorbent for oil spill clean-up applications, Lin *et al.* [78] prepared nanoporous polystyrene fibers with high specific surface area by employing electrospinning method. High sorption capacities of hydrophobic-oleophilic fibers were achieved for sunflower seed oil, bean oil and motor oil as 96.89, 111.80 and 113.87 g/g, respectively. In another study, hydrophobic-oleophilic fibrous mats were developed by Lin *et al.* [84]. The work demonstrated the fabrication of PS-PU fibrous mats by using co-axial electrospinning method, where PS and PU solutions were used as the shell and core solutions, respectively. By manipulating the factors such as solvent compositions in the shell solution, concentration of core solution and spinning voltages, the structure of fibrous mats could be controlled. The porous structure of the PS-PU fibers ranged between 20 and 80 nm, when investigated by synchrotron

Table 5.2 Various polymer based aerogels for cleaning of oil spills

Authors	Aerogels	Methods	Oils/ Organic solvents	Absorption capacity (g/g) or (times) or (w/w %)	WCA	OCA	Ref.
Chin et al.	Magnetic cellulose/TiO_2 aerogel	CO_2 supercritical point drying and Sol-gel process	Paraffin oil	28 times its own weight	-	-	[74]
Korhonen et al.	Nanocellulose/TiO_2 aerogel	Vacuum freeze drying	[Hexadecane, Octane, Mineral oil, Chloroform, Toluene, Paraffin oil, Octanol, Dodecane, Petroleum benzene, Hexane]	[20-40 g/g]	>90°	-	[17]
Nguyen et al.	MTMS-coated recycled cellulose aerogel	Simple alkaline/urea method and freeze drying	RB crude oil TGT crude oil RD crude oil	18.4 g/g 18.5 g/g 20.5 g/g	>140°	-	[73]
Jiang et al.	Triethoxyl[octyl] silane-coated cellulose nanofibril aerogel	Chemical vapor deposition and freeze drying	[Chloroform, DMSO, Ethylene glycol, DMF, Soybean oil, Toluene, Pump oil, Acetone, Cyclohexane, Decane, Octane, Hexane]	[139-356 g/g]	-	-	[82]

Table 5.2 (continued)

Authors	Aerogels	Methods	Oils/ Organic solvents	Absorption capacity (g/g) or (times) or (w/w %)	WCA	OCA	Ref.
Zheng et al.	Silane treated PVA-cellulose nanofibril hybrid aerogel	Simple thermal chemical vapor deposition pocess and freeze drying	[Diesel, Gasoline, Crude oil, Corn oil, Pump oil, Hexane, Chloro-form, Toluene]	[44-96 times its own weight]	150.3±1.2°	-	[83]
Sai et al.	Hydrophobic and oleophilic bacterial cellulose aerogel	Liquid phase reaction followed by freeze drying	[n-Hexane, Ace-tone, Paraffin liq-uid, Plant oil, Gasoline, Diesel, Toluene, Chloro-benzene, Di-chloromethane, Chloroform]	[Up to 185 g/g]	146.5°	-	[80]
Si et al.	Superhydrophobic-superoleophilic PAN/SiO$_2$ nanofibers based aerogel	Freeze drying method	Petroleum ether	-	162°	0°	[81]

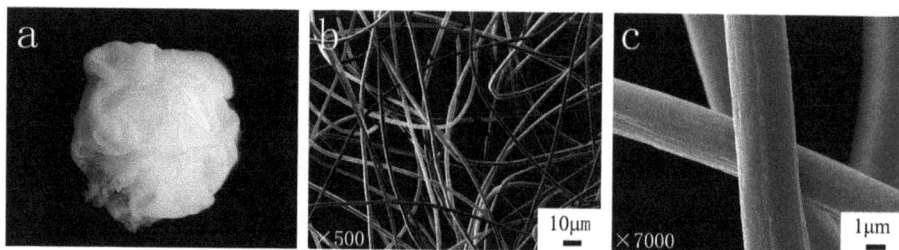

Figure 5.6 (a) Photo, (b) low and (c) high magnification SEM images of PVC/PS sorbent. Reproduced from reference 19 with permission from American Chemical Society.

radiation small-angle X-ray scattering (SAXS) and Brunauer-Emmett-Teller (BET). The sorption capacities of fibrous mats for sunflower seed oil and motor oil were 47.78 and 64.40 g/g, respectively and were strongly affected by the inter-fiber voids among the fibers. It was noticed that the sorption capacity of the sorbent was maintained even after five cycles of absorption-desorption. Lin *et al.* [85] also fabricated fibrous mats consisting of PS fibers with micro-nanostructures by using electrospinning technique, mainly focusing on the porous structures of the fibers. The authors investigated the properties of PS fibers by tuning the composition of solvent, concentration of solution and molecular weight of polystyrene. The absorption ability of fibrous mats for sunflower seed oil and motor oil were determined as 79.62 and 84.41 g/g, respectively. Thus, different studies confirmed that polystyrene fibers had better oil absorption capacities than commercial polypropylene non-woven fabric [78,84,85].

In another study, Wu *et al.* [79] reported the fabrication of PS fibrous films with pores formation on the fibers via electrospinning method. The diameter and surface morphology of fibers impacted the oil adsorption capacity as well as oil/water selectivity. With decreasing the diameter of the fibers, the uptake rate and adsorption capacity of the oil increased. The oil adsorption capacities of PS fibrous film for motor oil, peanut oil, silicon oil and diesel oil were around 131.63, 112.30, 81.40 and 7.13 g/g, respectively. The WCA and OCA (motor oil) were measured to be 151.3±1.6° and 0°, respectively, indicating that the PS fibrous film exhibited superhydrophobic and superoleophilic properties. Zhao *et al.* [86] employed melt blown method for the preparation of polypropylene (PP) and poly (butylmethacrylate-co- hydroxyethylmethacrylate) (PBMA-co-HEMA) non-woven oil absorbents. In this study, PP and PBMA-co-HEMA blends were prepared with different composition ratios by weight. It was observed that with increasing the content of PBMA-co-HEMA in the resultant non-woven material, water contact angle and absorption capacity of the material increased.

PBMA-co-HEMA (20 wt%) exhibited the highest WCA (136.8±1.19°) and absorption capacities for crude oil, diesel and toluene as 9.02, 8.88, and 7.74 g/g, respectively. After eight cycles of absorption-desorption, only 1.5% reduction in the absorption capacity of the non-woven material was noticed. For the continuous separation of oil-water mixture, Xue et al. [87] proposed superhydrophobic and superoleophilic poly(ethylene terephthalate) (PET) textiles prepared via sol-gel coating method. Firstly, PET textiles were treated with NaOH to make the fiber surface rough and also to improve compatibility between the fibers and silica sols. Subsequently, by coating the PET fibers with -Si(CH$_3$)$_3$ functionalized SiO$_2$ nanoparticles, superhydrophobic-superoleophilic PET textiles were obtained. The water contact angle of the PET textiles was observed to be more than 150° and gasoline drops spread immediately upon contacting the textile. PET textiles indicated excellent superhydrophobic property even after 700 cycles. Li et al. [88] reported polyester textile with superhydrophobic and superoleophilic properties and employed as an effective membrane for

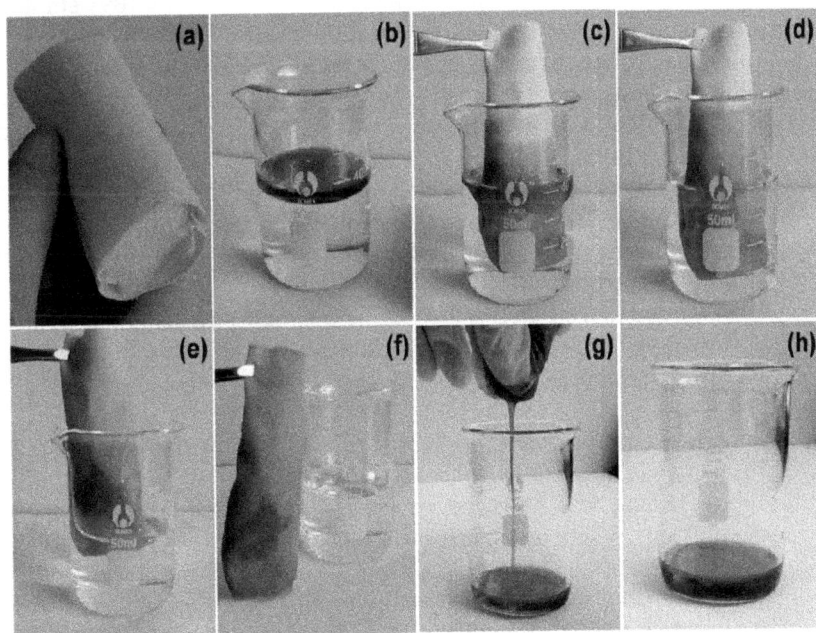

Figure 5.7 Photographs of (a) the textile bag, and (b-g) the oil adsorption and recycling process by using the superhydrophobic textile bag. Reproduced from Reference 88 with permission from American Chemical Society.

oil/water separation and selective removal of oil from water (Figure 5.7). Under hydrothermal conditions, fabrication of the textile was performed via direct *in-situ* growth of layered double hydroxides (LDH) micro-crystals on the surface of polyester textile and followed by further modification of the textile with sodium laurate to reduce the surface energy. TEM analysis revealed that LDH micro-crystals were perpendicularly aligned to the fiber surface of the textile, creating a nest-like microstructure that also led to the micro-nano sized spaces among LDH micro-crystals, contributing to the fiber surface roughness of the textile. Superhydrophobic and superoleophilic properties of the polyester textile were achieved with the combination of low surface energy molecules and hierarchical micro-nanostructure. Water contact angles were measured to be 144±1.2° and 154±1.6° for LDH coated textile and modified LDH coated textile with sodium laurate, respectively and in contrast, oils were spread easily on the resultant textile. The authors also subjected the as-prepared textile to 20 adsorption-squeezing cycles. No change in respect to the water contact angle was observed even after 20 cycles. In another study, Zhang *et al.* [89] employed chemical vapor deposition technique to conduct one-step growth of silicone nano-filaments onto the surface of polyester textile to produce superhydrophobic-superoleophilic polyester material. Wu *et al.* [90] also reported superhydrophobic-superoleophilic polyester material with high selectivity and recyclability for oil-water separation. The authors coated a commercial polyester material by dipping it into the solution of Fe_3O_4@hexadecyltriethoxysilane (HD)-silica/HD-polymer to produce superhydrophobic-superoleophilic material. The oil absorption capacities of coated polyester material for crude oil, diesel and petrol were determined as 396%, 406% and 355%, respectively. By using a magnet, the as-prepared material could be magnetically driven to clean the oil spills. Polymeric fibers for removal of oil or organic solvents from water are also listed in Table 5.3.

5.5 Polymeric Membranes for Oil-Water Mixture Separation

Membrane technology has been proved to a promising method to separate oil-in-water emulsions, which otherwise cannot be separated by applying the conventional techniques (flotation, gravity separation, coagulation, ultrasonic separation, skimming and centrifugation). Membrane based separation process is highly applicable for the separation of oil-in-water emulsions with promising methods such as reverse osmosis [8], ultrafiltration [9,10], microfiltration [11], dehydration of oil emulsion by pervaporation [12] and membrane distillation [13]. Li *et al.* [91] reported the aminated polyacrylonitrile (APAN) electrospun

Table 5.3 Polymer based fibers for oil spill clean up

Authors	Fibers/Mats/Textiles	Methods	Oils/ Organic solvents	Absorption capacity (g/g) or (times) or (w/w %)	WCA	OCA	Ref.
Zhu et al.	Polyvinyl chloride (PVC)/polystyrene (PS) fibers	Electrospinning process	Motor oil Peanut oil Diesel Ethylene glycol	146 g/g 119 g/g 38 g/g 81 g/g	-	-	[19]
Wu et al.	PS porous fibers	Electrospinning process	Motor oil Peanut oil Silicon oil Diesel oil	131.63 g/g 112.30 g/g 81.40 g/g 7.13 g/g	151.3±1.6°	0°	[79]
Lin et al.	PS-PU fibrous mats	Co-axial electrospinning	Motor oil Sunflower oil	64.40 g/g 47.48 g/g	Obtuse angle	-	[84]
Lin et al.	PS fibers	One step electrospinning	Motor oil Bean oil Sunflower oil	113.87 g/g 111.80 g/g 96.89 g/g	-	-	[78]
Lin et al.	PS fibers with micro - nanostructures	Electrospinning process	Motor oil Sunflower oil	84.41 g/g 79.62 g/g	Obtuse angle	-	[85]
Zhao et al.	PP/PBMA-co-HEMA	Melt blown method	Crude oil Diesel Toluene	9.02 g/g 8.88 g/g 7.74 g/g	136.8±1.19°	0°	[86]
Xue et al.	PET textiles	Sol-gel coating	Crude oil, Gasoline	-	>150°	-	[87]
Liu et al.	LDH coated polyester textile bag	Hydrothermal process	[n-Hexane, Chloroform, Diesel, Gasoline]	-	154±1.6°	-	[88]
Wu et al.	Fe$_3$O$_4$@HD-silica/HD polymer coated polyester	Dip coating method	Petrol Diesel Crude oil	355 % 406 % 396 %	-	-	[90]

nano-fibrous membranes exhibiting superhydrophobic, superoleophilic and excellent recyclable properties for oil-water separation. Firstly, the authors aminated the electrospun PAN nanofibers with diethylenetriamine (DETA) solution, followed by anchoring of silver (Ag) nano-clusters onto the APAN nanofibrous membranes via electroless plating technique. Subsequently, the modification of the membrane surface with long alkyl chain thiols was performed. The surface roughness of APAN nano-fibrous membranes could be controlled by adjusting the electroless plating time. The permeate fluxes of APAN-Ag-SR-50 nano-fibrous membrane with 100 μm thickness were found to be higher for low viscosity oils, and comparatively lower for the high viscosity oils such as vegetable oil (954.9±38.5 L/m^{2}h) and vacuum pump oil (447.5±100.3 L/m^{2}h). The water and oil contact angles of as-prepared membrane were determined as 162±1.9° and 0°, respectively. Tai et al. [92] employed electrospinning technique for the preparation of flexible electrospun SiO$_2$-carbon nanocompositesbased nano-fibrous membranes, where PAN was used as a carrying polymer for SiO$_2$ precursor and a precursor for carbon nanofiber (CNF). At 125 °C, ultra-hydrophobic (WCA = 144.2±1.2°) and superoleophilic (OCA = 0°) properties were attained after coating of silicone oil onto the surface of the membrane via vapor deposition method. The chemical and thermal stability tests suggested that the membrane was capable to maintain its ultra-hydrophobicity in a pH range 2-14. The wettability of the as-prepared membrane was resistant up to 350 °C. The authors also determined the average value of permeate flux of membrane (with thickness 213.3±21.8 μm) for hexane, iso-octane and petroleum spirit as 2648.8±89.7, 1719.1±36.2, and 3032.4±234.6 L/m^{2}h, respectively. These permeate flux values are equivalent or even higher than the flux values achieved in some of the advanced membranes such as superhydrophobic polyvinylidene fluoride membrane (PVDF) and mineral coated polypropylene micro-filtration membrane having permeate flux values of (700-3500 L/m^{2}h) [93] and (>2000 L/m^{2}h) [94], respectively. Chakrabarty et al. [95] also reported the modified polysulfone (PSF) membranes for the removal of oil from stable oil-in-water emulsion. In another study, Zhang et al. [96] reported the polymer of intrinsic microporosity (PIM-1) and polyhedral oligomeric silsesquioxane (POSS) based electrospun microfibrous membranes which could be used to separate immiscible oil-water mixtures and water-in-oil emulsions with efficiencies higher than 99.95% and 99.97%, respectively. Higher separation efficiencies were gained due to two reasons, i. e. superhydrophobic-superoleophilic surface of the membrane and pore sizes (1-5 μm) of membrane which were smaller than the emulsion droplet sizes (5-20 μm). It was observed that PIM-1 fibrous membrane exhibited the WCA of 132° but with an increase of POSS concentration to 40 wt%, hierarchical structures (surface roughness) were gained on the fiber surface to achieve the superhydrophobic membrane with WCA 155°. On immersing PIM-1/POSS fibrous membrane in water, the superhydrophobicity of membrane was confirmed by the presence of a silvery plastron layer on the surface of fibrous membrane which was caused by the formation of air cushion between the membrane surface and water [97]. The authors also employed the PIM-1/POSS fibrous membrane to separate the n-hexane-water mixture and determined the permeate flux of about 2890±100 L/m^2h. The as-prepared membranes were also capable to adsorb oil red O and blue 35 with corresponding adsorption capacities of 7.49 and 8.33 mg/g, respectively. In another study, superhydrophobic-superoleophilic nanofibrous membranes were reported by Huang *et al.* [98] for the effective separation of water-in-oil emulsions. The membrane was prepared by the combination of *in-situ* polymerized fluorinated benzoxazine (F-PBZ), silica nanofibrous membranes (SNF) and alumina (Al_2O_3) nanoparticles. The membranes (F-SNF/Al_2O_3 NPs) exhibited excellent antifouling properties, durability, reusability and the high flux of 892±50 L/m^2h.

Tang *et al.* [99] designed novel superhydrophobic-superoleophilic nanofibrous membranes for gravity driven oil-water separation. Firstly, multi-walled carbon nanotubes (MWCNTs) reinforced electrospun poly (m-phenylene isophthalamide) (PMIA) nano-fibrous membranes were prepared via electrospinning technique, followed by the fabrication of *in-situ* polymerized fluorinated-polybenzoxazine functional layer incorporated with SiO_2 nanoparticles (F-PBZ/SiO_2 NPs layer) on the surface of PMIA/MCNTs nano-fibrous membrane. Modification of hydrophilic PMIA nano-fibrous membrane with F-PBZ/SiO_2 NPs layer resulted into a superhydrophobic and superoleophilic membrane. The as-prepared membrane exhibited the water contact angle and oil contact angle of 161° and 0°, respectively. The permeate flux of membrane was determined as 3311 L/m^2h, when employed to separate the mixture of dichloromethane and water. The membranes exhibited high separation efficiency for oil-water mixtures along with excellent stability over the pH range of 2-12. In a similar study, Shang *et al.* [100] reported similar F-PBZ/SiO_2 NPs modified electrospun cellulose acetate nano-fibrous membrane with superhydrophobic (WCA = 161°) and superoleophilic (OCA = 3°) properties for the separation of dichloromethane from water. Chen *et al.* [101] reported hydrophilic cellulose acetate (CA)-graft-PAN membranes to achieve high permeation flux as well as excellent fouling resistance for oil-water mixture separation. The hydrophilic CA-graft-PAN membranes were fabricated by employing phase inversion method. Membrane material was synthesized by grafting PAN onto CA powder by using free-radical polymerization. Thus, by embedding the PAN segments into the matrix of membrane during phase inversion method, low surface energy of the membrane was achieved to enhance the hydrophilic property of the CA-graft-PAN membrane for the successful oil-water separation. Modified PSF electrospun nano-fiber mats as effective oil-water separation membranes were reported by Obaid *et al.* [102] The authors incorporated NaOH nanoparticles in the PSF nano-fibers, followed by the formation of thin polyamide layer via interfacial polymerization of m-phenylenediamine (MPD) and 1,3,5-benzenetricarbonyl chloride (TMC) on the surface of the electrospun mats. The water contact angle of PSF nanofiber membrane decreased from ~130° to 13° due to the combined effect of NaOH nanoparticles and thin polyamide layer. NaOH nanoparticles were observed to absorb water due to the formation of hydrogen bonds, which led to a decrease in the water contact angle. Similarly, the presence of amide content also reduced the water contact angle [103,104]. The as-prepared membrane was subjected to three successive cycles for separation of soybean oil-water mixture and water flux of the membrane reduced from 8 to 5.5 m^3/m^2 day due to membrane fouling. Zhu *et al.* [105] employed surface-initiated atom transfer radical polymerization (SI-ATRP) technique for the fabrication of a novel zwitterionic polyelectrolyte grafted PVDF membrane with ultrahigh efficiency to separate oil from water. Xiang *et al.* [106] also prepared superoleophobic PVDF membrane coated with polydopamine via facile solution-immersion method to separate both surfactant-stabilized and surfactant-free oil-water emulsions. After the hydrophilic coating of polydopamine onto the PVDF membrane surface, the water contact angle of membrane reduced from 118±1.5° to 53±2.3°. The oil contact angle was found to be 152±0.3°, when a 3 µl chloroform droplet was dropped on the membrane. The authors also studied the flux values of toluene-in-water (T/W), chloroform-in-water (C/W) and dichloroethane-in-water (DCE/W) emulsions (both surfactant stabilized and surfactant-free emulsions). Dry-wet spinning technique was employed by Li *et al.* [107] to prepare a hollow fiber ultrafiltration membrane for the separation of oil-water mixture. Another superhydrophilic and oleophobic nano-fibrous membranes for ultrafast oil-water separation was reported by Raza *et al.* [108]. The fabrication of the membrane was achieved by the deposition of *in-situ* crosslinked polyethylene glycol diacrylate (PEGDA) nano-fibers onto the polyacrylonitrile/polyethylene glycol (PG NF) membrane. The as-prepared PEGDA@PG NF membranes exhibited high flux values of 10975, 11565, 11976, and 12656 L/m^2h for soybean oil-water (S-W), crude oil-water (C-W), petrol-water (P-W) and hexane-water (H-W) mixtures, respectively. For separation of oil-water emulsion, Chen *et al.* [109] reported polyethersulfone membranes modified with amphiphilic copolymer Pluronic F127. Zhang *et al.* [110] employed the salt-induced phase-inversion approach for the preparation of the superhydrophilic-underwater superoleophilic polyacrylic acid-grafted-polyvinylidene fluoride (PAA-g-PVDF) membranes for effective separation of oil-in-water emulsions. Under applied pressure of 0.1 bar, PAA-g-PVDF membranes exhibited high flux values for surfactant-free and surfactant-stabilized (sodium dodecyl sulfate (SDS) and Tween 80) oil-in-water emulsions.

Various superhydrophobic-superoleophilic or superhydrophilic-superoleophobic polymeric membranes used for oil spill clean up are listed in Table 5.4.

5.6 Other Polymer Absorbents for Oil Spill Clean Up

Li *et al.* [111] reported a smart fiber membrane endowed with switchable wetting states, i.e. superhydrophilicity-underwater superoleophobicity and superhydrophobicity-superoleophilicity. With the combination of underwater oleophilicity/hydrophilicity of poly (methyl methacrylate) (PMMA) and pH responsive poly (4-vinylpyridine) (P4VP), the as-prepared membrane attained switchable surface wettability towards oil and water. The smart fiber membrane responded through switching the pH medium for on-demand oil/water separation. Due to the protonation and deprotonation of pyridyl groups, the wettability of the as-prepared membrane could be switchable at different pH [112-115]. At pH 3, P4VP (weak polybase) became protonated and due to extended conformation of P4VP segments caused by the electrostatic repulsion between pyridyl groups, the protonated P4VP segments stretched out. This promoted the water uptake into the membrane texture, leading to the supehydrophilic surface of the membrane and subsequently, water selectively passed through the membrane. At pH 7, the separation process got reversed after the deprotonation of P4VP (Figure 5.8). Zhang *et al.* [116] employed the solvothermal method to prepare superhydrophobic-superoleophilic nano-porous polydivinylbenzene (PDVB) adsorbents for the adsorption of various organic contaminants and volatile organic compounds such as n-heptane, kerosene, benzene and nitrobenzene. Large surface area, adjustable pore sizes and nano-porus structures were exhibited by the PDVB material. Atta *et al.* [117] reported the cross-linked poly (octadecene-alt-maleic anhydride) copolymers, which were used as crude oil sorbers. Similarly, Yuan *et al.* [28] reported cross-linked polyolefin terpolymer as an oil superabsorbent, having oil uptake capacity about 45 times its own weight. The tea-bag nanocomposite system as an oil absorbent material have been reported by Avila *et al.* [118]. This system is based on PS superhydrophobic membranes, surrounding the exfoliated graphite. The

Table 5.4 Polymeric membranes for oil-water mixture separation

Authors	Membranes	Methods	Oils/ Organic solvents	Permeate Flux (L/m²h) or (m³/m² day) or (L/ m²h bar)	WCA	OCA	Ref.
Li et al.	Superhydrophobic-superoleophilic APAN-Ag-SR nanofibrous membrane	Electrospinning and electroless plating technique	Vacuum pump oil Vegetable oil [Petroleum ether, n-hexane, chloroform]	447.5±100.3 L/m²h 954.9±38.5 L/m²h 2250-3375 L/m²h	162.4 ± 1.9°	0°	[91]
Tai et al.	Hydrophobic carbon-silica nanofibrous membrane	Electrospinning method	Petroleum spirit Iso-octane Hexane	3032.4±234.6 L/m²h 1719.1±36.2 L/m²h 2648.8±89.7 L/m²h	144.2 ± 1.2°	0°	[92]
Chen et al.	Superhydrophilic mineral coated PAA-grafted polypropylene membrane	Coating deposition	[Petroleum ether, Hexadecane, Gasoline, Diesel, Dichloroethane]	>2000 L/m²h	-	>150°	[94]
Zhang et al.	Superhydrophobic-superoleophilic PIM-1/POSS fibrous membrane	Electrospinning method	n-Hexane Cyclohexane n-Heptane Petroleum ether Soybean oil Motor oil	2890±100 L/m²h - - - - -	155°	- - - - - -	[96]
Tang et al.	Superhydrophobic-superoleophilic F-PBZ/SiO₂ NPs modified PMIA nanofibrous membrane	Electrospinning method and in-situ polymerization	Dichloromethane	3311 L/m²h	161°	0°	[99]
Shang et al.			Dichloromethane	-	161°	0°	[100]
Chen et al.	Hydrophilic cellulose acetate (CA)-graft-polyacrylonitrile (PAN) membrane	Phase inversion method	Vacuum pump oil (GS-1)	227.8-334.5 L/m²h (Pure water flux)	[55.4 ± 1.7) – [56.9 ± 0.9)	-	[101]

Table 5.4 (continued)

Authors	Membranes	Methods	Oils/ Organic solvents	Permeate Flux (L/m²h) or (m³/m² day) or (L/m²h bar)	WCA	OCA	Ref.
Obaid et al.	Hydrophilic modified polysulfone nanofibrous membrane	Electrospinning method	Soybean oil	5.5-8 (m³/m²day) (water flux)	13°	-	[102]
Huang et al.	Superhydrophobic-superoleophilic (F-SNF/ Al₂O₃ NPs) nanofibrous membrane	In-situ polymerization approach	Petroleum ether	892±50 L/m²h	161°	0°	[98]
Zhu et al.	Superhydrophilic-superoleophobic PMAPS-grafted-PVDF membrane	Surface-initiated atom transfer radical polymerical technique (SI-ATRP)	Diesel oil	11000-13500 (L/ m²h bar) water flux	-	>150°	[105]
			Petroleum ether	-		>150°	
			Soybean oil	-		>150°	
			Hexane	-		>150°	
			Isooctane	-		>150°	
Xiang et al.	Superoleophobic PVDF membrane coated with polydopamine	Facile solution-immersion method	Surfactant stabilized (C/W) emulsion	3969 L/m²h	53 ± 2.3°	152 ± 0.3°	[106]
			Surfactant-free (C/W) emulsion	14037 L/m²h		152 ± 0.3°	
			Surfactant stabilized (T/W) emulsion	1991 L/m²h		-	
			Surfactant-free (T/W) emulsion	15882 L/m²h		-	

Table 5.4 (continued)

Authors	Membranes	Methods	Oils/ Organic solvents	Permeate Flux (L/m²h) or (m³/m² day) or (L/ m²h bar)	WCA	OCA	Ref.
			Surfactant stabilized (DCE/W) emulsion	2351 L/m²h		-	
			Surfactant-free (DCE/W) emulsion	10183 L/m²h		-	
Raza et al.	Superhydrophilic and oleophobic PEGDA@PG NF membrane	Electrospinning method	Hexane	12656 L/m²h	-	-	[108]
			Petrol	11976 L/m²h		-	
			Crude oil	11565 L/m²h		-	
			Soybean oil	10975 L/m²h		-	
Zhang et al.	Superhydrophilic and underwater superoleophobic PAA-g-PVDF membrane	Salt induced phase-inversion method	1,2-Dichlorophenol	-	0°	160°	[110]
			Hexadecane	2320 L/m²h		-	
			Toluene	1680 L/m²h		-	
			Diesel	1550 L/m²h		-	
			SDS-Hexadecane	1140 L/m²h		-	
			Tween 80-Toulene	1020 L/m²h		-	
			SDS-Diesel	720 L/m²h		-	

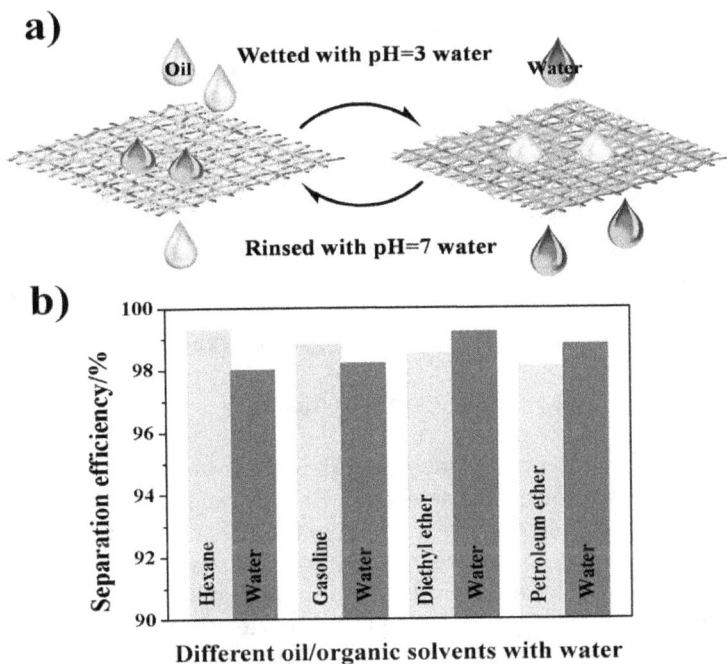

Figure 5.8 (a) Schematic diagrams of pH-switchable oil/water separation and (b) Oil/organic solvent separation efficiency (dry state) and water separation efficiency (after wetted with pH 3 water) of four different mixtures. Reproduced from Reference 111 with permission from American Chemical Society.

absorption rates of tea-bag system for vacuum pump oil, new and used motor oil and water-oil emulsions ranged between 2.5 and 40 g/g. Gu *et al.* [119] reported highly hydrophobic and superoleophilic magnetic polymer nanocomposites prepared via emulsion polymerization. Styrene (St), divinylbenzene (DVB) and methylmethacrylate (MMA) were used for the preparation of core-shell magnetic polymer nanocomposites. Fe_3O_4@poly(St/DVB) magnetic nanocomposites and poly(MMA/St/DVB) layer were employed as core and shell, respectively for preparing highly selective absorbent for the successful separation of oil from water. The absorption capacities of as-prepared magnetic polymer nanocomposites for different oils were found to be 3.63 times their own weight. Another highly hydrophobic and superoleophilic core-shell Fe_3O_4/polystyrene magnetic nanocomposites with absorption capacities of 3 times their own weight for lubricating oils were reported by Chen *et al.* [120].

β-Cyclodextrin based polymeric microspheres (POAms) as oil absorbent were reported by Song *et al.* [121]. Butyl acrylate and octadecyl acrylate were employed as co-monomers for the preparation of POAms via suspension polymerization. The absorption capacities of POAms for diesel, kerosene, gasoline, toluene, xylenes, $CHCl_3$, CCl_4 were determined as 18.2, 27.1, 30, 42.8, 48.7, 75.1, and 83.4 g/g, respectively. Mao *et al.* [122] reported the preparation of another porous magnetic poly (styrene-divinylbenzene)/Fe_3O_4 microspheres via seed swelling polymerization process. The absorption capacities of resultant microspheres for various oils were up to 8.42 times their own weight. The microspheres also exhibited good reusability for 10 cycles, though a slight reduction in oil absorption capacity was noticed after the 10th cycle.

5.7 Polymeric Adsorbents for Phenol Removal

A wide range of polymer based adsorbents have been employed for the adsorption of phenol from industrial wastewater. Huang *et al.* [123] reported the HJ-1 resin (hyper-crosslinked polymeric adsorbent) for adsorption of phenol and *p*-cresol from industrial wastewater. Adsorption kinetics was obsered to follow pseudo-second-order kinetics and langmuir model could be fitted with the adsorption isotherm data. Caetano *et al.* [124] reported the removal of phenol using three polymeric resins, i. e., MN 200 (crosslinked polystyrene), AuRIX 100 (styrene divinylbenzene with guanidine functional group) and Dowex XZ (styrene divinylbenzene with t-butylamine functional group). To define the controlling mechanism of adsorption process, the shell progressive model (SPM) and homogeneous particle diffusion model (HPDM) were used, which fitted well with the kinetic experimental data. Adsorption isotherm data was successfully described by the langmuir model. Senel *et al.* [125] reported the removal of 2,4,6-trichlorophenol, *o*-chlorophenol, phenol and *p*-chlorophenol by using recyclable polyamide hollow fibers. The maximum adsoprtion capacities of 2,4,6-trichlorophenol, *o*-chlorophenol, phenol and *p*-chlorophenol were determined as 179.2, 202.8, 145.9, and 194.5 μmol/g, respectively. The hollow fibers were also observed to be recyclable for more than 10 cycles without any observable decrease in the adsorption capacity. The kinetic studies indicated that the adsorption of chlorophenols onto hollow fibers fitted the pseudo-second-order model. An *et al.* [126] reported polyethyleneimine-grafted-silica gel particles (PEI/SiO_2) for the adsorption of phenol from aqueous solution. Due to the coupling effect of γ-chloropropyl trimethoxysilane, the surface of silica gel particle was grafted with the macromolecule polyethyleneimine. The adsorption ability of PEI/SiO_2 for phenol was determined to be 160 mg/g. The equilibrium adsorption data was well described by Freundlich isotherm model. In another study, An *et al.* [127] also employed molecular imprinting technique to prepare the imprinted polymer MIP-PEI/SiO$_2$ with excellent affinity and selectivity for phenol. Shao *et al.* [128] employed plasma-induced grafting technique to prepare the polyaniline (PANI)-grafted-multiwalled carbon nanotubes (MWCNTs) for the removal of phenol from contaminated water. It was also observed that due to the grafting of PANI onto MWCNTs, there was a strong conjugate effect between PANI and phenol, which jointly enhanced the adsorption capacity to adsorb phenol. Zeng *et al.* [129] reported polymeric adsorbents, PDM-1 (macroporous styrene-co-divinylbenzene copolymer with ester groups) and PDM-2 (PDM-1 copolymer was post-crosslinked by Friedel-Crafts reaction of pendant vinyl groups), which were employed for the adsorption of phenol from aqueous solution. Adsoprtion kinetics were observed to follow pseudo-second-order model and Freundlich isotherm model was well fitted with the equilibrium adsorption data. Another water-compatible hypercrosslinked NJ-8 (oxygen modified polystyrene) polymeric adsorbent was reported by Li *et al.* [130] for the adsorption of phenolic derivatives such as *p*-nitrophenol, *p*-cresol, *p*-chlorophenol and phenol. The equilibrium adsorption data was well described by using the Freundlich isotherm. Ming *et al.* [131] reported the comparison of NDA 103 (aminated polystyrene), amberlite XAD4 (polystyrene) and Amberlite IRA96C (weak base polystyrene) for the adsorption of phenol from water solution at temperatures 293-313 K. Langmuir and Freundlich isotherms for three different polymeric adsorbents were well fitted to the equilibrium adsorption data. It was observed that phenol adsorption capacity of NDA 103 (driven by hydrogen bonding and van der Waals, simultaneously) was higher than Amberlite XAD4 and Amberlite IRA96C (driven by van der Waals interaction only and hydrogen bonding only, respectively). Zhang *et al.* [132] also investigated the phenol adsorption performance of aminated polystyrene resins NDA-101 with surface area of 845 m^2/g and NDA-103 with surface area of 611 m^2/g. It was observed that due to the combination of π-π stacking interactions with phenol and multiple hydrogen bonding, aminated resins achieved better adsorption capacity. Both Freundlich and Langmuir equations fitted well to the adsorption isotherms. Pan *et al.* [133] prepared NDA-701 (styrene-divinylbenzene) as a hyper-crosslinked polymer adsorbent with a unique bimodal pore size distribution and high surface area (824 m^2/g) for adsorption of 4-nitrophenol from aqueous solution. 4-nitrophenol adsorption isotherm was well described by Freundlich model. Kinetic studies demonstrated the adsorption of 4-nitrophenol onto NDA-701 to follow the pseudo-second-order model. In another study, Pan *et al.* [134] also reported macroporous polystyrene, its chemically modified animated derivative and a weak anion exchanger (CHA-111, MCH-111 and ND-900, respectively) for the adsorption of phenolic compouds such as phenol, *p*-cresol, *p*-chlorophenol, *p*-nitrophenol, 2,4-dichlorophenol and 2,4-dinitrophenol. Li *et al.* [135] compared the adsorption capacity of Amberlite XAD-4 and MX-4 polymeric adsorbents for different phenolic compounds from aqueous solution. To obtain MX-4, which could be used directly without any wetting process, Amberlite XAD-4 was chemically modified with acetyl group. The equilibrium adsorption capacities for phenolic compounds were observed to increase by 20 % after the modification of Amberlite XAD-4. Isotherm data for phenol, *p*-cresol, *p*-chlorophenol and *p*-nitrophenol onto Amberlite XAD-4 and MX-4 well fitted to the Freundlich model. Effective removal of phenol from aqueous solution by employing Amberlite XAD-7 (porous acrylic ester polymer) was also reported by Pan *et al.* [136]. The kinetic analysis fitted with pseudo-second-order kinetic model and adsorption isotherms were well described by both Langmuir and Freundlich models. Thermodynamic parameters also revealed that phenol adsorption process was exothermic and spontaneous in nature. Juang *et al.* [137] reported non-ionic macroreticular resins for adsorption isotherms of phenol and 4-chlorophenol from aqueous solution. Zeng *et al.* [138] prepared a series of post-crosslinked methyl methacrylate/divinylbenzene (MMA-co-DVB) and ethylene glycol dimethacrylate/ divinylbenzene (EGDMA-co-DVB) copolymer adsorbents by modifying them with different amounts of MMA and EGDMA for the enhanced adsorption of phenol from aqueous solution. Phenol adsoprtion isotherms onto post-crosslinked copolymers were well desribed by both Frendlich and Langmuir models. Wagner *et al.* [139] used polystyrene and polymethacrylate resins which were crosslinked with divinylbenzene for the adsorption of different phenolic derivatives from polluted water. The reins were used for phenol and chlorophenols adsorption at temperatures ranging from 294.15 to 318.15 K. Adsorption isotherms for different phenols were well fitted to Redlich-Peterson model.

The adsorption of phenol from cyclohexane by using polymethylmethacrylate-co-divinylbenzene, polymethylmethacrylate-co-divinylbenzene/ polystyrene-co-divinylbenzene and poly (N-p-vinylbenzyl acetylamide) was reported by Li *et al.* [140]. Dursun *et al.* [141] reported chitin, a natural occuring polymer, for the adsorption of phenol from aqueous solution [141]. Freundlich model fitted well to the adsorption data and pseudo-second-order model correlated to the experimental data. Thermodynamics parameters revealed that phenol adsorption process onto chitin was endothermic and spontaneous in nature. Kawabata *et al.* [142] reported the adsorption of phenol from aqueous solution by using vinylpyridine-divinylbenzene copolymer. Fowkes *et al.* [143] reported

PMMA films for the adsorption of phenol from methylene iodide solution. Burleigh *et al.* [144] reported the adsorption of 4-methylphenol, 4-nitrophenol and 4-chlorophenol by using porous polysilsesquioxanes. Huang *et al.* [145] reported HJ-Z01 resin (a hypercrosslinked resin modified with N-methylacetamide) for the adsorption of phenol from aqueous solution. Isotherms for adsorption of phenol onto HJ-Z01 could be well correlated by Langmuir and Freundlich models. Experimental data fitted with pseudo-second-order model. Gupta *et al.* [146] reported the pervaporation performances of hydroxyterminated polybutadiene based polyurethaneurea membranes for the separation of phenol from water-phenol mixture. All the polymeric adsorbents for the adsorption of phenol and its derivatives from aqueous solution are also listed in Table 5.5.

5.8 Conclusion

Oil spills and phenol-contaminated water are serious concern for the ecosystem, which need to be addressed with responsible actions. In recent years, diverse polymeric materials have been developed to effectively deal with these issues.

In this chapter, different advanced polymer systems exhibiting functionalities for adsorption, absorption and separation of these contaminants from aqueous bodies have been reviewed. Firstly, diverse polymeric materials such as sponges, aerogels, fibers, membranes and other polymeric materials (polymer nanocomposites, textile bag, microspheres, tea bag, smart materials, etc.) have been focused for the effective removal of oil spill clean up. Such materials have been largely fabricated by either designing with superoleophobic-superhydrophilic or superhydrophobic-superoleophilic surfaces to separate the water-oil mixture. Secondly, various synthetic polymeric resins as adsorbents have been reviewed for the removal of phenol and its derivatives from industrial wastewater.

Table 5.5 Various polymeric adsorbents for the adsorption of phenol and its derivative

Authors	Adsorbent	Structure	Pollutant	Model	Kinetics order	Ref.
Huang et al.	HJ-1 resin	Formaldehyde carbonyl groups modified polystyrene	Phenol, p-Cresol	Langmuir	Pseudo-second order	[123]
Caetano et al.	MN 200 Dowex XZ	Crosslinked polystyrene Styrene divinylbenzene with t-butylamine functional group	Phenol	Langmuir	-	[124]
	AuRIX 100	Styrene divinylbenzene with guanidine				
Senel et al.	Polyamide hollow fibers carrying Re-active Green HE 4BD	Polyamide hollow fibers	2,4,6-Trichloro-phenol, o-Chlorophenol, Phenol and p-Chlorophenol	-	Pseudo-second order	[125]
An et al.	PEI/SiO$_2$	Macromolecule polyethylene-imine-grafted-silica gel parti-cles	Phenol	Freundlich	-	[126]
An et al.	MIP-PEI/SiO$_2$	Molecular imprinted PEI/SiO$_2$polymer	Phenol and p-Nitrophenol	Freundlich	-	[127]
Shao et al.	PANI/MWCNTs nanocompo-sites	Polyaniline-grafted-MWCNT nanocomposites	Phenol	-	-	[128]
Zeng et al.	PDM-1 PDM-2	macroporous styrene-co-DVB with ester groups PDM-1 copolymer post-cross-linked by Friedel –Crafts reaction of pendant vinyl groups	Phenol	Freundlich	Pseudo-second order	[129]
Li et al.	NJ-8	Oxygen modified polystyrene	[Phenol, p-Cresol, p-Chlorophenol and p-Nitrophe-nol]	Freundlich	-	[130]

Table 5.5 (continued)

Authors	Adsorbent	Structure	Pollutant	Model	Kinetics order	Ref.
Ming *et al.*	NDA-103 XAD4 Amberlite IRA96C	Aminated polystyrene Polystyrene Weak base polystyrene	Phenol	Langmuir, Freundlich	-	[131]
Zhang *et al.*	NDA-101 NDA-103	Aminated polystyrene with surface area of 845 m^2/g Aminated polystyrene with surface area of 611 m^2/g	Phenol	Langmuir, Freundlich	-	[132]
Pan *et al.*	NDA-701	Styrene-Divinylbenzene.	4-Nitrophenol	Freundlich	Pseudo-second order	[133]
Pan *et al.*	CHA-111 MCH-111 ND-900	Macroporous polystyrene Aminated derivative of CHA-111 Weakly anion exchanger (polystyrene)	[phenol, *p*-Cresol, *p*-Chlorophenol, *p*-Nitrophenol, 2,4-Dichlorophenol, 2,4-Dinitrophenol]	Freundlich	-	[134]
Li *et al.*	XAD-4 MX-4	Polystyrene Polystyrene modified with acetyl group	[phenol, *p*-Cresol, *p*-Chlorophenol, *p*-Nitrophenol]	Freundlich	-	[135]
Pan *et al.*	Amberlite XAD-7	Porous acrylic ester polymer	Phenol	Langmuir, Freundlich	Pseudo-second order	[136]
Zeng *et al.*	MMA-co-DVB EGDMA-co-DVB	Methylmethacrylate-divinylbenzene copolymer Ethylene glycol dimethacrylate- divinylbenzene co-polymer	Phenol	Langmuir, Freundlich	-	[138]

Table 5.5 (continued)

Authors	Adsorbent	Structure	Pollutant	Model	Kinetics order	Ref.
Wagner et al.	Amberlite XAD-4 Serdolite PAD I, II and III TRP-100 Lewatit VP-OC 1163	Polystyrene and polymeth-acrylate resins crosslinked with divinylbenzene	[Phenol, 4-Chlorophenol, 2- Chlorophenol, 2,4,6-trichloro-phenol]	Redlich-Peterson	-	[139]
Li et al.	Various copol-ymers	P[MMA-co-DVB] Interpenetrating polymer based on P(MMA-co-DVB) and P(S-co-DVB) Poly (N-p-vinylbenzyl acet-ylamide)	Phenol	Freundlich	-	[140]
Dursun et al.	Chitin	Natural occurring poly-mer	Phenol	Freundlich	Pseudo-second order	[141]
Kawabata et al.	Vinylpyridine-Divinylben-zene	Vinylpyridine-Divinylben-zene copolymer	Phenol	-	-	[142]
Fowkes et al.	PMMA films	PMMA films	Phenol	-	-	[143]
Burleigh et al.	Polysilsesqui-oxanes	Arylene- and ethylene-bridged Polysilsesquiox-anes	[4-Methylphenol, 4-Nitrophenol, 4-Chlorophenol]	-	-	[144]
Huang et al.	HJ-Z01	Hypercrosslinked resin modified with N-methyla-cetamide	Phenol	Langmuir and Freundlich	Pseudo-second order	[145]
Gupta et al.	PUUSD mem-brane	Hydroxyterminated poly-butadiene based polyure-thaneurea membranes	Phenol	-	-	[146]

References

1. Neff, J. M. (2002) *Bioaccumulation in Marine Organisms: Effect of Contaminants from Oil Well Produced Water*, Elsevier, USA.
2. Fakhru'l-Razi, A., Pendashteh, A., Abdullah, L. C., Biak, D. R. A., Madaeni, S. S., and Abidin, Z. Z. (2009) Review of technologies for oil and gas produced water treatment. *Journal of Hazardous Materials*, **170**, 530-551.
3. Fingas, M. (2010) *Oil Spill Science and Technology*, 2nd edition, Gulf Professional Publishing, USA.
4. Um, M.-J., Yoon, S.-H., Lee, C.-H., Chung, K.-Y., and Kim, J.-J. (2001) Flux enhancement with gas injection in crossflow ultrafiltration of oily wastewater. *Water Research*, **35**, 4095-4101.
5. Gaaseidnes, K., and Turbeville, J. (1999) Separation of oil and water in oil spill recovery operations. *Pure and Applied Chemistry*, **71**, 95-101.
6. Cheryan, M., and Rajagopalan, N. (1998) Membrane processing of oily streams. Wastewater treatment and waste reduction. *Journal of Membrane Science*, **151**, 13-28.
7. Xue, Z., Cao, Y., Liu, N., Feng, L., and Jiang, L. (2014) Special wettable materials for oil/water separation. *Journal of Materials Chemistry A*, **2**, 2445-2460.
8. Mohammadi, T., Kazemimoghadam, M., and Saadabadi, M. (2003) Modeling of membrane fouling and flux decline in reverse osmosis during separation of oil in water emulsions. *Desalination*, **157**, 369-375.
9. Karakulski, K., Kozlowski, A., and Morawski, A. (1995) Purification of oily wastewater by ultrafiltration. *Separations Technology*, **5**, 197-205.
10. Bodzek, M., and Konieczny, K. (1992) The use of ultrafiltration membranes made of various polymers in the treatment of oil-emulsion wastewaters. *Waste Management*, **12**, 75-84.
11. Ohya, H., Kim, J., Chinen, A., Aihara, M., Semenova, S., Negishi, Y., Mori, O., and Yasuda, M. (1998) Effects of pore size on separation mechanisms of microfiltration of oily water, using porous glass tubular membrane. *Journal of Membrane Science*, **145**, 1-14.
12. Deng, S., Sourirajan, S., Chan, K., Farnand, B., Okada, T., and Matsuura, T. (1991) Dehydration of oil-water emulsion by pervaporation using porous hydrophilic membranes. *Journal of Colloid and Interface Science*, **141**, 218-225.
13. Gryta, M., and Karakulski, K. (1999) The application of membrane distillation for the concentration of oil-water emulsions. *Desalination*, **121**, 23-29.
14. Zhong, J., Sun, X., and Wang, C. (2003) Treatment of oily wastewater produced from refinery processes using flocculation and ceramic membrane filtration. *Separation and Purification Technology*, **32**, 93-98.
15. Chu, Y., and Pan, Q. (2012) Three-dimensionally macroporous Fe/C nanocomposites as highly selective oil-absorption materials. *ACS Applied Materials & Interfaces*, **4**, 2420-2425.
16. Gui, X., Zeng, Z., Lin, Z., Gan, Q., Xiang, R., Zhu, Y., Cao, A., and Tang, Z. (2013) Magnetic and highly recyclable macroporous carbon nanotubes for spilled oil sorption and separation. *ACS Applied Materials & Interfaces*, **5**, 5845-5850.

17. Korhonen, J. T., Kettunen, M., Ras, R. H., and Ikkala, O. (2011) Hydrophobic nanocellulose aerogels as floating, sustainable, reusable, and recyclable oil absorbents. *ACS Applied Materials & Interfaces*, **3**, 1813-1816.
18. Choi, H. M., and Cloud, R. M. (1992) Natural sorbents in oil spill cleanup. *Environmental Science & Technology*, **26**, 772-776.
19. Zhu, H., Qiu, S., Jiang, W., Wu, D., and Zhang, C. (2011) Evaluation of electrospun polyvinyl chloride/polystyrene fibers as sorbent materials for oil spill cleanup. *Environmental Science & Technology*, **45**, 4527-4531.
20. Lessard, R. R., and DeMarco, G. (2000) The significance of oil spill dispersants. *Spill Science & Technology Bulletin*, **6**, 59-68.
21. Kujawinski, E. B., Kido Soule, M. C., Valentine, D. L., Boysen, A. K., Longnecker, K., and Redmond, M. C. (2011) Fate of dispersants associated with the Deepwater Horizon oil spill. *Environmental Science & Technology*, **45**, 1298-1306.
22. Yang, C., Kaipa, U., Mather, Q. Z., Wang, X., Nesterov, V., Venero, A. F., and Omary, M. A. (2011) Fluorous metal–organic frameworks with superior adsorption and hydrophobic properties toward oil spill cleanup and hydrocarbon storage. *Journal of the American Chemical Society*, **133**, 18094-18097.
23. Sakthivel, T., Reid, D. L., Goldstein, I., Hench, L., and Seal, S. (2013) Hydrophobic high surface area zeolites derived from fly ash for oil spill remediation. *Environmental Science & Technology*, **47**, 5843-5850.
24. Bastani, D., Safekordi, A., Alihosseini, A., and Taghikhani, V. (2006) Study of oil sorption by expanded perlite at 298.15 K. *Separation and Purification Technology*, **52**, 295-300.
25. Syed, S., Alhazzaa, M., and Asif, M. (2011) Treatment of oily water using hydrophobic nano-silica. *Chemical Engineering Journal*, **167**, 99-103.
26. Majano, G., and Mintova, S. (2010) Mineral oil regeneration using selective molecular sieves as sorbents. *Chemosphere*, **78**, 591-598.
27. Zhou, X. M., and Chuai, C. Z. (2010) Synthesis and characterization of a novel high-oil-absorbing resin. *Journal of Applied Polymer Science*, **115**, 3321-3325.
28. Yuan, X., and Chung, T. M. (2012) Novel solution to oil spill recovery: using thermodegradable polyolefin oil superabsorbent polymer (oil–SAP). *Energy & Fuels*, **26**, 4896-4902.
29. Sun, X.-F., Sun, R., and Sun, J.-X. (2002) Acetylation of rice straw with or without catalysts and its characterization as a natural sorbent in oil spill cleanup. *Journal of Agricultural and Food Chemistry*, **50**, 6428-6433.
30. Husseien, M., Amer, A., El-Maghraby, A., and Hamedallah, N. (2009) A comprehensive characterization of corn stalk and study of carbonized corn stalk in dye and gas oil sorption. *Journal of Analytical and Applied Pyrolysis*, **86**, 360-363.
31. Radetic, M. M., Jocic, D. M., Jovancic, P. M., Petrovic, Z. L., and Thomas, H. F. (2003) Recycled wool-based nonwoven material as an oil sorbent. *Environmental Science & Technology*, **37**, 1008-1012.
32. Ali, N., El-Harbawi, M., Jabal, A.A., and Yin, C.-Y. (2012) Characteristics and oil sorption effectiveness of kapok fibre, sugarcane bagasse and rice husks: oil removal suitability matrix. *Environmental Technology*, **33**, 481-486.

33. Nordvik, A.B. (1995) The technology windows-of-opportunity for marine oil spill response as related to oil weathering and operations. *Spill Science & Technology Bulletin*, **2**, 17-46.
34. Wayment, E., and Wagstaff, B. (1999) Appropriate technology for oil spill management in developing nations. *Pure and Applied Chemistry*, **71**, 203-208.
35. Burton, Z., and Bhushan, B. (2006) Surface characterization and adhesion and friction properties of hydrophobic leaf surfaces. *Ultramicroscopy*, **106**, 709-719.
36. Poynor, A., Hong, L., Robinson, I.K., Granick, S., Zhang, Z., and Fenter, P.A. (2006) How water meets a hydrophobic surface. *Physical Review Letters*, **97**, 266101.
37. Onda, T., Shibuichi, S., Satoh, N., and Tsujii, K. (1996) Super-water-repellent fractal surfaces. *Langmuir*, **12**, 2125-2127.
38. Bico, J., Marzolin, C., and Quéré, D. (1999) Pearl drops. *Europhysics Letters*, **47**, 220.
39. Bhushan, B., and Jung, Y. C. (2011) Natural and biomimetic artificial surfaces for superhydrophobicity, self-cleaning, low adhesion, and drag reduction. *Progress in Materials Science*, **56**, 1-108.
40. Darmanin, T., and Guittard, F. (2009) Super oil-repellent surfaces from conductive polymers. *Journal of Materials Chemistry*, **19**, 7130-7136.
41. Steele, A., Bayer, I., and Loth, E. (2008) Inherently superoleophobic nanocomposite coatings by spray atomization. *Nano Letters*, **9**, 501-505.
42. Zeng, J., Wang, B., Zhang, Y., Zhu, H., and Guo, Z. (2014) Strong amphiphobic porous films with oily-self-cleaning property beyond nature. *Chemistry Letters*, **43**, 1566-1568.
43. Drelich, J., and Chibowski, E. (2010) Superhydrophilic and superwetting surfaces: definition and mechanisms of control. *Langmuir*, **26**, 18621-18623.
44. Song, S., Jing, L., Li, S., Fu, H., and Luan, Y. (2008) Superhydrophilic anatase TiO2 film with the micro-and nanometer-scale hierarchical surface structure. *Materials Letters*, **62**, 3503-3505.
45. Hutton, S., Crowther, J., and Badyal, J. (2000) Complexation of fluorosurfactants to functionalized solid surfaces: smart behavior. *Chemistry of Materials*, **12**, 2282-2286.
46. Howarter, J. A., and Youngblood, J. P. (2007) Self-Cleaning and Anti-fog surfaces via stimuli-responsive polymer brushes. *Advanced Materials*, **19**, 3838-3843.
47. Howarter, J. A., Genson, K. L., and Youngblood, J. P. (2011) Wetting behavior of oleophobic polymer coatings synthesized from fluorosurfactant-macromers. *ACS Applied Materials & Interfaces*, **3**, 2022-2030.
48. Melian, E. P., Diaz, O. G., Arana, J., Rodriguez, J. D., Rendon, E. T., and Melian, J. H. (2007) Kinetics and adsorption comparative study on the photocatalytic degradation of o-, m-and p-cresol. *Catalysis Today*, **129**, 256-262.
49. Babic, K., Driessen, G., Van der Ham, A., and De Haan, A. (2007) Chiral separation of amino-alcohols using extractant impregnated resins. *Journal of Chromatography A*, **1142**, 84-92.
50. Arana, J., Melian, E. P., Lopez, V. R., Alonso, A. P., Rodriguez, J. D., Diaz, O. G., and Pena, J. P. (2007) Photocatalytic degradation of phenol and phenolic compounds: Part I. Adsorption and FTIR study. *Journal of Hazardous Materials*, **146**, 520-528.
51. Ahmaruzzaman, M., and Sharma, D. (2005) Adsorption of phenols from wastewater. *Journal of Colloid and Interface Science*, **287**, 14-24.

52. Bulut, Y., and Baysal, Z. (2006) Removal of Pb (II) from wastewater using wheat bran. *Journal of Environmental Management*, **78**, 107-113.
53. Kujawski, W., Warszawski, A., Ratajczak, W., Porebski, T., Capała, W., and Ostrowska, I. (2004) Removal of phenol from wastewater by different separation techniques. *Desalination*, **163**, 287-296.
54. Otero, M., Zabkova, M., and Rodrigues, A.E. (2005) Adsorptive purification of phenol wastewaters: Experimental basis and operation of a parametric pumping unit. *Chemical Engineering Journal*, **110**, 101-111.
55. Gupta, V. K., Mohan, D., Suhas, A., and Singh, K. P. (2006) Removal of 2-aminophenol using novel adsorbents. *Industrial & Engineering Chemistry Research*, **45**, 1113-1122.
56. Slein, M., and Sansone, E. (1981) Degradation of Chemical Carcinogens. *Journal of Occupational and Environmental Medicine*, **23**, 162.
57. Wang, C.-F., and Lin, S.-J. (2013) Robust superhydrophobic/superoleophilic sponge for effective continuous absorption and expulsion of oil pollutants from water. *ACS Applied Materials & Interfaces*, **5**, 8861-8864.
58. Jiang, G., Hu, R., Xi, X., Wang, X., and Wang, R. (2013) Facile preparation of superhydrophobic and superoleophilic sponge for fast removal of oils from water surface. *Journal of Materials Research*, **28**, 651-656.
59. Zhang, Z., Sebe, G., Rentsch, D., Zimmermann, T., and Tingaut, P. (2014) Ultralight-weight and flexible silylated nanocellulose sponges for the selective removal of oil from water. *Chemistry of Materials*, **26**, 2659-2668.
60. Wang, S., Peng, X., Zhong, L., Tan, J., Jing, S., Cao, X., Chen, W., Liu, C., and Sun, R. (2015) An ultralight, elastic, cost-effective, and highly recyclable superabsorbent from microfibrillated cellulose fibers for oil spillage cleanup. *Journal of Materials Chemistry A*, **3**, 8772-8781.
61. Wu, D., Fang, L., Qin, Y., Wu, W., Mao, C., and Zhu, H. (2014) Oil sorbents with high sorption capacity, oil/water selectivity and reusability for oil spill cleanup. *Marine Pollution Bulletin*, **84**, 263-267.
62. Liu, Y., Ma, J., Wu, T., Wang, X., Huang, G., Liu, Y., Qiu, H., Li, Y., Wang, W., and Gao, J. (2013) Cost-effective reduced graphene oxide-coated polyurethane sponge as a highly efficient and reusable oil-absorbent. *ACS Applied Materials & Interfaces*, **5**, 10018-10026.
63. Pan, Y., Shi, K., Peng, C., Wang, W., Liu, Z., and Ji, X. (2014) Evaluation of hydrophobic polyvinyl-alcohol formaldehyde sponges as absorbents for oil spill. *ACS Applied Materials & Interfaces*, **6**, 8651-8659.
64. Nguyen, D. D., Tai, N.-H., Lee, S.-B., and Kuo, W.-S. (2012) Superhydrophobic and superoleophilic properties of graphene-based sponges fabricated using a facile dip coating method. *Energy & Environmental Science*, **5**, 7908-7912.
65. Zhu, Q., Chu, Y., Wang, Z., Chen, N., Lin, L., Liu, F., and Pan, Q. (2013) Robust superhydrophobic polyurethane sponge as a highly reusable oil-absorption material. *Journal of Materials Chemistry A*, **1**, 5386-5393.
66. Turco, A., Malitesta, C., Barillaro, G., Greco, A., Maffezzoli, A., and Mazzotta, E. (2015) A magnetic and highly reusable macroporous superhydrophobic/superoleophilic PDMS/MWNT nanocomposite for oil sorption from water. *Journal of Materials Chemistry A*, **3**, 17685-17696.

67. Zhang, A., Chen, M., Du, C., Guo, H., Bai, H., and Li, L. (2013) Poly (dimethylsiloxane) oil absorbent with a three-dimensionally interconnected porous structure and swellable skeleton. *ACS Applied Materials & Interfaces*, 5, 10201-10206.

68. Nagappan, S., Park, J. J., Park, S. S., Lee, W.-K., and Ha, C.-S. (2013) Bio-inspired, multi-purpose and instant superhydrophobic–superoleophilic lotus leaf powder hybrid micro–nanocomposites for selective oil spill capture. *Journal of Materials Chemistry A*, 1, 6761-6769.

69. Zhu, Q., Pan, Q., and Liu, F. (2011) Facile removal and collection of oils from water surfaces through superhydrophobic and superoleophilic sponges. *The Journal of Physical Chemistry C*, 115, 17464-17470.

70. Li, H., Liu, L., and Yang, F. (2013) Covalent assembly of 3D graphene/polypyrrole foams for oil spill cleanup. *Journal of Materials Chemistry A*, 1, 3446-3453.

71. A Tanobe, V., Sydenstricker, T., Amico, S., Vargas, J., and Zawadzki, S. (2009) Evaluation of flexible postconsumed polyurethane foams modified by polystyrene grafting as sorbent material for oil spills. *Journal of Applied Polymer Science*, 111, 1842-1849.

72. Calcagnile, P., Fragouli, D., Bayer, I. S., Anyfantis, G. C., Martiradonna, L., Cozzoli, P. D., Cingolani, R., and Athanassiou, A. (2012) Magnetically driven floating foams for the removal of oil contaminants from water. *ACS Nano*, 6, 5413-5419.

73. Nguyen, S. T., Feng, J., Le, N. T., Le, A. T., Hoang, N., Tan, V. B., and Duong, H. M. (2013) Cellulose aerogel from paper waste for crude oil spill cleaning. *Industrial & Engineering Chemistry Research*, 52, 18386-18391.

74. Chin, S. F., Romainor, A. N. B., and Pang, S. C. (2014) Fabrication of hydrophobic and magnetic cellulose aerogel with high oil absorption capacity. *Materials Letters*, 115, 241-243.

75. Choi, H.-M., Kwon, H.-J., and Moreau, J. P. (1993) Cotton nonwovens as oil spill cleanup sorbents. *Textile Research Journal*, 63, 211-218.

76. Deschamps, G., Caruel, H., Borredon, M.-E., Bonnin, C., and Vignoles, C. (2003) Oil removal from water by selective sorption on hydrophobic cotton fibers. 1. Study of sorption properties and comparison with other cotton fiber-based sorbents. *Environmental Science & Technology*, 37, 1013-1015.

77. Wang, J., Zheng, Y., and Wang, A. (2013) Coated kapok fiber for removal of spilled oil. *Marine Pollution Bulletin*, 69, 91-96.

78. Lin, J., Shang, Y., Ding, B., Yang, J., Yu, J., and Al-Deyab, S. S. (2012) Nanoporous polystyrene fibers for oil spill cleanup. *Marine Pollution Bulletin*, 64, 347-352.

79. Wu, J., Wang, N., Wang, L., Dong, H., Zhao, Y., and Jiang, L. (2012) Electrospun porous structure fibrous film with high oil adsorption capacity. *ACS Applied Materials & Interfaces*, 4, 3207-3212.

80. Sai, H., Fu, R., Xing, L., Xiang, J., Li, Z., Li, F., and Zhang, T. (2015) Surface modification of bacterial cellulose aerogels' web-like skeleton for oil/water separation. *ACS Applied Materials & Interfaces*, 7, 7373-7381.

81. Si, Y., Fu, Q., Wang, X., Zhu, J., Yu, J., Sun, G., and Ding, B. (2015) Superelastic and superhydrophobic nanofiber-assembled cellular aerogels for effective separation of oil/water emulsions. *ACS Nano*, 9, 3791-3799.

82. Jiang, F., and Hsieh, Y.-L. (2014) Amphiphilic superabsorbent cellulose nanofibril aerogels. *Journal of Materials Chemistry A*, 2, 6337-6342.

83. Zheng, Q., Cai, Z., and Gong, S. (2014) Green synthesis of polyvinyl alcohol (PVA)–cellulose nanofibril (CNF) hybrid aerogels and their use as superabsorbents. *Journal of Materials Chemistry A*, **2**, 3110-3118.

84. Lin, J., Tian, F., Shang, Y., Wang, F., Ding, B., Yu, J., and Guo, Z. (2013) Co-axial electrospun polystyrene/polyurethane fibres for oil collection from water surface. *Nanoscale*, **5**, 2745-2755.

85. Lin, J., Ding, B., Yang, J., Yu, J., and Sun, G. (2012) Subtle regulation of the micro-and nanostructures of electrospun polystyrene fibers and their application in oil absorption. *Nanoscale*, **4**, 176-182.

86. Zhao, J., Xiao, C., and Xu, N. (2013) Evaluation of polypropylene and poly (butyl-methacrylate-co-hydroxyethylmethacrylate) nonwoven material as oil absorbent. *Environmental Science and Pollution Research*, **20**, 4137-4145.

87. Xue, C.-H., Ji, P.-T., Zhang, P., Li, Y.-R., and Jia, S.-T. (2013) Fabrication of superhydrophobic and superoleophilic textiles for oil–water separation. *Applied Surface Science*, **284**, 464-471.

88. Liu, X., Ge, L., Li, W., Wang, X., and Li, F. (2014) Layered double hydroxide function-alized textile for effective oil/water separation and selective oil adsorption. *ACS Applied Materials & Interfaces*, **7**, 791-800.

89. Zhang, J., and Seeger, S. (2011) Polyester materials with superwetting silicone nanofilaments for oil/water separation and selective oil absorption. *Advanced Functional Materials*, **21**, 4699-4704.

90. Wu, L., Zhang, J., Li, B., and Wang, A. (2014) Magnetically driven super durable superhydrophobic polyester materials for oil/water separation. *Polymer Chemistry*, **5**, 2382-2390.

91. Li, X., Wang, M., Wang, C., Cheng, C., and Wang, X. (2014) Facile immobilization of Ag nanocluster on nanofibrous membrane for oil/water separation. *ACS Applied Materials & Interfaces*, **6**, 15272-15282.

92. Tai, M. H., Gao, P., Tan, B. Y. L., Sun, D. D., and Leckie, J. O. (2014) Highly efficient and flexible electrospun carbon–silica nanofibrous membrane for ultrafast gravity-driven oil–water separation. *ACS Applied Materials & Interfaces*, **6**, 9393-9401.

93. Zhang, W., Shi, Z., Zhang, F., Liu, X., Jin, J., and Jiang, L. (2013) Superhydrophobic and superoleophilic PVDF membranes for effective separation of water-in-oil emulsions with high flux. *Advanced Materials*, **25**, 2071-2076.

94. Chen, P.-C., and Xu, Z.-K. (2013) Mineral-coated polymer membranes with superhydrophilicity and underwater superoleophobicity for effective oil/water separation. *Scientific Reports*, **3**.

95. Chakrabarty, B., Ghoshal, A., and Purkait, M. (2008) Ultrafiltration of stable oil-in-water emulsion by polysulfone membrane. *Journal of Membrane Science*, **325**, 427-437.

96. Zhang, C., Li, P., and Cao, B. (2015) Electrospun microfibrous membranes based on PIM-1/POSS with high oil wettability for separation of oil–water mixtures and cleanup of oil soluble contaminants. *Industrial & Engineering Chemistry Research*, **54**, 8772-8781.

97. Zhou, X., Zhang, Z., Xu, X., Guo, F., Zhu, X., Men, X., and Ge, B. (2013) Robust and durable superhydrophobic cotton fabrics for oil/water separation. *ACS Applied Materials & Interfaces*, **5**, 7208-7214.

98. Huang, M., Si, Y., Tang, X., Zhu, Z., Ding, B., Liu, L., Zheng, G., Luo, W., and Yu, J. (2013) Gravity driven separation of emulsified oil–water mixtures utilizing in situ polymerized superhydrophobic and superoleophilic nanofibrous membranes. *Journal of Materials Chemistry A*, **1**, 14071-14074.

99. Tang, X., Si, Y., Ge, J., Ding, B., Liu, L., Zheng, G., Luo, W., and Yu, J. (2013) In situ polymerized superhydrophobic and superoleophilic nanofibrous membranes for gravity driven oil–water separation. *Nanoscale*, **5**, 11657-11664.

100. Shang, Y., Si, Y., Raza, A., Yang, L., Mao, X., Ding, B., and Yu, J. (2012) An in situ polymerization approach for the synthesis of superhydrophobic and superoleophilic nanofibrous membranes for oil–water separation. Nanoscale 4, 7847-7854.

101. Chen, W., Su, Y., Zheng, L., Wang, L., and Jiang, Z. (2009) The improved oil/water separation performance of cellulose acetate-graft-polyacrylonitrile membranes. *Journal of Membrane Science*, **337**, 98-105.

102. Obaid, M., Barakat, N. A., Fadali, O., Motlak, M., Almajid, A. A., and Khalil, K. A. (2015) Effective and reusable oil/water separation membranes based on modified polysulfone electrospun nanofiber mats. *Chemical Engineering Journal*, **259**, 449-456.

103. Extrand, C. (2002) Water contact angles and hysteresis of polyamide surfaces. *Journal of Colloid and Interface Science*, **248**, 136-142.

104. Extrand, C. (2004) Contact angles and their hysteresis as a measure of liquid-solid adhesion. *Langmuir*, **20**, 4017-4021.

105. Zhu, Y., Zhang, F., Wang, D., Pei, X. F., Zhang, W., and Jin, J. (2013) A novel zwitterionic polyelectrolyte grafted PVDF membrane for thoroughly separating oil from water with ultrahigh efficiency. *Journal of Materials Chemistry A*, **1**, 5758-5765.

106. Xiang, Y., Liu, F., and Xue, L. (2015) Under seawater superoleophobic PVDF membrane inspired by polydopamine for efficient oil/seawater separation. Journal of Membrane Science 476, 321-329.

107. Li, H.-J., Cao, Y.-M., Qin, J.-J., Jie, X.-M., Wang, T.-H., Liu, J.-H., and Yuan, Q. (2006) Development and characterization of anti-fouling cellulose hollow fiber UF membranes for oil–water separation. *Journal of Membrane Science*, **279**, 328-335.

108. Raza, A., Ding, B., Zainab, G., El-Newehy, M., Al-Deyab, S.S., and Yu, J. (2014) In situ cross-linked superwetting nanofibrous membranes for ultrafast oil–water separation. *Journal of Materials Chemistry A*, **2**, 10137-10145.

109. Chen, W., Peng, J., Su, Y., Zheng, L., Wang, L., and Jiang, Z. (2009) Separation of oil/water emulsion using Pluronic F127 modified polyethersulfone ultrafiltration membranes. *Separation and Purification Technology*, **66**, 591-597.

110. Zhang, W., Zhu, Y., Liu, X., Wang, D., Li, J., Jiang, L., and Jin, J. (2014) Salt-induced fabrication of superhydrophilic and underwater superoleophobic PAA-g-PVDF membranes for effective separation of iil-in-water emulsions. *Angewandte Chemie International Edition*, **53**, 856-860.

111. Li, J.-J., Zhou, Y.-N., and Luo, Z.-H. (2015) Smart fiber membrane for pH-induced oil/water separation. *ACS Applied Materials & Interfaces*, **7**, 19643-19650.

112. Zhang, W., Shi, L., Ma, R., An, Y., Xu, Y., and Wu, K. (2005) Micellization of

thermo-and pH-responsive triblock copolymer of poly (ethylene glycol)-b-poly (4-vinylpyridine)-b-poly (N-isopropylacrylamide). *Macromolecules*, **38**, 8850-8852.

113. Mendrek, S., Mendrek, A., Adler, H. J., Dworak, A., and Kuckling, D. (2009) Synthesis and characterization of pH sensitive poly (glycidol)-b-poly (4-vinylpyridine) block copolymers. *Journal of Polymer Science Part A: Polymer Chemistry*, **47**, 1782-1794.

114. Escale, P., Rubatat, L., Derail, C., Save, M., and Billon, L. (2011) pH Sensitive Hierarchically Self-Organized Bioinspired Films. *Macromolecular Rapid Communications*, **32**, 1072-1076.

115. Geng, Z., Guan, S., Jiang, H.-M., Liu, Z.-W., and Jiang, L. (2014) pH-sensitive wettability induced by topological and chemical transition on the self assembled surface of block copolymer. *Chinese Journal of Polymer Science*, **32**, 92-97.

116. Zhang, Y., Wei, S., Liu, F., Du, Y., Liu, S., Ji, Y., Yokoi, T., Tatsumi, T., and Xiao, F.-S. (2009) Superhydrophobic nanoporous polymers as efficient adsorbents for organic compounds. *Nano Today*, **4**, 135-142.

117. Atta, A. M., El-Hamouly, S. H., AlSabagh, A. M., and Gabr, M. M. (2007) Crosslinked poly (octadecene-alt-maleic anhydride) copolymers as crude oil sorbers. *Journal of Applied Polymer Science*, **105**, 2113-2120.

118. Avila, A. F., Munhoz, V. C., de Oliveira, A. M., Santos, M. C., Lacerda, G. R., and Goncalves, C. P. (2014) Nano-based systems for oil spills control and cleanup. *Journal of Hazardous Materials*, **272**, 20-27.

119. Gu, J., Jiang, W., Wang, F., Chen, M., Mao, J., and Xie, T. (2014) Facile removal of oils from water surfaces through highly hydrophobic and magnetic polymer nanocomposites. *Applied Surface Science*, **301**, 492-499.

120. Chen, M., Jiang, W., Wang, F., Shen, P., Ma, P., Gu, J., Mao, J., and Li, F. (2013) Synthesis of highly hydrophobic floating magnetic polymer nanocomposites for the removal of oils from water surface. *Applied Surface Science*, **286**, 249-256.

121. Song, C., Ding, L., Yao, F., Deng, J., and Yang, W. (2013) β-Cyclodextrin-based oil-absorbent microspheres: Preparation and high oil absorbency. *Carbohydrate Polymers*, **91**, 217-223.

122. Mao, J., Jiang, W., Gu, J., Zhou, S., Lu, Y., and Xie, T. (2014) Synthesis of P (St-DVB)/Fe 3 0 4 microspheres and application for oil removal in aqueous environment. *Applied Surface Science*, **317**, 787-793.

123. Huang, J. (2009) Treatment of phenol and p-cresol in aqueous solution by adsorption using a carbonylated hypercrosslinked polymeric adsorbent. *Journal of Hazardous Materials*, **168**, 1028-1034.

124. Caetano, M., Valderrama, C., Farran, A., and Cortina, J. L. (2009) Phenol removal from aqueous solution by adsorption and ion exchange mechanisms onto polymeric resins. *Journal of Colloid and Interface Science*, **338**, 402-409.

125. Senel, S., Kara, A., Alsancak, G., and Denizli, A. (2006) Removal of phenol and chlorophenols from water with reusable dye-affinity hollow fibers. *Journal of Hazardous Materials*, **138**, 317-324.

126. An, F., and Gao, B. (2008) Adsorption of phenol on a novel adsorption material PEI/SiO$_2$. *Journal of Hazardous Materials*, **152**, 1186-1191.

127. An, F., Gao, B., and Feng, X. (2008) Adsorption and recognizing ability of molecular imprinted polymer MIP-PEI/SiO 2 towards phenol. *Journal of Hazardous Materials*, **157**, 286-292.

128. Shao, D., Hu, J., Chen, C., Sheng, G., Ren, X., and Wang, X. (2010) Polyaniline multi-walled carbon nanotube magnetic composite prepared by plasma-induced graft technique and its application for removal of aniline and phenol. *The Journal of Physical Chemistry C*, **114**, 21524-21530.
129. Zeng, X., Fan, Y., Wu, G., Wang, C., and Shi, R. (2009) Enhanced adsorption of phenol from water by a novel polar post-crosslinked polymeric adsorbent. *Journal of Hazardous Materials*, **169**, 1022-1028.
130. Li, A., Zhang, Q., Zhang, G., Chen, J., Fei, Z., and Liu, F. (2002) Adsorption of phenolic compounds from aqueous solutions by a water-compatible hypercrosslinked polymeric adsorbent. *Chemosphere*, **47**, 981-989.
131. Ming, Z. W., Long, C. J., Cai, P. B., Xing, Z. Q., and Zhang, B. (2006) Synergistic adsorption of phenol from aqueous solution onto polymeric adsorbents. *Journal of Hazardous Materials*, **128**, 123-129.
132. Zhang, W., Du, Q., Pan, B., Lv, L., Hong, C., Jiang, Z., and Kong, D. (2009) Adsorption equilibrium and heat of phenol onto aminated polymeric resins from aqueous solution. *Colloids and Surfaces A: Physicochemical and Engineering Aspects*, **346**, 34-38.
133. Pan, B., Du, W., Zhang, W., Zhang, X., Zhang, Q., Pan, B., Lv, L., Zhang, Q., and Chen, J. (2007) Improved adsorption of 4-nitrophenol onto a novel hyper-cross-linked polymer. *Environmental Science & Technology*, **41**, 5057-5062.
134. Pan, B., Zhang, X., Zhang, W., Zheng, J., Pan, B., Chen, J., and Zhang, Q. (2005) Adsorption of phenolic compounds from aqueous solution onto a macroporous polymer and its aminated derivative: isotherm analysis. *Journal of Hazardous Materials*, **121**, 233-241.
135. Li, A., Zhang, Q., Chen, J., Fei, Z., Long, C., and Li, W. (2001) Adsorption of phenolic compounds on Amberlite XAD-4 and its acetylated derivative MX-4. *Reactive and Functional Polymers*, **49**, 225-233.
136. Pan, B., Pan, B., Zhang, W., Zhang, Q., Zhang, Q., and Zheng, S. (2008) Adsorptive removal of phenol from aqueous phase by using a porous acrylic ester polymer. *Journal of Hazardous Materials*, **157**, 293-299.
137. Juang, R.-S., and Shiau, J.-Y. (1999) Adsorption isotherms of phenols from water onto macroreticular resins. *Journal of Hazardous Materials*, **70**, 171-183.
138. Zeng, X., Yu, T., Wang, P., Yuan, R., Wen, Q., Fan, Y., Wang, C., and Shi, R. (2010) Preparation and characterization of polar polymeric adsorbents with high surface area for the removal of phenol from water. *Journal of Hazardous Materials*, **177**, 773-780.
139. Wagner, K., and Schulz, S. (2001) Adsorption of phenol, chlorophenols, and dihydroxybenzenes onto unfunctionalized polymeric resins at temperatures from 294.15 K to 318.15 K. *Journal of Chemical & Engineering Data*, **46**, 322-330.
140. Li, H., Xu, M., Shi, Z., and He, B. (2004) Isotherm analysis of phenol adsorption on polymeric adsorbents from nonaqueous solution. *Journal of Colloid and Interface Science*, **271**, 47-54.
141. Dursun, A. Y., and Kalayci, C. S. (2005) Equilibrium, kinetic and thermodynamic studies on the adsorption of phenol onto chitin. *Journal of Hazardous Materials*, **123**, 151-157.

Polymers in Oil and Gas Industry

142. Kawabata, N., and Ohira, K. (1979) Removal and recovery of organic pollutants from aquatic environment. 1. Vinylpyridine-divinylbenzene copolymer as a polymeric adsorbent for removal and recovery of phenol from aqueous solution. *Environmental Science & Technology*, **13**, 1396-1402.

143. Fowkes, F. M., Kaczinski, M. B., and Dwight, D. W. (1991) Characterization of polymer surface sites with contact angles of test solutions. 1. Phenol and iodine adsorption from methylene iodide onto PMMA films. *Langmuir*, **7**, 2464-2470.

144. Burleigh, M. C., Markowitz, M. A., Spector, M. S., and Gaber, B. P. (2002) Porous polysilsesquioxanes for the adsorption of phenols. *Environmental Science & Technology*, **36**, 2515-2518.

145. Huang, J., Jin, X., and Deng, S. (2012) Phenol adsorption on an N-methylacetamide-modified hypercrosslinked resin from aqueous solutions. *Chemical Engineering Journal*, **192**, 192-200.

146. Gupta, T., Pradhan, N. C., and Adhikari, B. (2003) Separation of phenol from aqueous solution by pervaporation using HTPB-based polyurethaneurea membrane. *Journal of Membrane Science*, **217**, 43-53.

6

Resorcinol-Formaldehyde Cryogel Nanocomposites for Oil Spill Clean Up

6.1 Introduction

Oil spills have resulted in catastrophic effects on aquatic ecosystems over the past years [1]. Therefore, there is a continuous requirement to develop ideal absorbent materials exhibiting low cost, environment favorability, high efficiency, high absorption capacity, excellent selectivity and recyclability. Variety of methods have been employed to clean the oil spills such as bio-remediation [2], mechanical collection [3], in-situ burning [4], chemical dispersants [5] and absorbent materials [6]. Among these methods, sorbent materials have proven to be the most desirable option for the removal of oil spills from water bodies. In this category, some materials including wool fibers [7], organoclays [8], activated carbon [9], zeolites [10], and alkyl acrylate copolymers [11] have been employed to clean the oil spill. However, in practical applications, usage of these materials is limited due to their low absorption capacities, environmental incompatibilities and poor recyclability. To overcome these challenge, various advanced materials based on graphene [12], carbon nanotubes [13], mesh films [14], aerogels [15], sponges [16], foams [17], membranes [18] and fibers [19] have been developed in the past few years for the separation of oil from water.

Cryogels have large and well-defined interconnected porous heterogeneous structure [20,21]. Free-radical cryo-polymerization technique is generally employed to prepare cryogels from polymeric gel precursors below the solvent freezing point, as freeze-drying process maintains the interconnected structure of the resultant cryogel [22]. The properties of cryogels such as pore dimensions, pore wall thickness, mechanical toughness and elasticity depend on the conditions of preparation, amount of solvent and precursors, temperature gradients, cooling rate and other conditions, etc. [23,24]. In this study, the preparation of three-dimensional carbonaceous monolithic cryogel nanocomposites from resorcinol-formaldehyde (R-F) resin precursor with graphene oxide (GO), phosphazene nanotubes (PZTs) and cobalt ferrite nanoparticles is reported. The nanocomposites exhibited porosity and high surface area, pav-

Naman Arora, Muthukumaraswamy R. Vengatesan and Vikas Mittal*, The Petroleum Institute (part of Khalifa University of Science and Technology), Abu Dhabi, UAE
*Current address: Bletchington, Wellington County, Australia

ing the way to utilize the nanomaterials effectively as a sorbent material at the macroscopic level to remove oil from oil/water mixture. During the formation of R-F based cryogel, divalent metal ions (such as Co^{2+} and Fe^{2+}) perform as both cross-linker and catalyst due to their binding ability towards oxygen groups [25] which leads to crosslinking between graphene oxide and R-F skeletons, thus, enhancing the strength of cryogel and also substituting Na_2CO_3 as a catalyst. Subsequently, the cryogel is converted into carbonaceous monoliths with excellent porosity and hydrophobicity *via* pyrolysis. During the carbonization process, divalent metal ions (Co^{2+} and Fe^{2+}) are converted into cobalt ferrite nanoparticles [26] which help in increasing the surface roughness and surface area of the carbonaceous monolithic cryogel, thus, enhancing the sorption capacity [27].

Graphene has a two-dimensional structure with hydrophobic properties, and tends to aggregate due to van der Waals interaction between neighboring sheets. Reduction in specific surface area of graphene occurs due to the aggregation of sheets which is not beneficial for the adsorption of organic contaminants. The adsorption of molecules onto the sorbents is basically determined by hydrogen bonding, $\pi - \pi$ interactions, van der Waals interactions and electrostatic interactions [28,29]. Several methods have been adopted to reduce the sheet staking of the graphene. In particular, for this purpose, one-dimensional nanomaterials like carbon nanofibers [30], CNTs [31], sepiolite clay [32], etc. have been employed. Therefore, in this study, in order to attain high specific surface area of sorbent, graphene sheets have been modified with phosphazene nanotubes so as to avoid the stacking of graphene sheets. Phosphazene is a versatile class of hybrid organic-inorganic material with alternating phosphorus and nitrogen atoms. These materials have been used as membrane materials, biomaterials, optical materials, flame retardant material, etc. [33-37]. Phosphazene nanotubes containing hydroxyl groups are synthesized through one-pot reaction with simple process, high yield, controllable morphology, and low cost [38], which provides cost advantage for manufacturing of CNTs [39].

In specific, this work reports graphene-PZTs based carbonaceous monolithic cryogel containing cobalt ferrite nanoparticles as an efficient sorbent. Crude oil and vegetable oil were selected as model pollutants for evaluating the oil sorption performance under laboratory conditions.

6.2 Materials and Methods

6.2.1 Materials

Natural graphite powder (10 mesh), resorcinol, 2-nitrophenol (Merck), formaldehyde (37%, Merck), H_2SO_4 (95-97%, Merck), H_3PO_4 (85%, Merck), $KMnO_4$ (Eurolab), HCl (37%, Panreac), hydrogen peroxide (35%, Merck), $CoCl_2.6H_2O$ (Sigma-Aldrich), iron(III) nitrate nanohydrate (≥98%, Sigma-Aldrich), phosphonitrilic chloride trimer (99 %, Sigma-Aldrich), bisphenol S (Sigma-Aldrich), triethylamine (≥99%, Sigma Aldrich) and tetrahydrofuran were used for material synthesis. Crude oil and vegetable oil were procured locally.

6.2.2 GO Preparation

GO was prepared by using the improved Tour's synthesis method [40]. For this, graphite (10 g) was added into the mixture of H_2SO_4 (540 ml) and H_3PO_4 (60 ml) in a molar ratio of 9:1. The mixture was allowed to stir for 30 min, followed by gradual addition of $KMnO_4$ (56 g) and continuous stirring for 72 h. The reaction was initiated by pouring H_2O_2 slowly and allowed to stir for 8 h until the mixture color turned yellow. Produced graphene oxide was washed with 1 M HCl solution, until no sulfate ions were detected, followed by washing with excess of distilled water to remove chloride ions. The GO solution was then poured in Teflon petri-dish for freeze drying.

6.2.3 Preparation of PZTs

3 g of phosphonitrilic chloride trimer was dissolved in 550 ml of tetrahydrofuran (THF) and the mixture was sonicated for 20 min. Another mixture containing triethylamine (TEA) (7.8 ml) and THF (50 ml) was prepared, followed by its dropwise addition in the mixture of phosphonitrilic chloride trimer and THF under ultra-sonication at 40 °C. Bisphenol S (6.5 g) was added after 30 min to the same mixture followed by the sonication for another 7 h. The final solution was filtered and the precipitates were washed with the mixture of THF and distilled water (1:1). The resulting product was dried under vacuum at 50 °C overnight.

6.2.4 Synthesis of Graphene-PZT based R-F Cryogel Containing Cobalt Ferrite Nanoparticles

GO exfoliation was carried out under ultra-sonication (2 h) by adding GO (0.2 g) into deionized water (100 ml). Later, phosphazene nanotubes (0.2 g) were added into the GO suspension followed by stirring for 2 h. $CoCl_2.6H_2O$ (0.0025

M) and $Fe(NO_3)_3.9H_2O$ (0.0025 M) were added as Co^{2+} and Fe^{2+} sources. Resorcinol (2.48 g) and formaldehyde (3.65 g) were then added in the molar ratio of 1:2. After continuous stirring for 4 h to form a homogeneous solution, the mixture was cast into a glass mold. It was kept at 85 °C for curing until a red-brown colored hydrogel was obtained. The hydrogel was washed with excess of distilled water to remove the residual ions, followed by a lyophilization process for 5 d at a temperature of -90 °C under vacuum to obtain a GO-PZT based R-F cryogel. The cryogel was carbonized for 2 h at 700 °C at a heating rate of 10 °C/min under N_2 atmosphere and cooled naturally in a quartz tube furnace. Finally, GO-PZTs based R-F cryogel containing cobalt ferrite nanoparticles ($CoFe_2O_4$/graphene-PZTs/R-F cryogel or resultant cryogel) was obtained. To investigate the influence of PZTs and cobalt ferrite nanoparticles on the formation of 3-D carbonaceous monolith material, the graphene/R-F based cryogel containing cobalt ferrite nanoparticles ($CoFe_2O_4$/graphene/R-F) was prepared using the same methodology. Graphene-R-F based cryogel (graphene/R-F/Na_2CO_3) was also synthesized using Na_2CO_3 catalyst by replacing Co^{2+} and Fe^{2+} ions for the gelation process.

6.3 Adsorbent Characterization

Wide-angle X-ray diffraction (WAXRD) patterns of the samples were collected using analytical powder (X'Pert PRO) diffractometer with instrument parameters of 40 kV and $CuK\alpha$ (1.5406 Å) radiation in reflection mode. The samples were step-scanned between 5 and 60° at a step size of $2\theta = 0.02°$ s^{-1}.

Scanning electron microscopy (SEM) images and EDX analysis data were used to obtain morphology of the sample, using FEI Quanta, FEG250 (USA) microscope at accelerating voltages of 10-20 kV. FEI transmission electron microscope (TEM, Tecnai G20) at 200kV was used to obtain the morphology of $CoFe_2O_4$/graphene/PZTs/R-F nanocomposites. The samples for TEM analysis were prepared by dispersing 0.5 mg of adsorbent in 20 ml of dimethylformamide under sonication at room temperature for 10 min. Two drops of suspension were poured onto a 400 mesh copper grid covered with thin lacey carbon film and dried in air.

Nitrogen adsorption-desorption isotherms were obtained on a Micromeritics ASAP 2010 volumetric adsorption analyzer. Prior to each adsorption measurement, the samples were degassed for 16 h at 200 °C, under vacuum (P < 5-10 mbar). Specific surface area of the nanocomposites was determined from the linear part of the BET equation. The pore volume was calculated using the BET plot corresponding to the amount of nitrogen adsorbed at the last adsorption point (P/P_0 = 0.99). Pore size distribution was estimated using the Barrett-Joyner-Halenda (BJH) method.

6.4 Oil Sorption Studies by Cryogels

Oil can be removed from aqueous media by different phenomena such as adsorption, absorption or both (sorption). Herein, sorption process has been employed to remove the oil from water. In the oil sorption experiments, the sorption capacities of graphene/R-F/Na_2CO_3, $CoFe_2O_4$/graphene/R-F and $CoFe_2O_4$/graphene/PZTs/R-F for crude oil and vegetable oil were measured. Initial weight of different monolithic cryogels was measured as $M_\circ$ (before immersing them into 200 ml glass bowl containing 100 ml of oil/water mixture). Oil/ water mixture was prepared using 30 ml of oil and 70 ml of water. The cryogels were allowed to remove the oil in the oil/water mixture at room temperature for an interval of 15 min. After being saturated with oil, the soaked cryogels were taken out and the weight was recorded as M_e. The following equation was used to determine the oil sorption capacities of the cryogels:

$$Oil\ absorption\ capacity\left(\frac{g}{g}\right) = \frac{(M_e - M_\circ)}{M_\circ}$$

In order to study the recyclability of the cryogels, *n*-hexane was used to remove the oil from the monolithic cryogels by soaking them in *n*-hexane for several hours. After extracting the oil from the cryogels completely, the gels were taken out and heated to 65 °C for the evaporation of residual *n*-hexane. Subsequently, the cryogels were used for the next sorption-desorption cycles.

6.5 Results and Discussion

6.5.1 Structure and Morphology of Adsorbent

Figure 6.1a shows the image of GO-PZTs/R-F based cryogel containing divalent metal salts of Co^{2+} and Fe^{2+}. After the carbonization process, the GO-PZTs/R-F based cryogel turned into black colored carbonaceous monolithic cryogel ($CoFe_2O_4$/graphene-PZTs/R-F), represented in Figure 6.1b. The cryogels were synthesized by curing the initial suspensions and further lyophilization of the corresponding hydrogels in order to eliminate the water inside, which resulted in dried monoliths/cryogels with desirable shape, depending on the glass molds. It can be noticed that GO-PZTs/R-F was a homogeneous monolith with yellow-brownish color, which was obtained when the divalent metal salts of Co^{2+} and Fe^{2+} were used to promote the gelation process of GO/R-F. The Co^{2+} and Fe^{2+} ions were chelated by the functional groups (negatively charged) of GO sheets and R-F, acting as both catalyst and crosslinking agents at the same time. After the carbonization of GO-PZTs/R-F cryogel, divalent metal ions (Co^{2+} and Fe^{2+}) and GO sheets were converted into cobalt ferrite nanoparticles and graphene sheets, respectively, along with PZTs turning into carbon fibers. As a result, $CoFe_2O_4$/graphene/PZTs/R-F) carbonaceous monolith was obtained.

The diffraction pattern of the GO/PZTs/R-F is presented in Figure 6.2a. A diffraction hump was observed in the range of (2θ = 20° to 33°). The diffraction pattern of $CoFe_2O_4$/graphene/PZTs/R-F is depicted in Figure 6.2b, where the weak diffraction peak can be noticed at 2θ = 26.2°, attributed to the (002) plane of carbon. This indicated graphitization of amorphous carbon catalyzed

Figure 6.1 (a) Digital image of GO/PZTs/R-F based cryogel containing divalent metal salt of Co^{2+} and Fe^{2+}, and (b) digital image of $CoFe_2O_4$/graphene/PZTs/R-F based cryogel.

by transition metals [41] and changes in interlayer spacing [42]. Strong diffraction peaks at 2θ = 43° and 52.5° were attributed to (400) and (422) crystal planes of cubic spinel structure of cobalt ferrite nanoparticles (*JCPDS card No: 22-1086*) and a small diffraction peak at 2θ = 45° corresponded to (110) plane of bcc Fe-Co [43]. Thus, it can be suggested that after the calcination at 700 °C, Co^{2+} and Fe^{2+} were completely reduced to zero valence state and cobalt ferrite nanoparticles were generated. This confirmed the achievement of intended crystalline state of the materials, which could be effectively controlled through the employed experimental protocols.

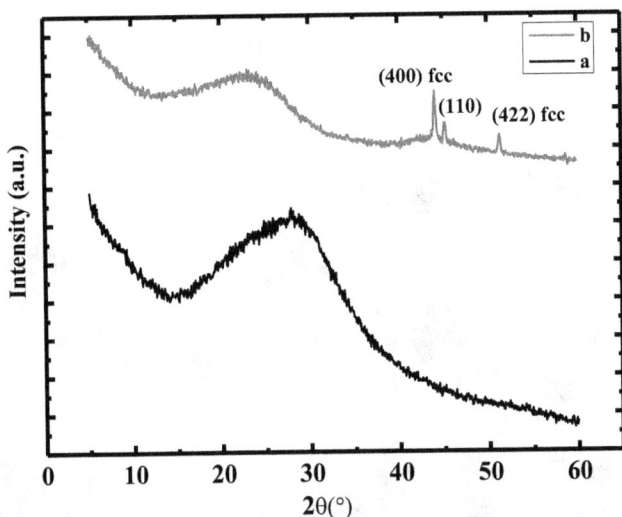

Figure 6.2 XRD patterns of (a) GO/PZTs/R-F, and (b) $CoFe_2O_4$/graphene/PZTs/R-F.

Figure 6.3 demonstrates the scanning electron micrographs of the resultant adsorbent ($CoFe_2O_4$/graphene/PZTs/R-F). It was observed that cobalt ferrite nanoparticles were present in a well-defined interconnected porous structure. As suggested earlier, the process of gelation was facilitated by the metal ions (Co^{2+} and Fe^{2+}) which acted as cross-linkers, provided active sites for the assembly of graphene layers [44], and facilitated the formation of corrugations in the resultant cryogel, therefore, forming a well-defined interconnected porous structure. For gaining further insights into the cryogel morphology, TEM image of the resultant adsorbent is shown in Figure 6.4. It was observed that carbonaceous matrix contained carbon nanofibers resulting from PZTs which helped in avoiding the stacking of graphene sheets, thus, enhancing the surface area available for sorption. Additionally, cobalt ferrite nanoparticles were observed to be distributed uniformly onto graphene sheets. The nanoparticle might improve nano-scale roughness of the graphene surface, which can further increase the effective surface area for the sorption process [27]. Determination of the surface area and pore volume of adsorbent was carried out by N_2 adsorption-desorption measurements at 73 K, along with pore size distribution determination based on BJH desorption method. The textural properties were studied for cryogel nanocomposites such as (i) graphene/R-F/Na_2CO_3, (ii) $CoFe_2O_4$/graphene/R-F and (iii) $CoFe_2O_4$/graphene/PZTs/R-F (Table 6.1).

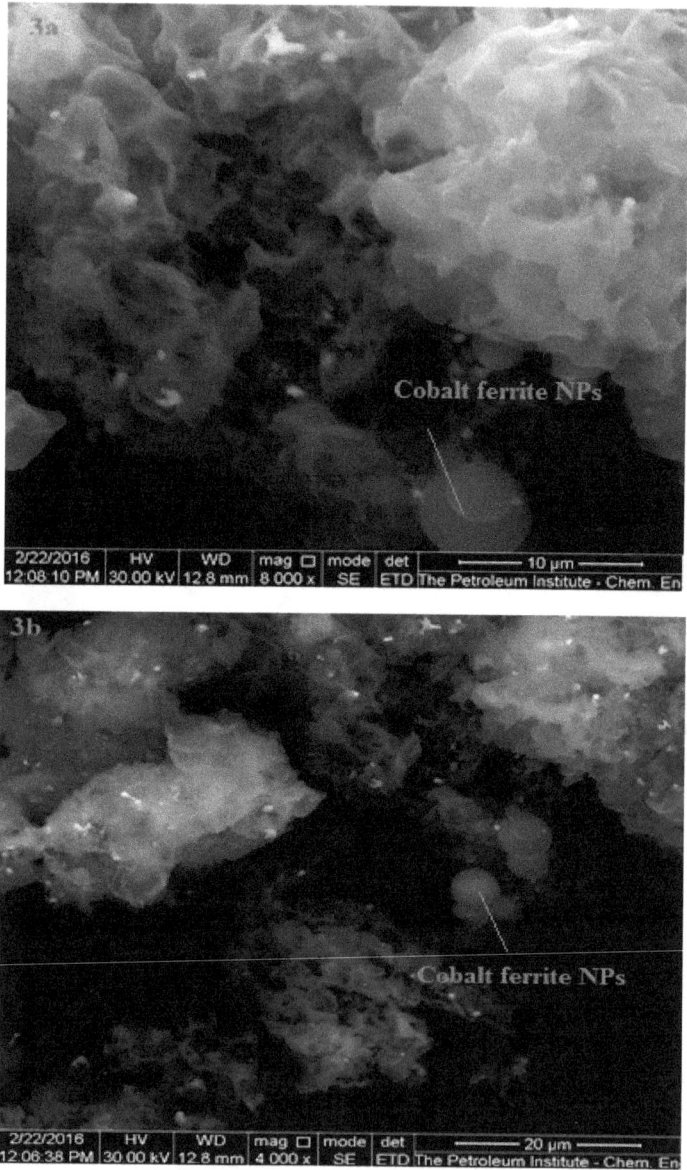

Figure 6.3 SEM image of CoFe$_2$O$_4$/graphene/PZTs/R-F cryogel at (a)10 μm and (b) 20 μm magnifications.

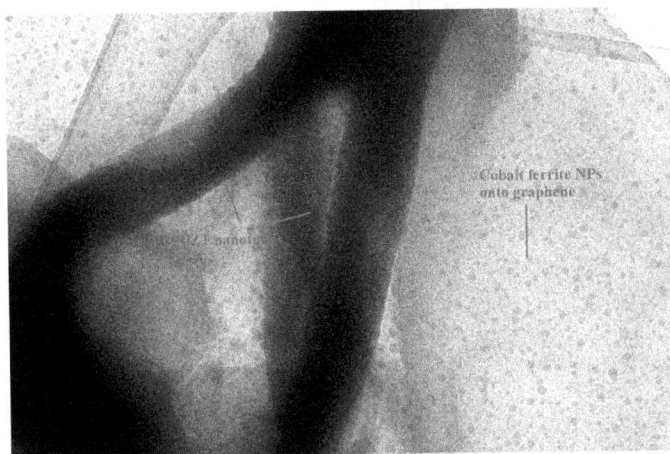

Figure 6.4 TEM image of $CoFe_2O_4$/graphene/PZTs/R-F cryogel.

From Table 6.1, it was observed that the presence of PZTs along with cobalt ferrite nanoparticles led to the achievement of higher BET surface area (629.5 m^2 g^{-1}) and pore radius (19.01 Å), as compared to the cryogels without PZTs and $CoFe_2O_4$ NPs, and cryogel with $CoFe_2O_4$ NPs but without PZTs.

Table 6.1 Textural properties of graphene/R-F/Na_2CO_3, $CoFe_2O_4$/graphene/R-F and $CoFe_2O_4$/graphene/PZTs/R-F based cryogels.

Sample	Surface area (m^2/g)	Pore volume (cc/g)	Pore radius (Å)
Graphene/R-F	324.3	0.54	7.05
$CoFe_2O_4$/graphene/R-F	478.7	0.42	7.12
$CoFe_2O_4$/graphene/PZTs/R-F	629.5	0.61	19.01

6.5.2 Hydrophobicity and Oil Sorption Capacity of Cryogels

With the combination of hydrophobic materials and multiple scale of surface roughness, enhanced hydrophobicity can be achieved [45]. The incorporation of intrinsic hydrophobic graphene sheets, carbon nanofibers of PZTs and cobalt ferrite nanoparticles into the R-F matrix, thus, provided the micro-nano scale roughness in combination with interconnected micro-porous structure of the R-F cryogel [45]. Other studies have also reported the use of nanoparticles for creating functional surfaces with hierarchical structures to improve the wettability of the substrate [27,46]. Figure 6.5 demonstrates the hydrophobic behavior of the $CoFe_2O_4$/graphene/PZTs/R-F cryogel. From the visual

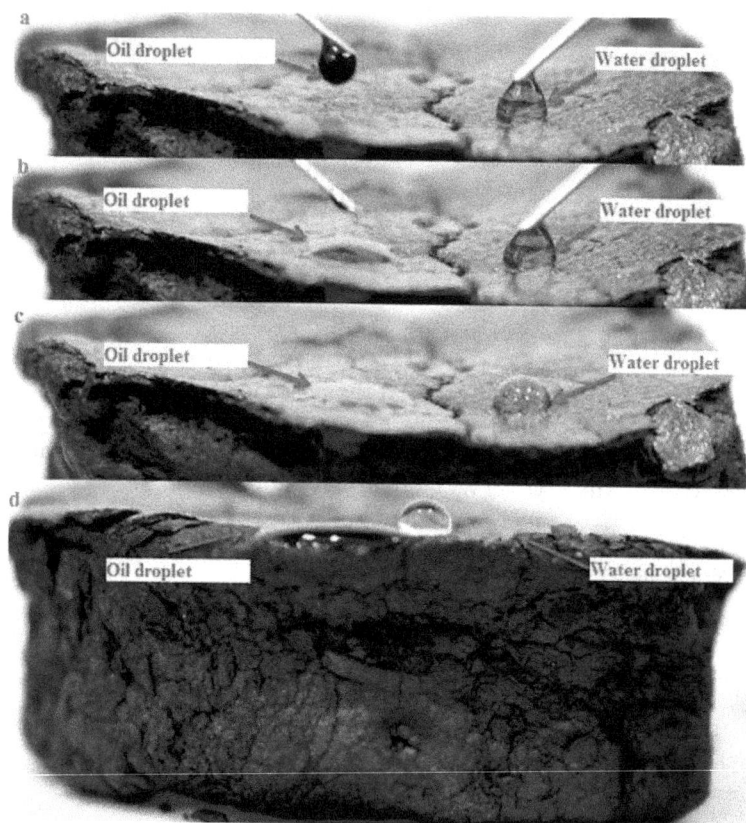

Figure 6.5 Hydrophobic behavior of the $CoFe_2O_4$/graphene/PZTs/R-F cryogel; water droplet gained quasi-spherical shape and oil droplet sorbed quickly onto the top surface of the cryogel.

images, it was observed that the oil droplet was quickly sorbed by the cryogel, while the water droplet achieved quasi-spherical shape on the surface of the cryogel with a water contact angle (WCA) of an obtuse angle. In a typical experiment for the measurement of oil sorption capacity, the monolithic carbonaceous cryogels were kept in contact with a layer of crude oil floating on the surface of water. Cryogels repelled the water phase and sorbed the oil within a time interval of 15 min due to their hydrophobic property, as shown in Figure 6.6, exhibiting excellent properties of oil sorbent for oil spill clean up.

Many studies have reported the oil sorption process to be driven by the van der Waals forces, hydrophobic interactions between the oil and sorbent, pore morphology and viscosity of oil [47-50]. To study these effects for the developed cryogels, the sorption capacities of three monolithic carbonaceous cryogels for crude oil with higher viscosity and vegetable oil with lower viscosity are compared in Figure 6.7. As mentioned above, oil viscosity played an important role in the sorption process. Thus, the penetration of the oil into the network of the cryogels is facilitated by the lower viscosity of oil, resulting in a higher sorption capacity. On the other hand, diffusion of oil into the pores of

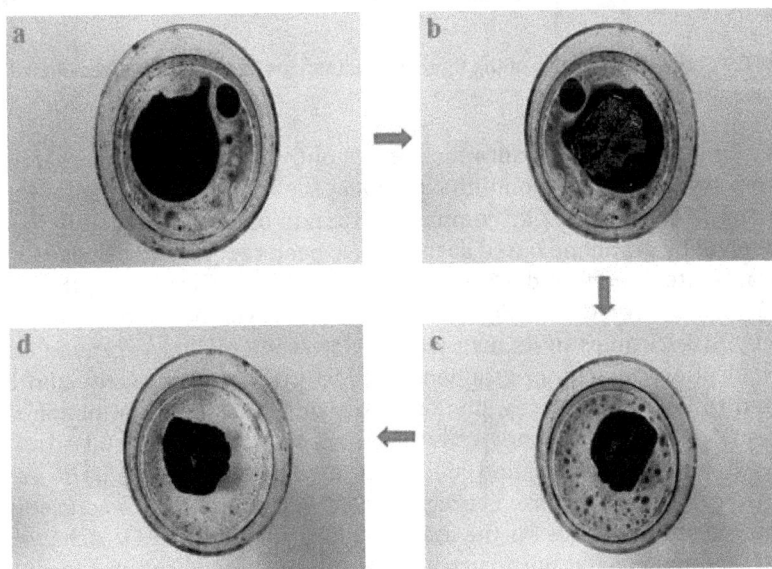

Figure 6.6 Oil sorption experiment by using $CoFe_2O_4$/graphene/PZTs/R-F cryogel for the time interval of 15 min.

the sorbent is inhibited by the high viscosity of oil, thus, resulting in a low sorption capacity [48]. In other words, this behavior can also be related as a function of time, indicating the sorption rate of the cryogel for vegetable oil is faster than the crude oil due to the high viscosity of the crude oil which delays

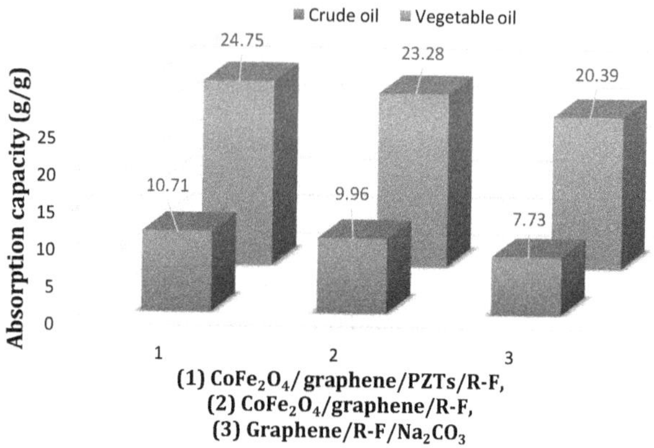

Figure 6.7 Sorption capacities of the three monolithic carbonaceous cryogels for crude oil and vegetable oil.

the wicking rate of the oil onto the surface of the sorbent [50]. Therefore, the sorption capacity of the monolithic carbonaceous cryogels was higher for vegetable oil (low viscosity) as compared to crude oil (high viscosity). The removal of oil by graphene based sorbents has been reported to be based on van der Waals interactions and is related to the specific surface area of the sorbent [50,51]. It is observed from Figure 6.7 that the cryogels achieved high sorption capacity (10-25 times of its own weight) for crude oil and vegetable oil. The lower sorption capacity of graphene/R-F/Na_2CO_3 based cryogel could be attributed to low surface area, poor porosity and stacking of graphene sheets (adsorption sites) which may reduce the adsorption sites for oil adsorption [26]. After the incorporation of metal ions (Co^{2+} and Fe^{2+}) in case of $CoFe_2O_4$/graphene/R-F based cryogel, where metal ions acted as cross-linkers and provided active sites for the assembly of graphene layers to achieve well-defined interconnected porous structure, improved oil sorption capacity was achieved. The conversion of metal ions into cobalt ferrite NPs after the calcination process helped in increasing the surface roughness of sorbent to enhance the hydrophobic property, which increased the oil sorption capacity. The high sorption capacity of $CoFe_2O_4$/graphene/PZTs/R-F based cryogel was attributed to high surface area, excellent porosity and the presence of cobalt ferrite NPs along with the carbon nanofibers of PZTs which avoided the stacking of graphene sheets by maintaining the considerable gap between two graphene sheets, hence, increasing the surface area to provide the maximum oil adsorption sites. The sorption capacities of $CoFe_2O_4$/graphene/PZTs/R-F, $CoFe_2O_4$/graphene/R-F and graphene/R-F/Na_2CO_3 cryogels for crude oil and vegetable oil were determined as 10.71, 9.96, and 7.73 g/g, respectively and 24.75, 23.28, and 20.39 g/g, respectively. The recyclability of the $CoFe_2O_4$/graphene/PZTs/R-F cryogel for crude oil and vegetable oil is also presented in Figure 6.8. It was noticed that the recyclability of the resultant cryogel did not deteriorate even after five cycles of sorption-desorption process, implying a good recycling performance.

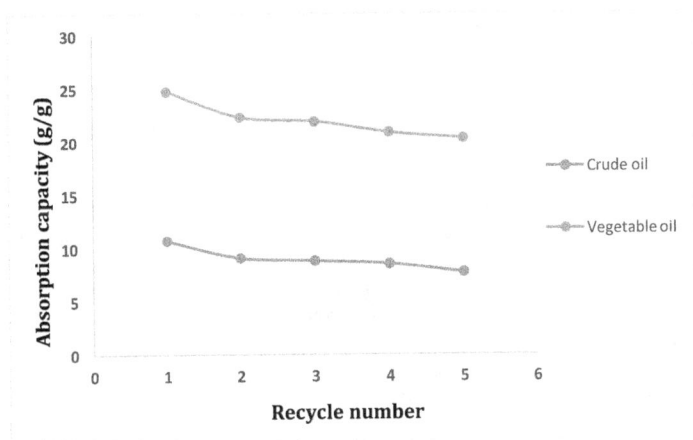

Figure 6.8 Recyclability of the $CoFe_2O_4$/graphene/PZTs/R-F cryogel for five cycles of sorption-desorption.

6.6 Conclusion

In this study, a well-defined interconnected porous structure cryogel ($CoFe_2O_4$/graphene/PZTs/R-F) has been prepared *via* freeze drying method, followed by carbonization in order to generate advanced materials to remove crude oil (high viscosity) and vegetable oil (low viscosity) from the oil/water mixture. The incorporation of PZTs into the monolithic carbonaceous cryogel enhanced the surface area of the sorbent by avoiding the stacking of the graphene sheets embedded in the R-F matrix. The combination of graphene, cobalt ferrite nanoparticles and carbon nanofibers of phosphazene tubes provided the micro-nano scale roughness in combination with a well-defined interconnected porous structure with good hydrophobic properties. The resultant cryogel exhibited good hydrophobic properties with water contact angle of obtuse angle. The sorption capacities of the resultant cryogel for crude oil and vegetable oil were higher than the sorption capacities of $CoFe_2O_4$/graphene/R-F and graphene/R-F/Na_2CO_3 cryogels. The recyclability of the $CoFe_2O_4$/graphene/PZTs/R-F cryogel did not deteriorate even after five cycles of the sorption-desorption process. Based on its promising performance, the developed cryogel is a promising sorbent for large quantity removal of various oils in different oil spill areas.

References

1. Chin, S. F., Romainor, A. N. B., and Pang, S. C. (2014) Fabrication of hydrophobic and magnetic cellulose aerogel with high oil absorption capacity. *Materials Letters,* **115**, 241-243.
2. Zahed, M. A., Aziz, H. A., Isa, M. H., Mohajeri, L., and Mohajeri, S. (2010) Optimal conditions for bioremediation of oily seawater. *Bioresource technology*, **101**(24), 9455-9460.
3. Broje, V., and Keller, A. A. (2006) Improved mechanical oil spill recovery using an optimized geometry for the skimmer surface. *Environmental Science & Technology*, **40**(24), 7914-7918.
4. Buist, I., Potter, S., Nedwed, T., and Mullin, J. (2011) Herding surfactants to contract and thicken oil spills in pack ice for in situ burning. *Cold Regions Science and Technology*, **67**(1), 3-23.
5. Kujawinski, E. B., Kido Soule, M. C., Valentine, D. L., Boysen, A. K., Longnecker, K., and Redmond, M. C. (2011) Fate of dispersants associated with the Deepwater Horizon oil spill. *Environmental Science & Technology*, **45**(4), 1298-1306.
6. Xue, Z., Cao, Y., Liu, N., Feng, L., and Jiang, L. (2014) Special wettable materials for oil/water separation. *Journal of Materials Chemistry A*, **2**(8), 2445-2460.
7. Radetic, M. M., Jocic, D. M., Jovancic, P. M., Petrovic, Z. L., and Thomas, H. F. (2003) Recycled wool-based nonwoven material as an oil sorbent. *Environmental Science & Technology*, **37**(5), 1008-1012.
8. Moazed, H., and Viraraghavan, T. (2005) Removal of oil from water by bentonite organoclay. *Practice Periodical of Hazardous, Toxic, and Radioactive Waste Management*, **9**(2), 130-134.
9. Ahmad, A., Sumathi, S., and Hameed, B. (2005) Residual oil and suspended solid removal using natural adsorbents chitosan, bentonite and activated carbon: A comparative study. *Chemical Engineering Journal*, **108**(1), 179-185.
10. Cui, J., Zhang, X., Liu, H., Liu, S., and Yeung, K. L. (2008), Preparation and application of zeolite/ceramic microfiltration membranes for treatment of oil contaminated water. *Journal of Membrane Science*, **325**(1), 420-426.

11. Jang, J., and Kim, B. S. (2000) Studies of crosslinked styrene–alkyl acrylate copolymers for oil absorbency application. I. Synthesis and characterization. *Journal of applied polymer Science*, **77**(4), 903-913.

12. Iqbal, M. Z., and Abdala, A. A. (2013) Oil spill cleanup using graphene. *Environmental Science and Pollution Research*, **20**(5), 3271-3279.

13. Gui, X., Zeng, Z., Lin, Z., Gan, Q., Xiang, R., Zhu, Y., Cao, A., and Tang, Z. (2013) Magnetic and highly recyclable macroporous carbon nanotubes for spilled oil sorption and separation. *ACS Applied Materials & Interfaces*, **5**(12), 5845-5850.

14. Wang, C.-F., Tzeng, F.-S., Chen, H.-G., and Chang, C.-J. (2012) Ultraviolet-durable superhydrophobic zinc oxide-coated mesh films for surface and underwater–oil capture and transportation. *Langmuir*, **28**(26), 10015-10019.

15. Reynolds, J. G., Coronado, P. R., and Hrubesh, L. W. (2001) Hydrophobic aerogels for oil-spill clean up–Synthesis and characterization. *Journal of Non-Crystalline Solids*, **292**(1), 127-137.

16. Zhu, Q., Pan, Q., and Liu, F., Facile removal and collection of oils from water surfaces through superhydrophobic and superoleophilic sponges. *The Journal of Physical Chemistry C*, **115**(35), 17464-17470.

17. Zhang, X., Li, Z., Liu, K., and Jiang, L. (2013) Bioinspired multifunctional foam with self-cleaning and oil/water separation. *Advanced Functional Materials*, **23**(22), 2881-2886.

18. Yuan, J., Liu, X., Akbulut, O., Hu, J., Suib, S. L., Kong, J., and Stellacci, F. (2008) Superwetting nanowire membranes for selective absorption. *Nature Nanotechnology*, **3**(6), 332-336.

19. Zhu, H., Qiu, S., Jiang, W., Wu, D., and Zhang, C. (2011) Evaluation of electrospun polyvinyl chloride/polystyrene fibers as sorbent materials for oil spill cleanup. *Environmental Science & Technology*, **45**(10), 4527-4531.

20. Bibi, N. S., Singh, N. K., Dsouza, R. N., Aasim, M., and Fernandez-Lahore, M. (2013) Synthesis and performance of megaporous immobilized metal-ion affinity cryogels for recombinant protein capture and purification. *Journal of Chromatography A*, **1272**, 145-149.

21. Zheng, S., Wang, T., Liu, D., Liu, X., Wang, C., and Tong, Z. (2013) Fast deswelling and highly extensible poly (N-isopropylacrylamide)-hectorite clay nanocomposite cryogels prepared by freezing polymerization. *Polymer*, **54**(7), 1846-1852.

22. Sahiner, N., Seven, F., and Al-lohedan, H. (2015) Superporous cryogel-M (Cu, Ni, and Co) composites in catalytic reduction of toxic phenolic compounds and dyes from wastewaters. *Water, Air, & Soil Pollution*, **226**(4), 1-12.

23. Reichelt, S., Becher, J., Weisser, J., Prager, A., Decker, U., Moeller, S., Berg, A., and Schnabelrauch, M. (2014) Biocompatible polysaccharide-based cryogels. *Materials Science and Engineering: C*, **35**, 164-170.

24. Petrov, P., Petrova, E., Tchorbanov, B., and Tsvetanov, C. B. (2007) Synthesis of biodegradable hydroxyethylcellulose cryogels by UV irradiation. *Polymer*, **48**(17), 4943-4949.

25. Park, S., Lee, K.-S., Bozoklu, G., Cai, W., Nguyen, S. T., and Ruoff, R. S. (2008) Graphene oxide papers modified by divalent ions-enhancing mechanical properties via chemical cross-linking. *ACS Nano*, **2**(3), 572-578.

26. Wei, G., Miao, Y.-E., Zhang, C., Yang, Z., Liu, Z., Tjiu, W. W., and Liu, T., Ni-doped graphene/carbon cryogels and their applications as versatile sorbents for water purification. *ACS Applied Materials & Interfaces*, **5**(15), 7584-7591.

27. Huang, M., Si, Y., Tang, X., Zhu, Z., Ding, B., Liu, L., Zheng, G., Luo, W., and Yu, J. (2013) Gravity driven separation of emulsified oil–water mixtures utilizing in situ polymerized superhydrophobic and superoleophilic nanofibrous membranes. *Journal of Materials Chemistry A*, **1**(45), 14071-14074.

28. Zhang, S., Shao, T., Bekaroglu, S. S. K., and Karanfil, T. (2009) The impacts of aggregation and surface chemistry of carbon nanotubes on the adsorption of synthetic organic compounds. *Environmental Science & Technology*, **43**(15), 5719-5725.

29. Pan, B., and Xing, B., Adsorption mechanisms of organic chemicals on carbon nanotubes. *Environmental Science & Technology*, **42**(24), 9005-9013.

30. Fan, Z.-J., Yan, J., Wei, T., Ning, G.-Q., Zhi, L.-J., Liu, J.-C., Cao, D.-X., Wang, G.-L., and Wei, F. (2011) Nanographene-constructed carbon nanofibers grown on graphene sheets by chemical vapor deposition: high-performance anode materials for lithium ion batteries. *ACS Nano*, **5**(4), 2787-2794.

31. Yang, S.-Y., Chang, K.-H., Tien, H.-W., Lee, Y.-F., Li, S.-M., Wang, Y.-S., Wang, J.-Y., Ma, C.-C. M., and Hu, C.-C. (2011) Design and tailoring of a hierarchical graphene-carbon nanotube architecture for supercapacitors. *Journal of Materials Chemistry*, **21**(7), 2374-2380.

32. Vengatesan, M., Singh, S., Stephen, S., Prasanna, K., Lee, C., and Mittal, V. (2017) Facile synthesis of thermally reduced graphene oxide-sepiolite nanohybrid via intercalation and thermal reduction method. *Applied Clay Science*, **135**, 510-515.

33. Allcock, H. R., and Ambrosio, A. M., Synthesis and characterization of pH-sensitive poly (organophosphazene) hydrogels. *Biomaterials*, **17**(23), 2295-2302.

34. Allcock, H. R., Bender, J. D., Chang, Y., McKenzie, M., and Fone, M. M. (2003) Controlled refractive index polymers: Polyphosphazenes with chlorinated-and fluorinated-, aryloxy-and alkoxy-side-groups. *Chemistry of Materials*, **15**(2), 473-477.

35. Allcock, H. R., Kellam, E. C., and Morford, R. V. (2001) Gel electrolytes from co-substituted oligoethyleneoxy/trifluoroethoxy linear polyphosphazenes. *Solid State Ionics*, **143**(3), 297-308.

36. Lu, S.-Y., and Hamerton, I. (2002) Recent developments in the chemistry of halogen-free flame retardant polymers. *Progress in Polymer Science*, **27**(8), 1661-1712.

37. Allcock, H. R. (2006) Recent developments in polyphosphazene materials science. *Current Opinion in Solid State and Materials Science*, **10**(5), 231-240.

38. Zhu, L., Xu, Y., Yuan, W., Xi, J., Huang, X., Tang, X., and Zheng, S. (2006) One-pot synthesis of poly(cyclotriphosphazene-co-4, 4'-sulfonyldiphenol) nanotubes via an in situ template approach. *Advanced Materials*, **18**(22), 2997-3000.

39. Wang, P., Cao, M., Wang, C., Ao, Y., Hou, J., and Qian, J. (2014) Kinetics and thermodynamics of adsorption of methylene blue by a magnetic graphene-carbon nanotube composite. *Applied Surface Science*, **290**, 116-124.

40. Marcano, D. C., Kosynkin, D. V., Berlin, J. M., Sinitskii, A., Sun, Z., Slesarev, A., Alemany, L. B., Lu, W., and Tour, J. M. (2010) Improved synthesis of graphene oxide. *ACS Nano*, **4**(8), 4806-4814.

41. Maldonado-Hodar, F., Moreno-Castilla, C., and Perez-Cadenas, A. (2004) Surface morphology, metal dispersion, and pore texture of transition metal-doped monolithic carbon aerogels and steam-activated derivatives. *Microporous and Mesoporous Materials*, **69**(1), 119-125.

42. Chen, W., and Yan, L. (2011) In situ self-assembly of mild chemical reduction graphene for three-dimensional architectures. *Nanoscale*, **3**(8), 3132-3137.

43. Wu, A., Yang, X., and Yang, H. (2013) Magnetic properties of carbon-encapsulated Fe–Co alloy nanoparticles. *Dalton Transactions*, **42**(14), 4978-4984.

44. Tang, Z., Shen, S., Zhuang, J., and Wang, X. (2010) Noble-metal-promoted three-dimensional macroassembly of single-layered graphene oxide. *Angewandte Chemie*, **122**(27), 4707-4711.

45. Nguyen, D. D., Tai, N.-H., Lee, S.-B., and Kuo, W.-S. (2012) Superhydrophobic and superoleophilic properties of graphene-based sponges fabricated using a facile dip coating method. *Energy & Environmental Science*, **5**(7), 7908-7912.

46. Rafiee, J., Rafiee, M. A., Yu, Z. Z., and Koratkar, N. (2010) Superhydrophobic to superhydrophilic wetting control in graphene films. *Advanced Materials*, **22**(19), 2151-2154.

47. Gui, X., Li, H., Wang, K., Wei, J., Jia, Y., Li, Z., Fan, L., Cao, A., Zhu, H., and Wu, D. (2011) Recyclable carbon nanotube sponges for oil absorption. *Acta Materialia*, **59**(12), 4798-4804.

48. Nguyen, S. T., Feng, J., Le, N. T., Le, A. T., Hoang, N., Tan, V. B., and Duong, H. M., Cellulose aerogel from paper waste for crude oil spill cleaning. *Industrial & Engineering Chemistry Research*, **52**(51), 18386-18391.

49. Wang, J., Zheng, Y., and Wang, A. (2013) Coated kapok fiber for removal of spilled oil. *Marine Pollution Bulletin*, **69**(1), 91-96.

50. Lin, J., Shang, Y., Ding, B., Yang, J., Yu, J., and Al-Deyab, S. S. (2012) Nanoporous polystyrene fibers for oil spill cleanup. *Marine Pollution Bulletin*, **64**(2), 347-352.

51. Zhao, J., Ren, W., and Cheng, H.-M. (2012) Graphene sponge for efficient and repeatable adsorption and desorption of water contaminations. *Journal of Materials Chemistry*, **22**(38), 20197-20202.

7

Polyurethane Membranes for Gas Separation

7.1 Introduction

Gas separation is one of the main processes in gas treatment plants in order to purify the gas streams or to recover useful gases. [1-4]. Sipek *et al.* [5] reviewed polymeric membranes as effective materials for separation of gases and vapors. A variety of glassy and rubbery polymers have been explored as membrane materials for separations of gaseous streams. Recently, Bernardo *et al.* [6] has also explored industrial applications and process intensification options for membrane gas separations. Membrane separation systems, being simpler in operation and maintenance, reliable, compact, efficient, are being widely used as an alternative to other cumbersome techniques such as absorption, adsorption or cryogenic distillation. Polymeric membranes for gas separation are widely used in diverse areas like carbon dioxide recovery, helium gas removal in natural gas purification, hydrogen recovery in ammonia plant purge streams, oxygen and nitrogen separations, CO_2 recovery from biogas, oxygen upgrading from air, etc. [7-9].

Polyurethanes (PU) are a versatile category of polymers possessing good physical and tensile strength, chemical resistance, bio-compatibility and mechanical properties. These properties make polyurethanes promising candidates for membrane separation. To underline the importance of polyurethane membranes for separation processes, a large number of reviews have been generated exploring the synthesis [10-12], versatility [13], and applications of polyurethanes as rigid foams [14], in scaffolds [15,16], in pharmacy [17], in composites [18], for separation applications in the form of foam sorbents [19], in medicine for bio-stability and carcinogenicity [20], for drug delivery [21], for cancer therapy [22], in spine surgery [23], for water purification [24], among others [25-27]. Polyurethanes consist of hard (glassy) and soft (elastomeric) segments [28]. The hard segments act as physical crosslinks as well as fillers and are in an amorphous glassy or crystalline state. On the other hand, the soft segments are rubbery which provide the polyurethanes both flexibility as well as elasticity. Polyurethanes are generally synthesized by the

Gigi George, Nidhika Bhoria and Vikas Mittal, The Petroleum Institute (part of Khalifa University of Science and Technology), Abu Dhabi, UAE*
*Current address: CMS College, Mahatma Gandhi University, Kerala, India; **Current address: Bletchington, Wellington County, Australia

reaction of polyfunctional isocyanates with polyols. By variation in reagents' molecular chain length, chemical nature and functionalities, a wide range of linear or cross-linked polyurethanes with different physiochemical properties can be obtained. Accordingly, polyurethane can vary in terms of chain length, free volume, density of polar groups, etc. Thus, various polyurethane grades though have the same chemical urethane group, however, behave differently depending on the reactants selected.

The soft segments in polyurethanes are generally made of polyether. By chain extending a terminal diisocyanate with a low molecular weight diol or diamine, the hard segment is usually prepared. Occasionally, polyurethanes undergo phase separation because of the hard and soft segment incompatibility. The intermolecular hydrogen bonding interactions between hard segments is the major reason for domain formation. A number of characterization techniques like dynamic mechanical analysis [29-31], thermal analysis [32-37], infrared spectroscopy [35-40] and scattering techniques [41] have been used for the study of polyurethanes morphology and hydrogen bonding. Phase separation into hard and soft segment domains is possible due to their type and process parameters used [31,42,43]. Howarth *et al.* [44] has also reviewed the various synthesis techniques of polyurethanes and suggested some modifications for the future. Overall, with different chemical characteristics and microstructures, polyurethanes are of high potential in gas separation [45]. Polyurethanes can behave as thermoplastic and thermosetting materials depending upon the chemical and morphological fabrication. Recently, George *et al.* [46] reviewed the polymer membranes for their acid gas separation applications including polyurethanes. For instance, Figure 7.1 also summarizes the permeability of various membranes for hydrogen sulfide, H_2S. As can be seen that there is still a large degree of advancement needed to generate high permeability PU membranes to match existing membrane systems.

In this review, various literature studies specifically reporting the developments in the synthesis and structure-property correlations of polyurethane membranes for specific applications as gas separation materials have been reviewed.

7.2 Theory of Membrane Gas Separation

In membrane separation performance, the essential characterizing features are the permeability coefficient, P_X, and the selectivity $\alpha_{A/B} = P_A/P_B$, where P_A is the more permeable gas's permeability while P_B is that of the lesser permeable gas [47]. The solution-diffusion phenomenon describes the gas transport

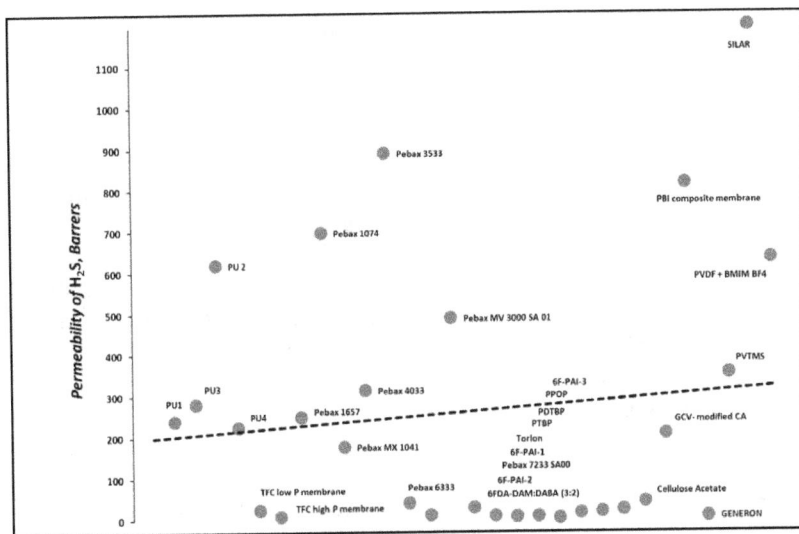

Figure 7.1 H_2S permeability of various membranes. Reproduced from Reference 46 with permission from Elsevier.

behavior of polymeric membranes. P_x is the contributed result of solubility and diffusivity through the membranes. Both solubility and diffusivity are polymer and penetrant dependent. Apart from polymer structure, a number of other parameters are responsible for its gas transport properties. The open volume available between polymer chains through which gas penetrant can pass through is an important parameter. Altering the polymer structure changes the T_g and the free volume which in turn can significantly influence the solubility and diffusivity. As the polymer chain packing efficiency increases, the free volume decreases thereby decreasing the permeability. The increasing penetrant size of gas molecules decreases the diffusion coefficient. Solubility is related to the condensability of the penetrant. The gas and vapor transport is strongly influenced by prevailing pressure and temperature conditions. According to the dual-mode sorption in glassy polymers and Henry's law in rubbery polymers, solubility increases with increase in pressure, while the changes in temperature have a reverse effect [48]. Many research studies have reported the gas permeation behavior through polyurethane membranes and established relationships between diffusion coefficient and the free volume using William-Landel-Ferry theory [49-51]. Schneider *et al.* [52] also observed that the changed glass transition temperature of polyurethanes due to variation in soft segment contents does not impact the gas and vapor permeation behavior [52].

7.3 Ideal Membrane Characteristics

As mentioned earlier, in polymeric membranes, the permeability can be evaluated based on diffusivity and solubility of the permeant into the polymer. Diffusivity is a kinetic term which is a function of the permeant molecules mobility inside the polymer. It depends on the free volume and mobile chain length of the polymer as well as the molecular diameter of the permeant. Solubility shows the affinity between the polymer and the permeant and is a thermodynamic term. Gas transport through membranes is strongly influenced by the polymer's state, i.e., glassy or rubbery. Generally, the gas sorption process in rubbery polymers, being in equilibrium state, $(T>T_g)$ is simpler than that of glassy polymers $(T<T_g)$ [53-55].

The selectivity, permeability and life-time are the evaluating parameters for polymeric membrane efficiency [56]. The extent of the recovery of the separation process is directly impacted by the selectivity and indirectly by the feed gas flow requirements and the total membrane area. The permeability directly affects the amount of membrane requirement, while the membrane durability affects the maintenance costs. For making membrane-based separations economical than other conservative processes, the polymeric membrane materials should have both high permeability and selectivity [57,58]. The efficiency and performance of the membrane systems also relies on the membrane configurations and modules, irrespective of the inherent properties of the polymeric material. An integration of all these features is important for generating a commercially viable product. High gas fluxes are characteristics of thin layers because of which packing of large membrane areas per unit volume can be achieved. The higher the degree of crosslinking in the membranes, the lower will be the permeability as the gas diffusion coefficients are much lower [59].

7.4 Influence of Polyurethane Structure on Permeability and Selectivity

7.4.1 Influence of Hard and Soft Segments

The majority of the permeability data so far for gas separation via polyurethane membranes are below Robeson's upper bound limit for gas pairs (CO_2/CH_4) and (O_2/N_2) [60]. There exists a tradeoff between the permeability and selectivity and a large number of studies have been carried out to overcome permeability-selectivity tradeoff relationship [61-63]. In polyurethanes, the gas permeation mainly occurs through the soft segments, while the hard segment act as physical crosslinks or impermeable filler in rubbery soft segment phase. An increase in the gas separation performance of polyurethane membranes has been a challenging task. Attempts have been made to enhance the permeability and selectivity of the membranes by optimizing the hard segment content [51]. Few of this include, studying the dependence on hard and soft segments in polyurethane membranes for gas permeability behavior [64], dependence on polymerization methods [65], using metal chelated polyurethane membranes [66-68], etc. Attempts to enhance the performance have also been made through the addition of amine and carboxyl functionalities [69] including the use of PDMS [70-72], epoxidation of hydroxyl terminated polybutadiene [73], polycarbonate–polyurethane membrane [74], etc.

Another attempt to advance the gas permeability and selectivity also included the usage of polymer blends [75-77]. In homogeneous blends, the interaction between the polymers influences the diffusion process, while in the case of heterogeneous blends, degree of heterogeneity influences the permeability significantly [78]. Different studies have reported on structure-morphology-property relationship of the thermoplastic polyurethanes by studying the synthesis, hard segment effect, hydrogen bonding effect, and properties [64,69,79-82]. Mohammadi *et al.* [83] studied the effects of temperature, pressure and stage cut on the gas transport properties through poly(ester urethane urea) for pure gases (CO_2, N_2, CH_4) as well as ternary gas mixtures of CO_2, CH_4, and H_2S. H_2S/CH_4 and CO_2/CH_4 selectivities of 43 and 16 and average permeabilities of 95 and 45 for H_2S and CO_2 were reported.

Wolinska-Grabczyk *et al.* [84] studied poly(acrylonitrile-*co*-butadiene)-based polyurethanes with varied extent of nitrile groups for CO_2 membrane gas separations. It was observed that with higher extent of polar nitrile group polymers, the permeability decreased, while the permselectivity increased. The polymer had less severe trade-off between the gas transport properties and exhibited much higher permeabilites. The polymer structure was analyzed and a relationship was established between the permeability and glass transition temperature.

Talakesh *et al.* [85] studied the polyether based polyurethanes with different hard and soft segments which were prepared by thermal phase inversion method. The soft segments were PEG (2000g/mol), PTMG (2000g/mol) and PTMG/PEG mixture, etc. It was reported that the chain mobility got restricted as the phase separation of hard and soft segments decreased. This led to a hike in the T_g values of the soft segment. By altering the physical conditions, the gas transport properties were studied using constant pressure method. With increase in the ether group content of the polymer structure, gas permeability decreased for pure gases, while CO_2/N_2 ideal selectivity increased. A drop in CO_2 permeability from nearly 130 Barrer in polyurethane containing PTMG in soft segment to 20 barrer in the PU containing PEG in soft segment was observed. An increase from 28 to 90 in CO_2/N_2 selectivity was reported. It was observed that highest selectivity for the polyurethane membranes with 75:25 wt% ratio of PEG:PTMG. Poreba *et al.* [86] also studied nanocomposites based on polycarbonate-based polyurethane with bentonite for thermal, mechanical and gas transport properties. Hexamethylene diisocyanate and butane-1,4-diol were used for hard segment formation. High degree of phase separation was observed for polyurethane based polymer and its nanocomposites. The gas permeation properties exhibited dependence on hard segment content, though not much significant change was observed after varying the polymer structures.

Khosravi *et al.* [87] studied the effect of polyurethane membrane with different polyol, diisocyanate, and chain extender on gas permeability. More phase interaction was observed by the changing of polyol type. The hydrogen bonding, which caused the packing density of the hard segments, was increased by the phase separation of hard and soft segments. This was achieved by changing the diisocyanate groups from cyclic aliphatic to linear aliphatic ones. The phase separation of hard and soft segments was increased on changing the chain extender from a diol one to a diamine. It was observed that more condensable gases permeate more in rubbery polymers and the solubility governs the selectivity of polyurethane. As the microphase separation increased, the permeability and rubbery behavior increased along with their selectivity. Due to the higher rubbery property, polypropylene glycol based polymers exhibited maximum permeability. The high C_3H_8/CH_4 selectivity of 5.47 and permeability of 200 Barrer was reported. Wang *et al.* [88] considered a series of polyurethane films consisting of hydroxyl terminated polybutadiene/acrylonitrile as soft segment, and consisting of TDI and butanediol as hard segments. It was observed that a direct relationship existed between the gas permeability and free-volume. This relation was based on the free-volume parameters and gas diffusivity. The free-volume played a significant part in determining the gas permeability. Marques *et al.* [89] also studied the free volume in polyurethane membranes using positron annihilation spectroscopy. Further, the authors studied the gas permeability and temperature–dependent free volume correlation in polyurethane membranes. Scholten *et al.* [90] also reported electrospun fibers based on polyurethane for removing volatile organic compounds from air. Polyurethanes had 4,4-methylene-bis(phenylisocyanate) (MDI) and aliphatic isophorone diisocyanate as hard segments, whereas butanediol and tetramethylene glycol were used to form the soft segments (Figure 7.2). The sorption performance and capacity of the generated polyurethane fibers was observed to be similar to activated carbon, thus, indicating advanced performance due to balance of hard and soft segments.

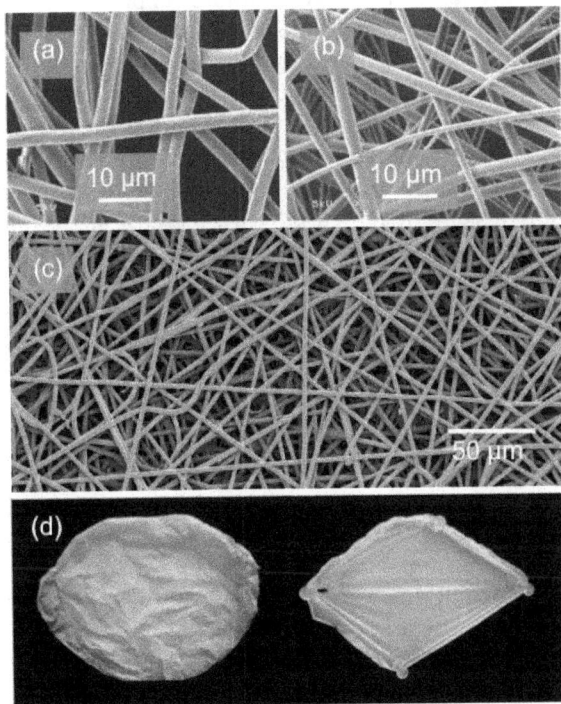

Figure 7.2 SEM images of (a) MDI-based and (b) isophorone-based non-woven fiber mats; (c) SEM image indicating the uniformity of MDI-based PU fiber diameter and mat density; (d) stretched fiber mat exhibiting lack of tearing or breaking. Reproduced from Reference 90 with permission from American Chemical Society.

Sadeghi *et al.* [91] studied the effect on the hard and soft segment microphase separation with the changes in diisocyanate from aromatic to linear aliphatic, and CO_2/N_2 selectivity of 45 and permeability of 186 Barrer were reported. Ruaan *et al.* [92] studied the microstructure behavior of hydroxyl terminated polybutadiene (HTPB) based polyurethane membranes. With increase in hard segment content, the oxygen and nitrogen permeability was observed to decrease. The authors suggested that the low O_2/N_2 selectivity was because of hydrogen bonding. It was indicated that the key to success for selectivity improvement is to avoid the hydrogen bond formation between the hard segments. Queiroz *et al.* [71] also studied the structural characteristics and gas permeation properties of PDMS/poly (propylene oxide) urethane urea bi-soft segment membranes. The authors reported increased CO_2, O_2, N_2 permeabilities with increase in PDMS membrane content. Lower degree of crosslinking and lower hydrogen bonding contribution between the hard segments exhibited higher permeability. Semsarzadeh *et al.* [93] investigated the effect of hard segments on the gas permeabilities for polyether based urethanes. As the hard segment content was increased, there was a decrease in the permeabilities of the gases. Gas permselectivity of the membranes with same hard segment content increased with increase in soft segment. These studies underlined the importance of optimization of hard and soft segment contents for optimum permeability and selectivity.

Galland *et al.* [94] studied the soft segment molecular weight as a major factor controlling the diffusion. It was reported the gas permeability depended on the chemical composition due to the nature of chain packing and degree of phase segregation. Huang *et al.* [95] recently studied thermoplastic polyurethane (TPU) films synthesized using layer-multiplying co-extrusion. The authors studied the morphological effect of confinement, along with gas barrier and mechanical properties. The soft TPU having the hard segment of 52% exhibited phase separation, whereas the rigid polymer having 100% hard-segment TPU exhibited amorphous structure. A multilayer structure with elasticity ratio of 100 and the viscosity ratio of 10 was genertaed. A significant reduction in oxygen permeability was observed when stretched at 75% which was due to the micro-confinement occurring during orientation. Park *et al.* [70] studied urethane urea membranes based on polysiloxane/polyether mixed soft segment for gas separation properties. The authors reported that small addition of PDMS into polyurethane matrices based on polyether increased the N_2, O_2, CO_2 permeabilities and N_2 selectivity. Also, the small addition of polyethers like PPO, PEO, PTMO and PEO-PPO-PEO inside the PDMS based polyurethane urea matrices decreased the gas permeabilities, but had no effect on the gas selectivities. In other studies, the authors studied segmented PU and PUU membranes with different soft segments for the separation of toluene and nitrogen. The poly(tetramethylene oxide) (PTMO)/PDMS mixed soft segment based polyurethane membranes exhibited good performance for toluene separation. The authors reported toluene/N_2 selectivity of about 70-140 and permeability as high as 17,500 Barrer. Increased gas permeability was observed with an increase of polysiloxane molecular weight [70,96,97]. Wolinska-Grabczyk *et al.* [98] studied segmented polyurethanes for gas properties with variation in the soft segments. PU membranes exhibited lower permeability when synthesized with more polar or shorter macro-diol segments. The authors observed that the diffusivity and solubility of oxygen and nitrogen were in correlation with glass transition temperature of macro-diol segments. Gomes *et al.* [72] generated poly(ether siloxane urethane urea) with varying content of polysiloxane. Gas permeation properties for O_2, N_2, CO_2, CH_4, n-C_4H_{10} were analyzed. Soft segments of polysiloxane and permeation properties were observed to have strong correlation. Gnanasekaran *et al.* [99] studied the structure-transport and microstructure of mixed soft-segmented poly(urethane-imide) membranes. Polycaprolactone diol, polypropylene glycol, and bis(3-aminopropyl)-terminated polydimethylsiloxane were used as soft segments for membrane synthesis. The membranes exhibited the potential application for n-C_4H_{10}/CH_4 separation because of higher selectivity for a mixture of gases as compared to single gases. Li *et al.* [100] studied a series of polyurethane ureas synthesized using various polyether diols. The polyethers were terathane (R) 2000, terathane (R) 2900, PEG 2000, PPG 2700, and a mixture of PEG 2000 and terathane (R) 2000. The fractional free volume increased with increase in soft segment content and the increase in polyether molecular weight, which increased the gas permeability.

McBride *et al.* [51] studied linear polyurethane membranes and the relation between the aromatic content present in diisocyanates and gas diffusion. The authors observed that the motion of the soft segment chains could be controlled, as hard domains at temperatures below the T_g acted as crosslinks. As the aromatic content increased, the motion of the soft segments was further restricted because of the increased crosslinking effectiveness. On the other hand, on increasing the length of soft segment, it increased the soft segment mobility. With increase in hard segment content, the activation energy increased which further reduced the permeability of the gas. Matsunaga *et al.* [101] studied the influence of chemical structure of thermoplastic polyurethane elastomers on CO_2 and O_2 gas permeation. Both the soft segment as well hard segments' chemical structure affected the gas permeabilities. The authors reported that there was an increase in gas permeabilities with soft segment chain lengths for membranes containing poly(oxytetramethylene)glycol. The diffusion process predominated for dissolution-diffusion gas permeation process.

7.4.2 Effect of Urethane/Urea Content

Teo *et al.* [102] studied the effect of chain extenders and PEG molecular weight on urethane moiety association. Polymeric membranes prepared from PEG with molecular weight of 600 exhibited promising performance for permeation of gas pairs like CO_2/CH_4, He/CH_4, H_2/N_2 and O_2/N_2. Sadeghi *et al.* [103] reported that the content of urea influenced the gas transport characteristics of PU membranes. Polytetramethyleneglycol (PTMG) and isophorone diisocyanate (IPDI) prepolymers were used for the synthesis. Butanediol (BDO) and butanediamine (BDA) were used as chain extenders to synthesize PUU with urethane/urea linkage. With increase in the urea linkages, the hard and soft segment micro-phase separation increased. Increasing urethane content in the polymers decreased the permeability, while with increasing urea content the gas selectivity decreased. The CO_2 permeability was reported to be 128 Barrer and CO_2/N_2 selectivity of 27. Molecular dynamics studies were carried out by Amani *et al.* [104] for understanding the effect of urethane and urea contents on gas separation properties of poly(urethane–urea) membranes. The membranes were synthesized from PTMG, IPDI and designed ratios of 1,4-butanediamine to 1,4-butanediol as chain extenders. For nanostructure characterizations of the membranes, the fractional free volume, X-ray diffraction patterns, glass transition temperature, density, and radial distribution function (RDF) were calculated. The gas permeations studied were done for O_2, N_2, CO_2, CH_4, and H_2S. Phase separation of hard and soft segments increased with increasing urea contents in the membranes and, thus, the *d*-spacing and fractional free volume. The gas permeability of the membranes proportionally increased with increasing urea linkages in the polymer. These studies underlined the importance of attaining balance of urethane and urea contents for effective gas permeation performance.

7.4.3 Effect of Temperature and Pressure

With temperature variations, the permeability and diffusivity change which allows the determination of activation energy. From this, the temperature dependence of the selectivity can be calculated. In case of large difference in permeation activation energies, the selectivity is also higher for those gas pairs [105,106]. For polyurethane membranes, the temperature variation effects on the transport properties have been studied by many researchers [107,108]. For polyurethane and polyurethane blends [77,109], the observed permeability order is $CO_2 > H_2 > O_2 > CH_4 > N_2$. Kinetic diameter [105], critical temperature and the solubility for the gas molecules have been reported to be responsible for this order. Lee *et al.* [75] studied the effect of synthesis temperature and molecular structure variation on the polymer network and gas transport properties. In specific, the gas permeability dependence of synthesis temperature, composition, aromatic content of diisocyanate, molecular weight of the polyol was analyzed. The increase in synthesis temperature and aromatic content decreased the permeability coefficient. There was an increase in tensile strength of the membrane due to the decreasing synthesis temperature and crosslinking density. The authors also studied further the effect of crosslinked state and annealing for gas transport in interpenetrating polymer network membranes.

The reduction in gas permeability due to the increase in pressure has been predicted by dual-sorption model [110], however, this has not been always observed [111,112]. In most of the glassy polymers [105,113-117], there is a decrease in permeability with pressure, when the permeating gas has high critical temperatures, like CO_2. In case of permeant gases like N_2 and O_2 with lower critical temperature, there is no influence of pressure variation on permeabilities, both for rubbery and glassy polymers [105,112,116,118-120]. However, there are some exceptions as well, which reported decrease in permeability of O_2, N_2, CH_4 with increasing pressures like poly(urethane urea) [121], polyimides [115], polyvinyl pyridine ethyl cellulose blend [114]. Madhavan *et al.* [119] studied a series of poly (dimethylsiloxane –urethane) membranes for gas transport properties. Pressure dependence on gas permeation was studied for oxygen, nitrogen and carbon dioxide gases. CO_2 permeability exhibited dependency on pressure, while O_2 and N_2 permeabilities didn't. The reported value of the O_2/N_2 permselectivity was 2.3 and CO_2/N_2 was 8.5. Thus, mixed effects of pressure have been observed on the gas transport properties of various polymers, depending on the type of permeant gas.

7.4.4 Effect of Molecular Chain Extension

Carboxyl or hydroxyl groups as organic functional groups in polyurethane membranes have direct influence on the gas transport properties of the membranes, molecular crystallinity, density and glass transition temperature. The separation coefficient for oxygen-nitrogen separation increases with increased content of functional groups, and is strongly subjective to the character of the functional group. The carboxyl group-containing poly(butylene glycol adipate) (PBA)-type polyurethane membrane was observed to exhibit higher density, glass transition temperature, crystallinity, higher gas permeability, separation coefficient [69].

Knight and Lyman [122] studied block copolyether membranes for the effect of chemical structure and fabrication variables on gas permeation. The authors observed that for copolyether-urethane-urea, a linear relation existed between the gas permeability and propylene glycol segment molecular weight. Copolymers of polypropylene segment were more permeable than those with polyethylene segment due to the crystallizing ability of polyethylene glycol. Chain packing nature of different chain extenders affected the permeability, higher the packing, the lower was the gas permeability. The addition of salts like LiBr or urea and variations in casting solvent did not impact gas permeability. Semsarzadeh *et al.* [123] studied polyether-based polyurethanes for the effects of chain extender length on the gas permeabilities. Toluene diisocyanate with 1000 and 2000 g/mol molecular weight and polytetramethylene glycol were used for polyurethane synthesis. Different chain extenders (1,6-hexane diol, 1,4-butane diol, ethylene glycol and 1,10-decane diol) were used. The permeability and diffusivity studies were performed for N_2, O_2, CH_4, CO_2. The glass transition temperature of the polymers decreased on increasing the chain extender's length. The phase separation was more probable with increased chain extender's length. With increasing the length of the chain extenders, the permeability and diffusivity of gases increased. Selectivity of CO_2/N_2 gets changed by chain extender length, while selectivity of CO_2/CH_4 and O_2/N_2 did not show any remarkable change. Damian *et al.* [124] studied the various hybrid membrane networks based on isocyanate chemistry. The permeability coefficients and the morphology depended upon the soft segment's polarity and chain length along with the composition of the networks.

7.4.5 Effect of Polymer Blending

Polymer blending is an attractive approach for enhancing the performance of polymeric membranes as it is both time and cost effective method for tuning the properties. In recent years, a variety of polymer blends have been explored for gas separation membranes. Mannaan *et al.* [125] recently reviewed polymer blend membranes for permeability, selectivity and phase behavior. For the enhancement of the transport properties, the morphology of the phase separated polymer blends was reported to play a major role. The fundamentals of polymer blends in transport processes (like gas barrier and separation) has also been discussed by Robeson *et al.* [126]. Kim *et al.* [127] studied the blend membranes of polyurethane with polyetherimides and poly (amide-imide) for CO_2 separations. As the blend ratio was increased, there was a decrease in gas permeability. The selectivity of CO_2/N_2 was improved with reduction in polyurethane content. The volume fraction of the dispersion component is an important factor to be taken care of for gas separation properties [127]. Patricio studied the gas transport behavior of gases such as H_2, N_2, O_2, CH_4, and CO_2 using polyurethane and PU/PMMA blend membranes [77]. In correlation with the decrease in average free volume size, the blends with higher wt% PU demonstrated lower permeabilities when compared with the PU membranes. For H_2/N_2 gas pair, the selectivity increased with increase in PMMA content in blends, while for other gas pairs, no marked changes were observed. The H_2 selectivity improvement could be associated with increased rigidity of amorphous phases. Ghalei *et al.* [128] studied PU/PVAc blend membranes for permeation of O_2, N_2, CO_2 and CH_4. The membranes had higher CO_2 permeability as compared to other gases, and higher CO_2/N_2 and CO_2/CH_4 selectivity. De Sales *et al.* [109] studied the polyurethane and PMMA blend membranes for temperature and pressure dependence of gas permeabilities for CO_2, H_2, O_2, CH_4, N_2 gases. With the addition of 30 wt % PMMA, the gas permeabilities decreased approximately 55%. It was observed that CO_2 had lowest permeation activation energy value (28 kJ/mol) with variation in temperature. For low temperatures, the gas pair selectivity increased, and the selectivity was higher for gas pairs having permeation activation energy value difference of about 15 kJ/mol. For pressure variation study, it was concluded that the permeabilities for CO_2 and H_2 gases through PU and the blend membrane increased by 35% at elevated pressure. CO_2 permeability increased by around 35% at higher pressure for the polyurethane and the blend membranes. Also, the O_2/N_2 selectivity increased with pressure, while the permeability to nitrogen decreased in the case of the 30% PMMA blend. Similar to earlier study, Semsarzadeh *et al.* [129] also studied the PU-PVAc blend membranes synthesized in the presence of various pluronic copolymer contents. Blends with 5 wt% PVAc exhibited higher CO_2 permeability in comparison to the PU membrane. Domain size of the dispersed PVAc was controlled by the addition of pluronic and had a positive effect on the permeability. Saedi *et al.* [130] investigated the blend membranes of polyethersulfone/polyurethane for CO_2 and CH_4 separations. The relative affinity of CO_2, CH_4 and H_2O for PES and PU were obtained using density functional theory calculations. Gas sorption by PU, viscosity of solution and PES membrane's mechanical properties were also analyzed. The fractional free volume and *d*-spacing of casting solution decreased the membrane porosity, glass transition temperature, thermal properties, gas sorption and plasticization. This was attributed to decreased Langmuir capacity and increased PES membrane strength at yield and elongation at yield. This resulted in the decrease of CO_2 permeability, a boost in CO_2/CH_4 selectivity and plasticization pressure of the PES membrane due to the presence of polyurethane. Also, the feed temperature affected the membrane behavior against pressure and the mixed gas composition.

7.4.6 Effect of Modifiers and Fillers

Another functional way to improve the gas transport behaviors of membranes can be improved by incorporating fillers. These fillers can be salts, metals, metal oxides, ions, inorganic silica, layered silicate, zeolites, carbon nanotube, graphene, etc. As the polymeric membranes are mixed with inorganic fillers, these membranes are thus termed as mixed matrix membranes. Incorporation of nanoporous filler particles is important breakthrough for boosting the gas separation capabilities of polymeric membranes. Out of a large variety of filler, metal organic frameworks (MOF) have become popular as a new group of nanoporous materials for enhanced membrane characteristics. Remarkable developments in gas permeability and selectivity have been reported for membranes based on MOFs. Erucar *et al.* [131] reviewed the recent developments in membranes incorporated with MOFs. The authors studied the experimental and computational methods to generate the polymer and MOF selection criteria for efficient gas separation membranes. Following sections summarize studies reporting the incorporation of polyurethane with a variety of reinforcements for generating effective gas separation performance of the membranes,

Zeolites

Zeolite incorporated nanocomposite membranes combine the advantages of both the polymer and the zeolite, thus, overcoming the individual shortcomings of the two materials [132]. Tirouni *et al.* [133] investigated polyurethane/zeolite mixed matrix membranes for the separation of C_2H_6 and C_3H_8 from CH_4. The hard and soft segment phase separation increased with the addition of butanediamine chain extender. With increase in urea groups of the polymer structure, both permeability and selectivity were observed to increase. Gas permeation data of polyurethane-zeolite 4Å membranes exhibited an increase in methane permeability and decrease in C_2H_6/CH_4 and C_3H_8/CH_4 selectivity, as the amount of Zeolite 4Å was increased up to 10 wt%. Polyurethane–zeolite (ZSM-5) membranes also exhibited significant improvements in selectivity and permeability of all hydrocarbons. With 20 wt% filled PU–ZSM 5 membranes, the propane permeability increased from 64.8 to 117.2 Barrer and C_3H_8/CH_4 selectivity increased from 2.6 to 3.64. Ciobanu *et al.* [134] studied the use of zeolite SAPO-5 nanocrystals for the synthesis of polyurethane composite membranes. Zeolite content was ranged from 10 to 70%. The zeolite nanocrystals were observed to act as a cross-linker for the polyurethane matrix.

Silica

A number of PU-silica based composite membranes have been reported in the literatiure. Khudyakov *et al.* [135] reviewed the status of UV-curable polyurethane nanocomposites incorporating nanosilica and organically-modified clay. A large number of studies have been reported related to the structure and characteristics of polyurethane nanocomposites formed through dark reactions as well as by UV-curing of urethane acrylate oligomers. Polyurethane nanocomposites with low loadings (less than 5%) of fillers have been reported to have dramatic property improvements [135,136].

In another study, Petrovic *et al.* [137] studied the polymerization of polypropylene glycol/hexamethylenediisocyanate/1,4-butanediol to gain insights about the gas transport properties. Tetraethoxysilane was used to prepare silica nanoparticles through sol-gel method, while the nanocomposites were synthesized by solution mixing technique. Various characterization techniques were used to confirm the desired nano-scale distribution of silica nanoparticles. Gas permeation studies of the membranes revealed the enhancement in CO/N selectivities with increasing amount of silica nanoparticles. It was also observed that the nanocomposites exhibited nearly two fold increment in the selectivity when compared with pristine polyurethanes membranes. However, the CO permeability exhibited a reduction of nearly 35% for the composite membranes while comparing against pure PU ones. Higuchi model may be used to predict the gas transport properties in polyurethane-silica membranes [138]. In this study, properties such as dielectric permeability of nanocomposite membranes were studied. Studies on the ether-based PU and ester-based PU exhibited that the amalgamation of silica nanoparticles helped the improvement of CO_2/CH_4 selectivity. By taking an account of the nanoparticles interfacial layer leading to the formation of void volumes, a model was introduced to help predicting the nanocomposite membrane performance [139].

Polycaprolactone-based polyurethane membranes were studied for the effect of the addition of silica nanoparticles [140]. The membranes were prepared using conventional methods such as solution mixing and casting. Polycaprolactone/hexamethylene diisocyanate/1,4-butanediol based polymer was produced by polymerization involving a two-step methodology. Membrane performance was predicted using modified Higuchi model to obtain a fair agreement in theoretical and experimental values. In another study, hybrid polyurethane/silica based composite membranes were prepared using tetraethoxysilane, cetyltrimethyl ammonium bromide and polyvinyl alcohol [141]. Spectroscopic techniques were used to verify the presence of silica in the polymer network and SEM for the nanoscale distribution of silica particles. It was observed that with increase in the silica content, the diffusivity of gases and CO_2 gas permeability were reduced. With silica content >10 wt%, an enhancement of CO_2/CH_4 selectivity was observed.

In another study, the transport performance of carbon dioxide and methane gases was investigated in polyesterurethane mixed matrix membranes (MMMs), in separate tests, containing different fumed silica nanoparticles [142]. Non-modified and commercially-modified silica (with octylsilane and polydimethylsiloxane) were used as fillers. The structural features were studied by various microscopic, spectroscopic and calorimetric techniques to confirm the interfaces induced in PU microphase when silica was present. Surface treatment of silica filler with long hydrophobic chains condensed the accumulation of nanoparticles and enhanced dispersion in MMMs. Regardless of silica type, both separation factor and CO_2 permeation were increased with the presence of silica nanoparticles, which was attributed to the interrupted chain packing and improved dynamic free volume. The results revealed that among silica nanoparticles, the unmodified particles with OH groups on the surface exhibited better performance for CO_2/CH_4 separation. A new model of gas permeation through PU/silica membranes was proposed considering the presence of filler aggregation in the matrix as well as free volume at the interface layer [142].

Another study revealed the gas transport of two types of polyurethane membranes which are synthesized from PCL225 and PPG polyether [143]. It was observed that by using silica content up to 2.5%, the permeability increased. However, the permeability exhibited a downward effect on increasing the silica content father. It was observed that the selectivity of propane over methane increased with the increase in the amount of silica particles. The studied membranes were the ones with 12.5% silica and the permeation tests were conducted at 2 bar pressure.

With the addition of nanofillers, the flexural properties of the polymer matrix are required to be retained. Ideally, the hybrid polymer materials are expected to exhibit enhancement in properties such as mechanical strength and polymer flexibility. Enhanced material properties are possible with more homogeneous distributions of the inorganic components leading to the formation of high performance functional membranes [144]. In another study, polyurethane (PU) based mixed matrix membranes were generated with polytetramethylene glycol and polyvinyl alcohol along with silica nanoparticles [145]. The dispersion in the nanocomposite membranes was confirmed using microscopic, spectroscopic techniques. Gas permeation studies for pure CH_4, O_2, CO_2, N_2 and He gases were performed through the composites with varying amounts of silica particles. It was observed that upon increasing the silica content, the permeability of the CO_2 gas was enhanced, while the transport properties of other gases decreased. The permeability of CO_2 increased from 68.4 to 96.7 Barrer, where the membrane contained 10 wt% of nanoparticles.

Layered Silicates

Osman *et al.* [146] studied the gas permeation properties of modified montmorillonite based nanocomposites with polyurethane adhesives. The nanocomposites were generated to act as barrier against oxygen and water vapor. Thus, their application in gas industries can be envisaged as membranes which allow permeation of hydrocarbons, but do not allow any other gases like O_2, CO_2 and water vapor to pass through. A correlation between the gas permeation and volume fraction was established. With the incorporation of small volume fractions of nanoparticles, the permeation rate of O_2 and water vapor was decreased (Figure 7.3 for oxygen permeation). A 30% reduction was observed with 3 vol% of filler, when bis(2-hydroxyethyl) hydrogenated tallow ammonium or alkylbenzyldimethylammonium ions were used for clay modification. The clay modified with dimethyl dihydrogenated tallow ammonium ions increased the O_2 permeation rate with increase in inorganic fraction. The above variation was attributed to the phase separation between the pure hydrocarbon modification and polar polyurethane. This underlined the need of generating compatibility between the filler and polymer phases in order to achieve optimum performance. Increase in oxygen permeation for dimethyl dihydrogenated tallow ammonium ions modified clay composites could still be used for O_2, water vapor separation as the membranes were impermeable to water vapor, whereas highly permeable to O_2 molecules. In a similar study, Mittal [147] also reported the gas permeation performance of polyurethane-clay nanocomposites and underlined the impact of synthesis procedure as well as filler-matrix compatibility. As shown in Figure 7.4, the composite with Tixogel VZ, the clay surface modified with a benzyl(hydrogenated tallow alkyl)dimethyl ammonium ions, exhibited a decrease in oxygen permeation. On the other hand, the composite with Tixogel VP, the clay surface modified with bis(hydrogenated tallow alkyl)dimethyl ammonium ions, exhibited an increase in oxygen permeation as a function of filler fraction. Similar to the previous study, the water vapor permeation in both the composites was reduced as a function of filler fraction. Thus, the chemical

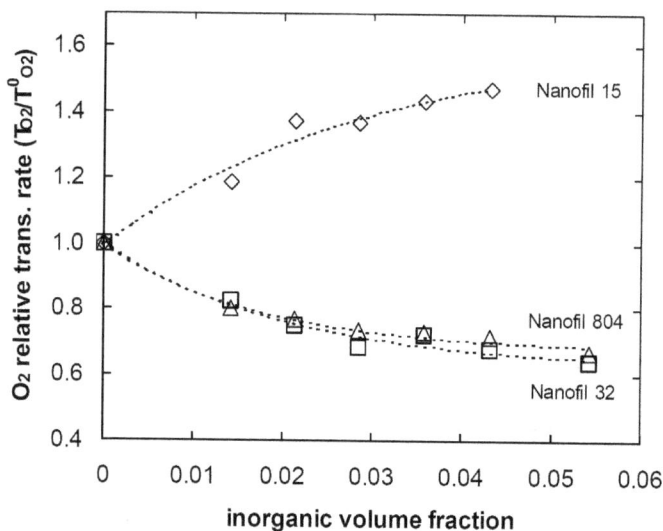

Figure 7.3 Dependence of the oxygen transmission rate through the PU-nanocomposites on the inorganic volume fraction. The dotted lines are guides for the eye. Reproduced from Reference 146 with permission from American Chemical Society.

nature of the surface coating of the fillers resulted in specific interactions with the polymer matrix, which consequently affected the nanocomposite properties. The authors also compared the oxygen permeation through polyurethane-clay nanocomposites with molecular dynamics predictions in order to gain more insights about the average aspect ratio of the filler platelets in the composites [148]. In polyurethane composites, the clay platelets were observed to be present with an average aspect ratio of 100. In addition, the good agreement with the experimental and predicted values was retained till 3 vol% filler fraction. Afterwards, the experimental values were observed to level off.

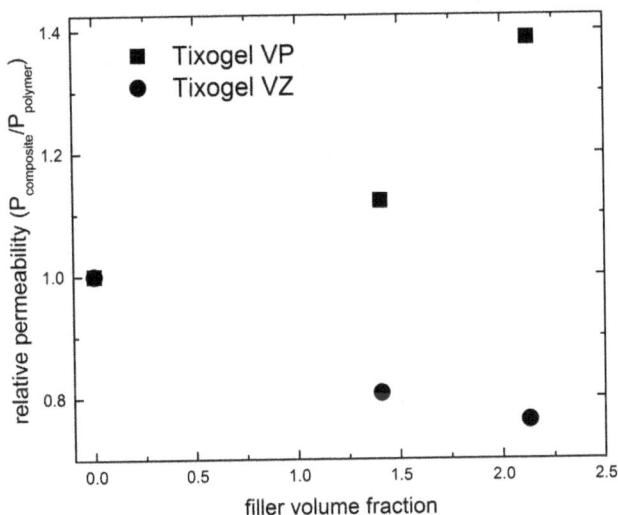

Figure 7.4 Oxygen permeation through the PU and PU nanocomposites as a function of filler volume fraction. Reproduced from reference 147 with permission from Wiley.

Also, as shown in Figure 7.5, the authors generated the relation between the reduction of oxygen permeation through the nanocomposites with filler fraction for both aligned and misaligned filler parties. The aligned filler particles were observed o be significantly effective in tuning the gas transport behavior of the nanocomposites. These studies indicated the successful development of polyurethane-clay nanocomposites with tuned reduction in water vapor permeation as well as increased or decreased O_2 permeation, based on the chosen filler system. Such systems can result in high selectivity membranes for the separation of hydrocarbons from other contaminant gases.

Dense polyurethane-based membranes containing hydrophilic clay were synthesized by Barboza *et al.* [149] and the permeability of carbon dioxide (CO_2) through the membranes was studied. Ethylene diamine was used as the chain extender, which led to the urea linkage network. The nanofiller fraction was optimized as 0.5 and 1% relative to the amount of poly(ethylene glycol).

The CO_2 permeability was observed to improve with the amount of PEG, whereas it was noticed to decrease with the amount of clay as the higher clay content contributed to the tortuous pathways for gas diffusion. A novel nano-composite was reported by Shamini and Yusoh, which consisted of Na^+ montmorillonite which was further modified using transition metal ions such as copper (II) chloride and iron (III) chloride [150]. The presence of metal ions reportedly contributed to better dispersion of nano-filler and also reduced clay agglomeration. The gas transport properties exhibited remarkable decrease. Polyurethane film containing 1% filler modified with iron chloride exhibited the permeability to decrease four fold. Iron and copper exhibited different patterns for the reduction in permeability.

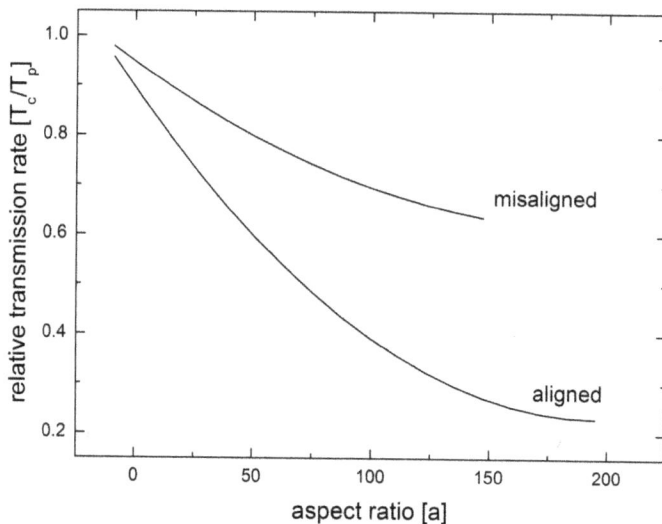

Figure 7.5 Numerical prediction of the effect of filler platelet misalignment on the gas permeation performance at 3 vol% loading. Reproduced from reference 148 with permission from Wiley.

It has to be mentioned that the reduction in permeability of different gases in the composite membranes does not indicate the reduction in the membrane performance, it only signifies the probable enhancement in the selectivity by blocking the passage of certain penetrants and possibly enhancing the permeation of others. In addition, the use of fillers also enhances the mechanical and thermal properties, along with enhancing environmental stability. Most of these developed composite membranes, thus, have strong potential of application in various industries to separate gas mixtures.

Metals, Metal Oxide and Metal Ions

Ameri *et al.* [151] used alumina (Al_2O_3) at different concentrations to prepare polyurethane nanocomposites through a bulk two-step polymerization. Different chain extenders were used to complete the urethane polymer network. Alumina ensured improved O_2/N_2, CO_2/CH_4, and CO_2/N_2 selectivity, but with reduced permeability values. In another study, cobalt chelated plasma treated membranes were generated and the effect of subsequent formamide formation in the membrane matrix on the permeability properties was evaluated [152]. The chelation was obtained by treatment with cobalt(II)/formamide solution. With ethylenediamine plasma treatment, selectivity increased from 2.6 to 3.1 GPU whereas the CoCl •6HO/formamide treatment improved the value to 4.4 GPU. The enhanced O/N selectivity was attributed to the improvement in oxygen affinity achieved through chelation and size sieving effect. Chen *et al.* [153] incorporated TiO_2 nanoparticles at varying concentrations to the polymer to prepare thermo-sensitive polyurethane (TSPU). Membrane formation temperature was optimized in the study to tune the transport behaviors of the nanocomposites. It was observed that the permeability coefficients altered with different membrane formation temperatures. Increasing nano-TiO_2 concentration also favorably helped the transport phenomenon. Conventionally seen tortuous diffusion pathway based low permeation mechanism failed to explain the counter-intuitive phenomenon observed in the study. The authors suggested that the soft segment of the thermo-sensitive polyurethane packed around the TiO_2 nanoparticles, as if these were in the bulk polymer, thus, leading to the observed results.

In another study, PU membranes were synthesized through thermal phase inversion method [154] Polyol:diisocyanate:chain extender was blended at 1:2:1 molar ratio and the interactions at intramolecular level were analyzed. TiO_2 was varied up to 30 wt% and transport behaviors were studied for N_2, O_2, CH_4 and CO_2 gases at varying temperatures. It was observed that an increased TiO_2 content led to a proportional rise in selectivity and a proportional fall in permeability values. Membranes of ionic polyurethane were also reported using N-methyldiethanolamine as chain extender, which was later complexed with cupric ions [155]. Different polymerization approaches such as single and double step were employed for the composite preparation. It was observed that the incorporation of CuCl enhanced hard segment aggregation, whereas incorporation of crosslinker hindered the cluster formation by the hard segment. The soft segment aggregation helped to increase the permeability.

Carbon Nanoparticles

The formation of polyurethane nanocomposites have proven the ability to exhibit advanced performance as compared to pure polymer. In one such study, poly(ether urethane) membranes containing multi-walled carbon nanotubes (MWCNTs) were studied for improvements in mechanical strength and permeation capabilities [156]. The MWCNTs were grafted with various functional groups to obtain three types of membranes with fillers such as MWCNT-COOH, MWCNT-OH, MWCNT-IPDI. Fourier-transform infrared spectroscopy was used to confirm the successful grafting of functional groups on to the surface of the MWCNTs. Techniques like SEM, mechanical testing and thermal analysis confirmed the superior properties of MWCNT-IPDI based membranes (Figure 7.6). Despite the development towards the generation of strong, dura-

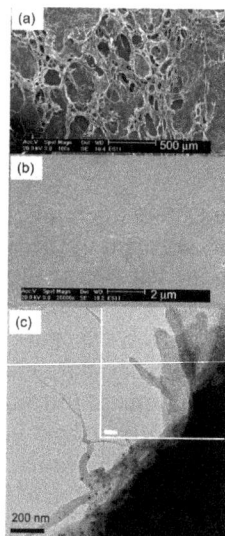

Figure 7.6 SEM and TEM images of PU nanocomposite membrane obtained by the electro spinning technique. Reproduced from Reference 156 with the permission from American Chemical Society.

ble, and cost-efficient carbon nanoparticles based polymer membranes, there is a need for the evolution of membrane materials with high gas transport properties [157,158].

Graphene

A large number of studies has reported the generation of polyurethane nano-composites by incorporating graphene/graphene oxide/modified or function-alized graphene [158-173]. Most of the studies have confirmed these nano-composites to have significantly superior thermal and mechanical properties. However, only a few studies report the gas permeation data for PU-graphene nanocomposites. Kim *et al.* [174] reported 90% decrease in nitrogen permea-tion with 3 wt% isocyanate treated graphene oxide filled thermoplastic polyu-rethanes. Figure 7.7 shows the mechanism of dispersion of functionalized

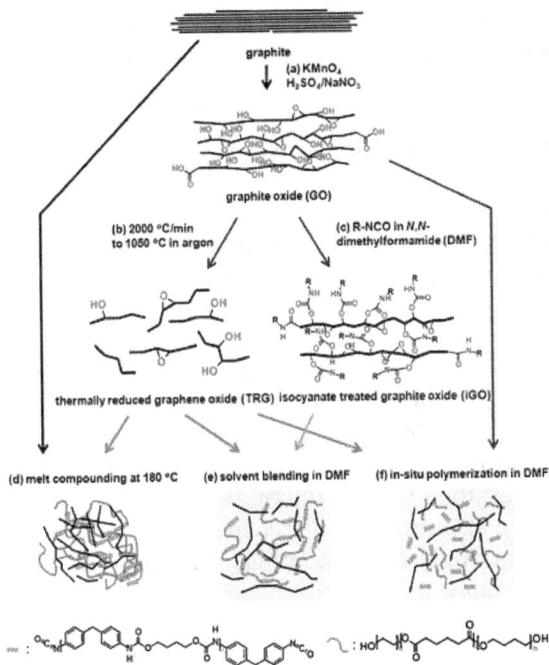

Figure 7.7 Scheme showing the mechanism of functionalized graphene dispersed in polyurethane. Reproduced from Reference 174 with the permission from American Chemical Society.

graphene in polyurethane. In another study, Thermoplastic composite films of polyurethane containing hexadecyl-functionalized low-defect graphene nanoribbons were studied by Xiang *et al.* [175] for improved gas barrier properties. A decrease in nitrogen gas diffusivity by 3 orders of magnitude was observed with only 0.5 wt% of filler. The films were reported to have potential applications in food packaging and light weight gas storage containers. Figure 7.8 also shows the nitrogen pressure drop across the PU membranes and PU composite membranes. The composite with 0.5 wt% filler exhibited strong resistance against nitrogen for a longer period of time, indicating its barrier towards nitrogen permeation. Kaveh *et al.* [176] reported excellent gas barrier properties with an 80% decrease in permeability of Helium for 1 % graphene oxide incorporated thermoplastic polyurethane films. Thus, until now, the graphene nanocomposites have been largely developed to generate potential candidates for gas barrier properties, however, optimization of the graphene based composites for enhanced permeation is still needed. However, as mentioned earlier, decrease permeability of a certain permeant still opens the potential for these membranes to be selective towards the separation of certain gas mixtures.

7.5 Other Natural Gas Specific Applications

Polyether-polyurethane interactions as well as water vapor and methane separation selectivity were studied by Di Landro *et al.* [177]. Water absorption was shown to be higher for urethane rich polymers. Water vapor permeability was 10^3-10^4 times more than methane as well as other permeant gases. This makes these materials a potential membrane material for natural gas dehydration [9,177]. High H_2S/CH_4 selectivities of poly(ether urethane) and poly(ether urethane urea) membranes were reported by Chatterjee *et al.* [178]. The authors reported polymer PU4 as a favorable membrane material for H_2S separation from CO_2 and CH_4 mixtures. H_2S/CH_4 selectivity was reported to be greater than 100 at 20 °C in the pressure range from 4-13.6 atm. Other studies have also reported high permeability, permselectivity, durability for acid gas separtiob from natural gas and carbon dioxide removal from synthesis gas for polyether-urethane or polyether-urea block copolymers [179,180].

Ponangi *et al.* [181] studied for the split-up of volatile organic compounds from nitrogen either in dry or in humidified form using PU membranes. The membranes were examined for the vapor phase separation of a variety of gases such as benzene, hexane, p-xylene, and benzene/toluene/xylene mixtures

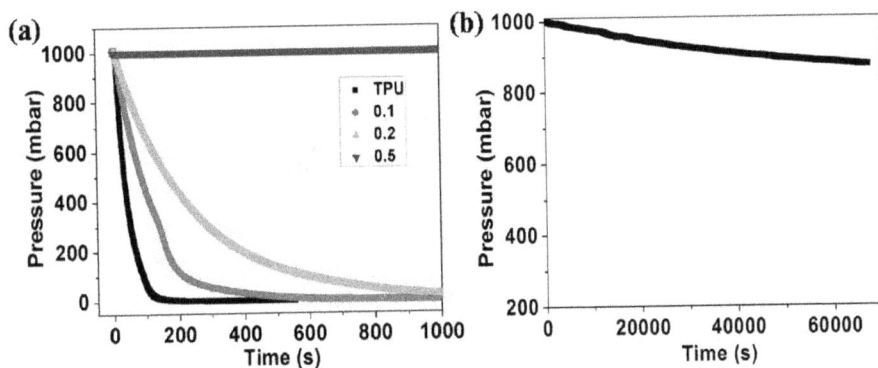

Figure 7.8 (a) Nitrogen pressure drop through PU and PU composite films; (b) nitrogen pressure drop through the composite wit 0.5 wt% filler over a longer period of time. Reproduced from Reference 175 with permission from American Chemical Society.

from nitrogen. The selectivities of organic/dry N_2 ranged from 30-210 and pressure normalized permeabilities of around 1.25×10^{-3} cm³ (STP)/(cm² s cmHg). The nitrogen permeability was increased with swelling of the polyether membranes, but the organic/N_2 selectivity was lowered. Water permeability was observed to be low and did not depend on the organic feed component. The authors also further studied the free volume in polyurethane membranes for organic vapor diffusion [182].

7.6 Conclusion

In this review, various advancements for enhancing the applications of polyurethane membranes for gas separation have been explored. Polyurethanes represent useful class of materials with properties such as mechanical strength, dimensional stability, thermal resistance, resistance towards chemicals and tunable gas transport behavior. Various modifications in the molecular microstructure of polyurethanes have been achieved in order to enhance the permeability as well as selectivity of various gases through these materials. These modifications include optimization of soft and hard segments, tuning of polarity, blending with other polymers, adjusting urea urethane interactions, incorporation of fillers, use of a wide variety of chain extenders and cross-linkers, etc. It is evident that the tunability of gas permeability as well as selectivity in polyurethane membranes would further enhance the industrial application of these materials for gas separation processes.

References

1. Clarizia, G. (2009) Polymer-based membranes applied to gas separation: material and engineering aspects. *Desalination*, **245**(1-3), 763-768.
2. Takht Ravanchi, M., Kaghazchi, T., and Kargari, A. (2009) Application of membrane separation processes in petrochemical industry: a review. *Desalination*, **235**(1-3), 199-244.
3. Yampolskii, Y. (2012) Polymeric gas separation membranes. *Macromolecules*, **45**(8), 3298-3311.
4. Sanders, D. F., Smith, Z. P., Guo, R., Robeson, L. M., McGrath, J. E., Paul, D. R., and Freeman, B. D. (2-13), Energy-efficient polymeric gas separation membranes for a sustainable future: A review. *Polymer*, **54**(18), 4729-4761.
5. Sipek, M., Friess, K., and Hynek, V. (2004) Membrane separation of mixtures of gases and vapors in practice. *Chemicke Listy*, **98**(1), 4-9.
6. Bernardo, P., Drioli, E., and Golemme, G. (2009) Membrane gas separation: a review/state of the art. *Industrial & Engineering Chemistry Research*, **48**(10), 4638-4663.
7. Maier, G. (1998) Gas separation with polymer membranes. *Angewandte Chemie, International Edition*, **37**(21), 2960-2974.
8. Matsuura, T. (1993) *Synthetic Membranes and Membrane Separation Processes*, Taylor & Francis.
9. George, G., Bhoria, N., and Mittal. V. (2015) Improved Polymer Membranes for Sour Gas Filtration and Separation. *Abu Dhabi International Petroleum Exhibition and Conference*, UAE. Online: https://www.onepetro.org/download/conference-paper/SPE-177805-MS?id=conference-paper%2FSPE-177805-MS (assessed 1st March 2017).
10. Krol, P. (2009) Polyurethanes - A review of 60 years of their syntheses and applications. *Polimery*, **54**(7-8), 489-500.
11. Hahn, C., Keul, H., and Moeller, M. (2012) Hydroxyl-functional polyurethanes and polyesters: synthesis, properties and potential biomedical application. *Polymer International*, **61**(7), 1048-1060.
12. Brzeska, J., Dacko, P., Janeczek, H., Janik, H., Sikorska, W., Rutkowska, M., and Kowalczuk, M. (2014) Synthesis, properties and applications of new (bio)degradable polyester urethanes. *Polimery*, **59**(5), 365-371.
13. Engels, H.-W., Pirkl, H.-G., Albers, R., Albach, R. W., Krause, J., Hoffmann, A., Casselmann, H., and Dormish, J. (2013) Polyurethanes: versatile materials and sustainable problem solvers for today's challenges. *Angewandte Chemie, International Edition*, **52**(36), 9422-9441.
14. Yang, C., Fischer, L., Maranda, S., and Worlitschek, J. (2015) Rigid polyurethane foams incorporated with phase change materials: A state-of-the-art review and future research pathways. *Energy and Buildings*, **87**, 25-36.
15. Yao, X., Tuo, X., and Wang, X. (2009) Preparing biodegradable polyurethane porous scaffold for tissue engineering application. *Progress in Chemistry*, **21**(7-8), 1546-1552.

16. Janik, H., and Marzec, M. (2015) A review: Fabrication of porous polyurethane scaffolds. *Materials Science & Engineering C*, **48**, 586-591.
17. Zoltowska, K., Sobczak, M., and Oledzka, E. (2014) Polyurethanes in pharmacy - current state and perspectives of the development. *Polimery*, **59**(10), 689-698.
18. Zia, K. M., Bhatti, H. N., and Bhatti, I. A. (2007) Methods for polyurethane and polyurethane composites, recycling and recovery: A review. *Reactive & Functional Polymers*, **67**(8), 675-692.
19. Braun, T. (1989) Quasi-spherical solid polymer membranes in separation chemistry: polyurethane foams as sorbents. Recent advances. *Fresenius' Zeitschrift fuer Analytische Chemie*, **333**(8), 785-792.
20. Pinchuk, L. (1994) A review of the biostability and carcinogenicity of polyurethanes in medicine and the new generation of 'biostable' polyurethanes. *Journal of Biomaterials Science, Polymer Edition*, **6**(3), 225-67.
21. Mishra, A., Singh, S. K., Dash, D., Aswal, V. K., Maiti, B., Misra, M., and Maiti, P. (2014) Self-assembled aliphatic chain extended polyurethane nanobiohybrids: Emerging hemocompatible biomaterials for sustained drug delivery. *Acta Biomaterialia*, **10**(5), 2133-2146.
22. Morral-Ruiz, G., Melgar-Lesmes, P., Solans, C., and Garcia-Celma, M. J. (2013) Multifunctional polyurethane-urea nanoparticles to target and arrest inflamed vascular environment: a potential tool for cancer therapy and diagnosis. *Journal of Controlled Release*, **171**(2), 163-171.
23. St. John, K. R. (2014) The use of polyurethane materials in the surgery of the spine: a review. *Spine Journal*, **14**(12), 3038-3047.
24. Teodosiu, C., Wenkert, R., Tofan, L., and Paduraru, C. (2014) Advances in preconcentration/removal of environmentally relevant heavy metal ions from water and wastewater by sorbents based on polyurethane foam. *Reviews in Chemical Engineering*, **30**(4), 403-420.
25. Boretos, J. W., and Pierce, W. S. (1968) Segmented polyurethane: A polyether polymer. An initial evalution for biomedical applications. *Journal of Biomedical Materials Research*, **2**(1), 121-130.
26. Boretos, J. W., Detmer, D. E., and Donachy, J. H. (1971) Segmented polyurethane: A polyether polymer, II. Two years experience. *Journal of Biomedical Materials Research*, **5**(4), 373-387.
27. Hepburn, C. (1982) *Polyurethane* Elastomers, Applied Science Publishers, USA.
28. Chang, Y. J. P., and Wilkes, G. L. (1975) Superstructure in segmented polyether–urethanes. *Journal of Polymer Science: Polymer Physics Edition*, **13**(3), 455-476.
29. Huh, D. S., and Cooper, S. L. (1971) Dynamic mechanical properties of polyurethane block polymers. *Polymer Engineering & Science*, **11**(5), 369-376.
30. Camargo, R., Macosko, C. W., Tirrell, M., and Wellinghoff, S. T. (1985) Phase separation studies in RIM polyurethanes catalyst and hard segment crystallinity effects. *Polymer*, **26**(8), 1145-1154.
31. Cooper, S. L., and Tobolsky, A. V. (1966) Properties of linear elastomeric polyurethanes. *Journal of Applied Polymer Science*, **10**(12), 1837-1844.
32. Clough, S., and Schneider, N. (1968) Structural studies on urethane elastomers. *Journal of Macromolecular Science, Part B: Physics*, **2**(4), 553-566.

33. Seymour, R. W., and Cooper, S. L. (1971) DSC studies of polyurethane block polymers. *Journal of Polymer Science, Part B: Polymer Letters*, 9(9), 689-694.

34. Seymour, R. W., and Cooper, S. L. (1973) Thermal analysis of polyurethane block polymers. *Macromolecules*, 6(1), 48-53.

35. Srichatrapimuk, V. W., and Cooper, S. L. (1978) Infrared thermal analysis of polyurethane block polymers. *Journal of Macromolecular Science, Part B*, 15(2), 267-311.

36. Sung, C. P., and Hu, C. (1981) Orientation studies of segmented polyether poly (urethaneurea) elastomers by infrared dichroism. *Macromolecules*, 14(1), 212-215.

37. Wang, C. B., and Cooper, S. L. (1983) Morphology and properties of segmented polyether polyurethaneureas. *Macromolecules*, 16(5), 775-786.

38. Seymour, R., Estes, G., and Cooper, S. (1970) Infrared studies of segmented polyurethan elastomers. I. Hydrogen bonding. *Macromolecules*, 3(5), 579-583.

39. Ishihara, H., Kimura, I., Saito, K., and Ono, H. (1974) Infrared studies on segmented polyurethane-urea elastomers. *Journal of Macromolecular Science, Part B: Physics*, 10(4), 591-618.

40. Brunette, C., Hsu, S., and MacKnight, W. (1982) Hydrogen-bonding properties of hard-segment model compounds in polyurethane block copolymers. *Macromolecules*, 15(1), 71-77.

41. Bonart, R., Morbitzer, L., and Hentze, G. (1969) X-ray investigations concerning the physical structure of cross-linking in urethane elastomers. II. Butanediol as chain extender. *Journal of Macromolecular Science, Part B: Physics*, 3(2), 337-356.

42. Blackwell, J., and Gardner, K. H. (1979) Structure of the hard segments in polyurethane elastomers. *Polymer*, 20(1), 13-17.

43. Fridman, I. D., and Thomas, E. L. (1980) Morphology of crystalline polyurethane hard segment domains and spherulites. *Polymer*, 21(4), 388-392.

44. Howarth, G. A. (2003) Polyurethanes, polyurethane dispersions and polyureas: Past, present and future. *Surface Coatings International, Part B: Coatings Transactions*, 86(2), 111-118.

45. Huang, S. L., and Lai, J. Y. (1995) Gas permeability of crosslinked HTPB–H12MDI-based polyurethane membrane. *Journal of Applied Polymer Science*, 58(10), 1913-1923.

46. George, G., Bhoria, N., Alhallaq, S., Abdala, A., and Mittal, V. (2016) Polymer membranes for acid gas removal from natural gas. *Separation and Purification Technology*, 158, 333-356.

47. Stern, S. A. (1994) Polymers for gas separations - The next decade. *Journal of Membrane Science*, 94, 1-65.

48. Matteucci, S., Yampolskii, Y., Freeman, B. D., and Pinnau, I. (2006) Transport of gases and vapors in glassy and rubbery polymers. In: *Materials Science of Membranes for Gas and Vapor Separation*, Yampolskii, Y., Pinnau. I., and Freeman, B. (eds.), John Wiley & Sons, UK, doi: 10.1002/047002903X.ch1.

49. Ziegel, K. (1971) Gas transport in segmented block copolymers. *Journal of Macromolecular Science, Part B: Physics*, 5(1), 11-21.

50. Pegoraro, M., Penati, A., and Zanderighi, L. (1986) Polyurethane membrane for gas fractionation. *Journal of Membrane Science*, 27(2), 203-214.

51. McBride, J. S., Massaro, T. A., and Cooper, S. L. (1979) Diffusion of gases through polyurethane block polymers. *Journal of Applied Polymer Science*, **23**(1), 201-214.

52. Schneider, N., S., Dusablon, L. V., Snell, E. W., and Prosser, R. A. (1969) Water vapor transport in structurally varied polyurethans. *Journal of Macromolecular Science, Part B: Physics*, **3**(4), 623-644.

53. Barrer, R., Barrie, J., and Slater, J. (1958) Sorption and diffusion in ethyl cellulose. Part III. Comparison between ethyl cellulose and rubber. *Journal of Polymer Science*, **27**(115), 177-197.

54. Vieth, W., Howell, J., and Hsieh, J. (1976) Dual sorption theory. *Journal of Membrane Science*, **1**, 177-220.

55. Chern, R. T., Koros, W. J., Sanders, E. S., and Yui, R. (1983) "Second component" effects in sorption and permeation of gases in glassy polymers. *Journal of Membrane Science*, **15**(2), 157-169.

56. Spillman, R. W. (1989) Economics of gas separation membranes. *Chemical Engineering Progress*, **85**(1), 41-62.

57. Ghosal, K., and Freeman, B. D. (1994) Gas separation using polymer membranes: an overview. *Polymers for Advanced Technologies*, **5**(11), 673-697.

58. Koros, W. J., and Fleming, G. K. (1993) Membrane-based gas separation. *Journal of Membrane Science*, **83**(1), 1-80.

59. Roualdes, S., Sanchez, J., and Durand, J. (2002) Gas diffusion and sorption properties of polysiloxane membranes prepared by PECVD. *Journal of Membrane Science*, **198**(2), 299-310.

60. Robeson, L. M. (1991) Correlation of separation factor versus permeability for polymeric membranes. *Journal of Membrane Science*, **62**(2), 165-185.

61. Mohr, J. M., and Paul, D. R. (1991) Surface fluorination of composite membranes. Part I. Transport properties. *Journal of Membrane Science*, **55**(1), 131-148.

62. Lin, X., Qiu, X., Zheng, G., and Xu, J. (1995) Gas permeabilities of poly (trimethylsilylpropyne) membranes surface modified with CCl4 plasma. *Journal of Applied Polymer Science*, **58**(11), 2137-2139.

63. Robeson, L. M. (2008) The upper bound revisited. *Journal of Membrane Science*, **320**(1-2), 390-400.

64. Hsieh, K. H., Tsai, C. C., and Tseng, S. M. (1990) Vapor and gas permeability of polyurethane membranes. Part I. Structure-property relationship. *Journal of Membrane Science*, **49**(3), 341-350.

65. Huang, S.-L., and Lai, J.-Y. (1995) On the gas permeability of hydroxyl terminated polybutadiene based polyurethane membranes. *Journal of Membrane Science*, **105**(1), 137-145.

66. Johnson, B. M., Baker, R. W., Matson, S. L., Smith, K. L., Roman, I. C., Tuttle, M. E., and Lonsdale, H. K. (1987) Liquid membranes for the production of oxygen-enriched air: II. Facilitated-transport membranes. *Journal of Membrane Science*, **31**(1), 31-67.

67. Bicak, N., Koza, G., and Atay, T. (1996) Metal chelating resins by condensation of ethylene diamine with p-dichloromethyl benzene. *Journal of Applied Polymer Science*, **61**(5), 799-804.

68. Park, C., Choi, M., and Lee, Y. M. (1995) Chelate membrane from poly (vinyl alcohol)/poly (N-salicylidene allyl amine) blend. I: Synthesis and characterization of Co (II) chelate membrane. *Polymer*, **36**(7), 1507-1512.

69. Hsieh, K. H., Tsai, C. C., and Chang, D. M. (1991) Vapor and gas permeability of polyurethane membranes. Part II. Effect of functional group. *Journal of Membrane Science*, **56**(3), 279-287.

70. Park, H. B., Kim, C. K., and Lee, Y. M. (2002) Gas separation properties of polysiloxane/polyether mixed soft segment urethane urea membranes. *Journal of Membrane Science*, **204**(1-2), 257-269.

71. Queiroz, D. P., and De Pinho, M. N. (2005) Structural characteristics and gas permeation properties of polydimethylsiloxane/poly(propylene oxide) urethane/urea bi-soft segment membranes. *Polymer*, **46**(7), 2346-2353.

72. Gomes, D., Peinemann, K.-V., Nunes, S. P., Kujawski, W., and Kozakiewicz, J. (2006) Gas transport properties of segmented poly(ether siloxane urethane urea) membranes. *Journal of Membrane Science*, **281**(1-2), 747-753.

73. Yang, J. M., Lai, W. C., and Lin, H. T. (2001) Properties of HTPB based polyurethane membrane prepared by epoxidation method. *Journal of Membrane Science*, **183**(1), 37-47.

74. Cao, N., Pegoraro, M., Bianchi, F., Di Landro, L., and Zanderighi, L. (1993) Gas transport properties of polycarbonate–polyurethane membrance. *Journal of Applied Polymer Science*, **48**(10), 1831-1842.

75. Doo, S. L., Dae, S. J., Tae, H. K., and Sung, C. K. (1991) Gas transport in polyurethane-polystyrene interpenetrating polymer network membranes. I. Effect of synthesis temperature and molecular structure variation. *Journal of Membrane Science*, **60**(2-3), 233-252.

76. Kumar, H. (2005) A study of sorption/desorption and diffusion of substituted aromatic probe molecules into semi interpenetrating polymer network of polyurethane/polymethyl methacrylate. *Polymer*, **46**(18), 7140-7155.

77. Patricio, P. S. O., de Sales, J. A., Silva, G. G., Windmoller, D., and Machado, J. C. (2006) Effect of blend composition on microstructure, morphology, and gas permeability in PU/PMMA blends. *Journal of Membrane Science*, **271**(1-2), 177-185.

78. George, S. C., and Thomas, S. (2001) Transport phenomena through polymeric systems. *Progress in Polymer Science*, **26**(6), 985-1017.

79. He, Y., Xie, D., and Zhang, X. (2014) The structure, microphase-separated morphology, and property of polyurethanes and polyureas. *Journal of Materials Science*, **49**(21), 7339-7352.

80. Ohst, H., Hildenbrand, K., and Dhein. R. (1991) Polymer Structure/Properties Correlation of Polyurethane PV-membranes for Aromatic Aliphatic Separation. *Proceedings 5th International Conference on Pervaporation Processes in the Chemical Industry*, Germany.

81. Pegoraro, M., Zanderight, L., Penati, A., Severini, F., Bianchi, F., Cao, N, Sisto, R., and Valentini, C. (1991) Polyurethane membranes from polyether and polyester diols for gas fractionation. *Journal of Applied Polymer Science*, **43**(4), 687-697.

82. Xiao, H., Ping, Z. H., Xie, J. W., and Yu, T. Y. (1990) Permeation of CO_2 through polyurethane. *Journal of Applied Polymer Science*, **40**(7-8),1131-1139.

83. Mohammadi, T., Tavakol Moghadam, M., Saeidi, M., and Mahdyarfar, M. (2008) Acid gas permeation behavior through poly (ester urethane urea) membrane. *Industrial & Engineering Chemistry Research*, **47**(19), 7361-7367.

84. Wolinska-Grabczyk, A., and Jankowski, A. (2011) CO_2/N_2 separation ability and structural characteristics of poly(butadiene-co-acrylonitrile)-based polyurethanes and hydrogenated nitrile rubbers. *Journal of Applied Polymer Science*, **122**(4), 2690-2696.

85. Talakesh, M. M., Sadeghi, M., Chenar, M. P., and Khosravi, A. (2012) Gas separation properties of poly(ethylene glycol)/poly(tetramethylene glycol) based polyurethane membranes. *Journal of Membrane Science*, **415-416**, 469-477.

86. Poreba, R., Spirkova, M., Brozova, L., Lazic, N., Pavlicevic, J., and Strachota, A. (2013) Aliphatic polycarbonate-based polyurethane elastomers and nanocomposites. II. Mechanical, thermal, and gas transport properties. *Journal of Applied Polymer Science*, **27**(1), 329-341.

87. Khosravi, A., and Sadeghi, M. (2013) Separation performance of poly(urethane-urea) membranes in the separation of C2 and C3 hydrocarbons from methane. *Journal of Membrane Science*, **434**, 171-183.

88. Wang, Z. F., Wang, B., Yang, Y. R., and Hu, C. P. (2003) Correlations between gas permeation and free-volume hole properties of polyurethane membranes. *European Polymer Journal*, **39**(12), 2345-2349.

89. Ferreira Marques, M. F., Lopes Gil, C., Gardo, P. M., Kajcsos, Z., de Lima, A. P., Queiroz, D. P., and de Pinho, M. N. (2003) Free-volume studies in polyurethane membranes by positron annihilation spectroscopy. *Radiation Physics and Chemistry*, **68**(3-4), 573-576.

90. Scholten, E., Bromberg, L., Rutledge, G. C., and Alan Hatton, T. (2011) Electrospun polyurethane fibers for absorption of volatile organic compounds from air. *ACS Applied Materials and Interfaces*, **3**(10), 3902-3909.

91. Sadeghi, M., Semsarzadeh, M. A., Barikani, M., and Ghalei, B. (2011) Study on the morphology and gas permeation property of polyurethane membranes. *Journal of Membrane Science*, **385-386**(1), 76-85.

92. Ruaan, R.-C., Ma, W.-C., Chen, S.-H., and Lai, J.-Y. (2001) Microstructure of HTPB-based polyurethane membranes and explanation of their low O_2/N_2 selectivity. *Journal of Applied Polymer Science*, **82**(6), 1307-1314.

93. Semsarzadeh, M. A., Sadeghi, M., Barkani, M., and Moadel, H. (2007) The effect of hard segments on the gas separation properties of polyurethane membranes. *Iranian Polymer Journal*, **16**(12), 819-827.

94. Galland, G., and Lam, T. (1993) Permeability and diffusion of gases in segmented polyurethanes: structure–properties relations. *Journal of Applied Polymer Science*, **50**(6), 1041-1058.

95. Huang, R., Chari, P., Tseng, J.-K., Zhang, G., Cox, M., and Maia, J. M. (2015) Microconfinement effect on gas barrier and mechanical properties of multilayer rigid/soft thermoplastic polyurethane films. *Journal of Applied Polymer Science*, **132**(18), DOI: 10.1002/app.41849.

96. Ho, B. P., Choon, K. K., and Young, M. L. (2002) Gas separation properties of polysiloxane/polyether mixed soft segment urethane urea membranes. *Journal of Membrane Science*, **204**(1-2), 257-269.

97. Park, H. B., and Lee, Y. M. (2002) Separation of toluene/nitrogen through segmented polyurethane and polyurethane urea membranes with different soft segments. *Journal of Membrane Science*, **197**(1-2), 283-296.

98. Wolinska-Grabczyk, A., and Jankowski, A. (2007) Gas transport properties of segmented polyurethanes varying in the kind of soft segments. *Separation and Purification Technology*, **57**(3), 413-417.

99. Gnanasekaran, D., and Reddy, B. S. (2014) A facile synthesis of mixed soft-segmented poly(urethane-imide) - polyhedral oligomeric silsesquioxone hybrid nanocomposites and study of their structure-transport properties. *Polymer International*, **63**(3), 507-513.

100. Li, H., Freeman, B. D., and Ekiner, O. M. (2011) Gas permeation properties of poly(urethane-urea)s containing different polyethers. *Journal of Membrane Science*, **369**(1-2), 49-58.

101. Matsunaga, K., Sato, K., Tajima, M., and Yoshida, Y. (2005) Gas permeability of thermoplastic polyurethane elastomers. *Polymer Journal*, **37**(6), 413-417.

102. Teo, L. S., Kuo, J. F., and Chen, C. Y. (1998) Study on the morphology and permeation property of amine group-contained polyurethanes. *Polymer*, **39**(15), 3355-3364.

103. Sadeghi, M., Semsarzadeh, M. A., Barikani, M., and Ghalei, B. (2010) The effect of urethane and urea content on the gas permeation properties of poly(urethane-urea) membranes. *Journal of Membrane Science*, **354**(1-2), 40-47.

104. Amani, M., Amjad-Iranagh, S., Golzar, K., Sadeghi, G. M. M., and Modarress, H. (2014) Study of nanostructure characterizations and gas separation properties of poly (urethane–urea) s membranes by molecular dynamics simulation. *Journal of Membrane Science*, **462**, 28-41.

105. Lin, W.-H., and Chung, T.-S. (2001) Gas permeability, diffusivity, solubility, and aging characteristics of 6FDA-durene polyimide membranes. *Journal of Membrane Science*, **186**(2), 183-193.

106. Marchese, J., Garis, E., Anson, M., Ochoa, N. A., and Pagliero, C. (2003) Gas sorption, permeation and separation of ABS copolymer membrane. *Journal of Membrane Science*, **221**(1), 185-197.

107. Wang, Z. F., Wang, B., Ding, X. M., Zhang, M., Liu, L. M., Qi, N., and Hu, J. L. (2004) Effect of temperature and structure on the free volume and water vapor permeability in hydrophilic polyurethanes. *Journal of Membrane Science*, **241**(2), 355-361.

108. Yang, S. J., Yang, J. M., and Lin, H. T. (2005) Evaluation of poly(N-isopropylacrylamide) modified hydroxyl-terminated polybutadiene based polyurethane membrane. *Journal of Membrane Science*, **258**(1), 97-105.

109. de Sales, J. A., Patricio, P. S. O., Machado, J. C., Silca, G. G., and Windmoller, D. (2008) Systematic investigation of the effects of temperature and pressure on gas transport through polyurethane/poly(methylmethacrylate) phase-separated blends. *Journal of Membrane Science*, **310**(1-2), 129-140.

110. Kanehashi, S., and Nagai, K. (2005) Analysis of dual-mode model parameters for gas sorption in glassy polymers. *Journal of Membrane Science*, **253**(1), 117-138.

111. Chiou, J., and Paul, D. (1986) Sorption and transport of inert gases in PVF2/PMMA blends. *Journal of Applied Polymer Science*, **32**(5), 4793-4814.

112. Li, X.-G., Kresse, I., Xu, Z.-K., and Springer, J. (2001) Effect of temperature and pressure on gas transport in ethyl cellulose membrane. *Polymer*, **42**(16), 6801-6810.

113.Mokdad, A., and Dubault, A. (2000) Transport properties of carbon dioxide through single-phase polystyrene/poly (vinylmethylether) blends. *Journal of Membrane Science*, **172**(1), 1-8.

114.Li, X.-G., Kresse, I., Springer, J., Nissen, J., and Yang, Y.-L. (2001) Morphology and gas permselectivity of blend membranes of polyvinylpyridine with ethylcellulose. *Polymer*, **42**(16), 6859-6869.

115.Chung, T. S., Cao, C., and Wang, R. (2004) Pressure and temperature dependence of the gas-transport properties of dense poly [2, 6-toluene-2, 2-bis (3, 4-dicarboxylphenyl) hexafluoropropane diimide] membranes. *Journal of Polymer Science, Part B: Polymer Physics*, **42**(2), 354-364.

116.Lopez-Gonzalez, M. M., Compan, V., Saiz, E., Riande, E., and Guzman, J. (2005) Effect of the upstream pressure on gas transport in poly (ether-imide) films. *Journal of Membrane Science*, **253**(1), 175-181.

117.Hu, C.-C., Lee, K.-R., Ruaan, R.-C., Jean, Y. C., and Lai, J.-Y. (2006) Gas separation properties in cyclic olefin copolymer membrane studied by positron annihilation, sorption, and gas permeation. *Journal of Membrane Science*, **274**(1), 192-199.

118.Lin, H., and Freeman, B. D. (2004) Gas solubility, diffusivity and permeability in poly (ethylene oxide). *Journal of Membrane Science*, **239**(1), 105-117.

119.Madhavan, K., and Reddy, B. S. R. (2006) Poly(dimethylsiloxane-urethane) membranes: Effect of hard segment in urethane on gas transport properties. *Journal of Membrane Science*, **283**(1-2), 357-365.

120.Raharjo, R. D., Lin, H., Sanders, D. F., Freeman, B. D., Kalakkunnath, S., and Kalika, D. S. (2006) Relation between network structure and gas transport in crosslinked poly (propylene glycol diacrylate). *Journal of Membrane Science*, **283**(1), 253-265.

121.Teo, L. S., Chen, C. Y., and Kuo, J. F. (1998) The gas transport properties of amine-containing polyurethane and poly(urethane-urea) membranes. *Journal of Membrane Science*, **141**(1), 91-99.

122.Knight, P., and Lyman, D. (1984) Gas permeability of various block copolyether-urethanes. *Journal of Membrane Science*, **17**(3), 245-254.

123.Semsarzadeh, M. A., Sadeghi, M., and Barikani, M. (2008) Effect of chain extender length on gas permeation properties of polyurethane membranes. *Iranian Polymer Journal*, **17**(6), 431-440.

124.Damian, C., Espuche, E., Escoubes, M., Cuney, S., and Pascault, J. P. (1997) Gas permeability of model polyurethane networks and hybrid organic-inorganic materials: Relations with morphology. *Journal of Applied Polymer Science*, **65**(12), 2579-2587.

125.Mannan, H. A., Mukhtar, H., Murugesan, T., Nasir, R., Mohshim, D. F., and Mushtaq, A. (2013) Recent applications of polymer blends in gas separation membranes. *Chemical engineering & Technology*, **36**(11), 1838-1846.

126.Robeson, L. M. (2010) Polymer blends in membrane transport processes. *Industrial & Engineering Chemistry Research*, **49**(23), 11859-11865.

127.Kim, M. J., Sea, B., Youm, K.-H., and Lee, K.-H. (2006) Morphology and carbon dioxide transport properties of polyurethane blend membranes. *Desalination*, **193**(1-3), 43-50.

128.Ghalei, B., and Semsarzadeh, M.-A. (2007) A novel nano structured blend membrane for gas separation. *Macromolecular Symposia*, **249-250**, 330-335.

129. Semsarzadeh, M. A., and Ghalei, B. (2012) Characterization and gas permeability of polyurethane and polyvinyl acetate blend membranes with polyethylene oxide-polypropylene oxide block copolymer. *Journal of Membrane Science*, **401-402**, 97-108.

130. Saedi, S., Madaeni, S. S., Hassanzadeh, K., Shamsabadi, A. A., and Laki. S. (2014) The effect of polyurethane on the structure and performance of PES membrane for separation of carbon dioxide from methane. *Journal of Industrial and Engineering Chemistry*, **20**(4), 1916-1929.

131. Erucar, I., Yilmaz, G., and Keskin, S. (2013) Recent advances in metal-organic framework-based mixed matrix membranes. *Chemistry-an Asian Journal*, **8**(8), 1692-1704.

132. Lue, S. J., Su, I.-M., Lee, D.-T., Chen, H.-Y., Shih, C.-M., Hu, C.-C., Jean, Y. C., and Lai. J.-Y. (2011) Correlation between free-volume properties and pervaporative flux of polyurethane– zeolite composites on organic solvent mixtures. *The Journal of Physical Chemistry B*, **115**(12), 2947-2958.

133. Tirouni, I., Sadeghi, M., and Pakizeh, M. (2015) Separation of C_3H_8 and C_2H_6 from CH_4 in polyurethane–zeolite 4Å and ZSM-5 mixed matrix membranes. *Separation and Purification Technology*, **141**, 394-402.

134. Ciobanu, G., Carja, G., and Ciobanu, O. (2007) Preparation and characterization of polymer-zeolite nanocomposite membranes. *Materials Science & Engineering C*, **27**(5-8), 1138-1140.

135. Khudyakov, I. V., Zopf, R. D., and Turro, N. J. (2009) Polyurethane nanocomposites. *Designed Monomers & Polymers*, **12**(4), 279-290.

136. George, G., Bhoria, N., and Mittal, V. (2016) *Supported UV polymerized ionic liquid membranes with block copolymer.* Journal of Membrane Science and Technology, **6**(166), DOI: 10.4172/2155-9589.1000166.

137. Petrovic, Z. S., Javni. I., Waddon, A., and Banhegyi, G. (2000) Structure and properties of polyurethane–silica nanocomposites. *Journal of Applied Polymer Science*, **76**(2), 133-151.

138. Sadeghi, M., Semsarzadeh, M. A., Barikani, M., and Chenar, M. P. Gas separation properties of polyether-based polyurethane–silica nanocomposite membranes. *Journal of Membrane Science*, **376**(1), 188-195.

139. Hassanajili, S., Masoudi, E., Karimi, G., and Khademi, M. (2013) Mixed matrix membranes based on polyetherurethane and polyesterurethane containing silica nanoparticles for separation of CO_2/CH_4 gases. *Separation and Purification Technology*, **116**, 1-12.

140. Sadeghi, M., Talakesh, M. M., Ghalei, B., and Shafiei, M. (2013) Preparation, characterization and gas permeation properties of a polycaprolactone based polyurethane-silica nanocomposite membrane. *Journal of Membrane Science*, **427**, 21-29.

141. Semsarzadeh, M. A., and Ghalei, B. (2013) Preparation, characterization and gas permeation properties of polyurethane-silica/polyvinyl alcohol mixed matrix membranes. *Journal of Membrane Science*, **432**, 115-125.

142. Hassanajili, S., Khademi, M., and Keshavarz, P. (2014) Influence of various types of silica nanoparticles on permeation properties of polyurethane/silica mixed matrix membranes. *Journal of Membrane Science*, **453**, 369-383.

143. Khosravi, A., Sadeghi, M., Banadkohi, H. Z., and Talakesh, M. M. (2014) Polyurethane-silica nanocomposite membranes for separation of propane/methane and ethane/methane. *Industrial and Engineering Chemistry Research*, **53**(5), 2011-2021.

144. Ribeiro, T., Baleizao, C., and Farinha, J. P. S. (2014) Functional films from silica/polymer manoparticles. *Materials*, **7**(5), 3881-3900.

145. Semsarzadeh, M. A., Ghalei, B., Fardi, M., Esmaeeli, M., and Vakili, E. (2014) Structural and transport properties of polydimethylsiloxane based polyurethane/silica particles mixed matrix membranes for gas separation. *Korean Journal of Chemical Engineering*, **31**(5), 841-848.

146. Osman, M. A., Mittal, V., Morbidelli, M., and Suter, U. W. (2003) Polyurethane adhesive nanocomposites as gas permeation barrier. *Macromolecules*, **36**(26), 9851-9858.

147. Mittal, V. (2014) Polyurethane-bentonite nanocomposites: morphology and oxygen permeation. *Advances in Polymer Technology*, **33**(3), DOI:10.1002/adv.21416.

148. Osman, M. A., Mittal, V., and Lusti, H. R. (2004) The aspect ratio and gas permeation in polymer-layered silicate nanocomposites. *Macromolecular Rapid Communications*, **25**, 1145-1149.

149. Barboza, E. M., Delpech, M. C., Garcia, M. E. F., and Pimenta, F. D. (2014) Evaluation of carbon dioxide gas barrier properties of membranes obtained from aqueous dispersions based on polyurethane and clay. *Polimeros*, **24**(1), 94-100.

150. Shamini, G., and Yusoh, K. (2014) Gas permeability properties of thermoplastic polyurethane modified clay nanocomposites. *International Journal of Chemical Engineering and Applications*, **5**(1), 64-68.

151. Ameri, E., Sadeghi, M., Zarei, N., and Pournaghshband, A. (2015) Enhancement of the gas separation properties of polyurethane membranes by alumina nanoparticles. *Journal of Membrane Science*, **479**, 11-19.

152. Chen, S.-H., Wu, T.-H., Ruaan, R.-C., and Lai, J.-Y. (1998) Effect of top layer swelling on the oxygen/nitrogen separation by surface modified polyurethane membranes. *Journal of Membrane Science*, **141**(2), 255-264.

153. Chen, Y., Wang, R., Zhou, J., Fan, H., Shi, B. (2011) Membrane formation temperature-dependent gas transport through thermo-sensitive polyurethane containing in situ-generated TiO_2 nanoparticles. *Polymer*, **52**(8), 1856-1867.

154. Sadeghi, M., Afarani, H. T., and Tarashi, Z. (2015) Preparation and investigation of the gas separation properties of polyurethane-TiO_2 nanocomposite membranes. *Korean Journal of Chemical Engineering*, **32**(1), 97-103.

155. Huang, S.-L., Ruaan, R.-C., and Lai, J.-Y. (1997) Gas permeability of cupric ion containing HTPB based polyurethane membranes. *Journal of Membrane Science*, **123**(1), 71-79.

156. Sen, R., Zhao, B., Perea, D., Itkis, M. E., Hu, H., Love. J., Bekyarova, E., and Haddon, R. C. (2004) Preparation of single-walled carbon nanotube reinforced polystyrene and polyurethane nanofibers and membranes by electrospinning. *Nano Letters*, **4**(3), 459-464.

157. Deng, J., Zhang, X., Wang, K., Zou, H., Zhang, Q., and Fu, Q. (2007) Synthesis and properties of poly (ether urethane) membranes filled with isophorone diisocyanate-grafted carbon nanotubes. *Journal of Membrane Science*, **288**(1), 261-267.

158.Yi, D. H., Yoo, H. J., Mahapatra, S. S., Kim, Y. A., and Cho, J. W. (2014) The synergistic effect of the combined thin multi-walled carbon nanotubes and reduced graphene oxides on photothermally actuated shape memory polyurethane composites. *Journal of Colloid and Interface Science*, **432**, 128-134.
159.Kumar, M., Chung, J. S., Kong, B.-S., Kim, E. J., and Hui, S. H. (2013) Synthesis of graphene-polyurethane nanocomposite using highly functionalized graphene oxide as pseudo-crosslinker. *Materials Letters*, **106**, 319-321.
160.Rana, S., Cho, J. W., and Tan, L. P. (2013) Graphene-crosslinked polyurethane block copolymer nanocomposites with enhanced mechanical, electrical, and shape memory properties. *RSC Advances*, **3**(33), 13796-13803.
161.Wang, X., Xing, W., Song, L., Yu, B., Hu, Y., and Yeoh, G. H. (2013) Preparation of UV-curable functionalized graphene/polyurethane acrylate nanocomposite with enhanced thermal and mechanical behaviors. *Reactive & Functional Polymers*, **73**(6), 854-858.
162.Dai, Y. T., Qiu, F. X., Xu., J. C., Yu, Z. P., Yang, P. F., Xu, B. B., Jiang, Y., and Yang, D. Y. (2014) Preparation and properties of UV-curable waterborne graphene oxide/polyurethane-acrylate composites. *Plastics Rubber and Composites*, **43**(2), 54-62.
163.Han, S., and Chun, B. C. (2014) Preparation of polyurethane nanocomposites via covalent incorporation of functionalized graphene and its shape memory effect. *Composites, Part A -Applied Science and Manufacturing*, **58**, 65-72.
164.Ma, W.-S., Wu, L., Yang, F., Wang, S.-F. (2014) Non-covalently modified reduced graphene oxide/polyurethane nanocomposites with good mechanical and thermal properties. *Journal of Materials Science*, **49**(2), 562-571.
165.Pokharel, P., and Lee, D. S. (2014) Thermal and mechanical properties of reduced graphene oxide/polyurethane nanocomposite. *Journal of Nanoscience and Nanotechnology*, **14**(8), 5718-5721.
166.Sadasivuni, K. K., Ponnamma, D., Kumar, B., Strankowski, M., Cardinaels, R., Moldenaers, P., Thomas, S., and Grohens, Y. (2014) Dielectric properties of modified graphene oxide filled polyurethane nanocomposites and its correlation with rheology. *Composites Science and Technology*, **104**, 18-25.
167.Khanna, S. K., and Phan, H. T. T. (2015) High strain rate behavior of graphene reinforced polyurethane composites. *Journal of Engineering Materials and Technology*, **137**(2), 021005.
168.Li, X., Deng, H., Li, Z., Xiu, H., Qi, X., Zhang, Q., Wang, K., Chen, F., and Fu, Q. (2015) Graphene/thermoplastic polyurethane nanocomposites: Surface modification of graphene through oxidation, polyvinyl pyrrolidone coating and reduction. *Composites, Part A -Applied Science and Manufacturing*, **68**, 264-275.
169.Luo, X., Zhang, P., Ren, J., Liu, R., Feng, J., and Ge, B. (2015) Preparation and properties of functionalized graphene/waterborne polyurethane composites with highly hydrophobic. *Journal of Applied Polymer Science*, **132**(23), DOI:10.1002/app.42005.
170.Pokharel, P., Choi, S., and Lee, D. S. (2015) The effect of hard segment length on the thermal and mechanical properties of polyurethane/graphene oxide nanocomposites. *Composites, Part A - Applied Science and Manufacturing*, **69**, 168-177.

171.Pokharel, P., Lee, S. H., and Lee, D. S. (2015) Thermal, mechanical, and electrical properties of graphene manoplatelet/graphene oxide/polyurethane hybrid nanocomposite. *Journal of Nanoscience and Nanotechnology*, **15**(1), 211-214.

172.Thakur, S., and Karak, N. (2015) A tough, smart elastomeric bio-based hyperbranched polyurethane nanocomposite. *New Journal of Chemistry*, **39**(3), 2146-2154.

173.Yu, B., Shi, Y., Yuan, B., Liu, L., Yang, H., Tai, Q., Lo, S., Song, L., and Hu, Y. (2015) Click-chemistry approach for graphene modification: effective reinforcement of UV-curable functionalized graphene/polyurethane acrylate nanocomposites. *RSC Advances*, **5**(18), 13502-13506.

174.Kim, H., Miura, Y., and Macosko, C. W. Graphene/polyurethane nanocomposites for improved gas barrier and electrical conductivity. *Chemistry of Materials*, **22**(11), 3441-3450.

175.Xiang, C., Cox, P. J., Kukovecz, A., Genorio, B., Hashim, D. P., Yan, Z., Peng, Z., Hwang, C.-C., Ruan, G., Samuel, E. L. G., Sudeep, P. M., Konya, Z., Vajtai, R., Ajayan, P. M., and Tour, J. M. (2013) Functionalized low defect graphene nanoribbons and polyurethane composite film for improved gas barrier and mechanical performances. *ACS Nano*, **7**(11), 10380-10386.

176.Kaveh, P., Mortezaei, M., Barikani, M., and Khanbabaei, G. (2014) Low-temperature flexible polyurethane/graphene oxide nanocomposites: effect of polyols and graphene oxide on physicomechanical properties and gas permeability. *Polymer-Plastics Technology and Engineering*, **53**(3), 278-289.

177.Di Landro, L., Pegoraro, M., and Bordogna, L. (1991) Interactions of polyether-polyurethanes with water vapour and water-methane separation selectivity. *Journal of Membrane Science*, **64**(3), 229-236.

178.Chatterjee, G., Houde, A. A., and Stern, S. A. (1997) Poly(ether urethane) and poly(ether urethane urea) membranes with high H_2S/CH_4 selectivity. *Journal of Membrane Science*, **135**(1), 99-106.

179.Simmons, J. W. (2003) Novel Block Polyurethane-ether and Polyurea-ether Gas Separation Membranes. US Patent US 6843829 B2.

180.George, G., and Mittal, V. (2015) CO_2 & H_2S removal from natural gas using polymer membranes. Virtual Special Issue, Journal of Membrane Science. Online: https://www.journals.elsevier.com/journal-of-membrane-science/virtual-special-issues/co2-h2s-removal-from-natural-gas-using-polymer-membranes (assessed 27th February 2017).

181.Ponangi, R. P., and Pintauro, P. N. (1996) Separation of volatile organic compounds from dry and humidified nitrogen using polyurethane membranes. *Industrial and Engineering Chemistry Research*, **35**(8), 2756-2765.

182.Ponangi, R., Pintauro, P. N., and De Kee, D. (2000) Free volume analysis of organic vapor diffusion in polyurethane membranes. *Journal of Membrane Science*, **178**(1), 151-164.

8

Polymer Modified/Enhanced Adsorbents for Gas Adsorption and Sweetening

8.1 Introduction

In order to satisfy the increase in global energy demand along with the need to reduce the environmental impacts of global greenhouse gas emissions, it is continuously strived to develop promising energy substitutes. Thus, some of the concerns of the present day energy technologies are production of adequate energy with good quality, environmental feasibility and economical viability. Natural gas, which is predominantly hydrocarbon in nature, is one of the prominent energy sources for many decades, with its usage as a fuel progressively increasing worldwide [1]. Commercial natural gas has several advantages such as high heating value and low carbon content per unit mass. However, along with methane, raw natural gas contains contaminants such as higher hydrocarbons, acidic gases (hydrogen sulfide (H_2S) and carbon dioxide (CO_2)), water and several other minute impurities. The existence of acid gases in the natural gas leads to shrinkage in natural gas heating value and also causes corrosion in facilities [2]. Hence, the removal of acid gases from natural gas is necessary prior to its transportation through pipelines [3, 4]. The ultimate level of CO_2 and H_2S are generally limited to 2 mol% and 4 ppm respectively by the regulations [5]. Thus, in order to meet the sales gas specifications and to keep the values of acid gases within the limits, the natural gas streams are treated using various processes. The acid gas removal process, which is generally termed as gas sweetening process, is employed for natural gas purification from acid gases like H_2S and CO_2. In this stage, the acid gases are separated using various technologies, depending on the operating conditions and levels of H_2S and CO_2 present in the natural gas [2].

Separation of acid gases from natural gas streams is very challenging and demands the use of special materials which can resist corrosive environments. Different technologies commonly used for natural gas purification are adsorption and absorption processes, membrane separation and cryogenic condensation. Various factors to be considered for the selection of appropriate technology are nature and quantity of contaminants present in feed gas,

Haleema Saleem and Vikas Mittal*, The Petroleum Institute (part of Khalifa University of Science and Technology), Abu Dhabi, UAE
*Current address: Bletchinton, Wellington County, Australia

quantity of hydrocarbons in the gas, targeted acid gas removal capacity, pipe-line specification, desired selectivity, operating cost and feed gas processing conditions [6]. Conventional liquid amine scrubbing process (chemical absorption) is one of the mostly used acid gas removal processes, where the separation of CO_2 and H_2S from gas mixtures like natural gas takes place through chemical reactions of amines with acid gases [7]. In this process, the sour gas is introduced into an absorber unit where it comes in contact with an aqueous solution of primary or secondary alkylamine e. g. monoethanolamine (MEA). As a result, CO_2 reacts with amine to form a solution containing corresponding alkyl-ammonium carbamate salt [8]. MEA has the ability to separate both H_2S as well as CO_2 from gas streams. In spite of the fact that amine scrubbing offers high selectivity of CO_2 and H_2S, the major limitation of this process is very high energy demand because of the need of regeneration of amine solution. This requirement is compounded further due to the strong interactions between acid gas molecules and amine. The energy penalty caused by MEA absorption process is approximately 15-37% of energy production of an ordinary power plant [9]. Due to the high capital and operational cost of this technology, the development of alternate low cost and environmentally effective technologies has become very important for the effective removal of acid gases.

Adsorption is regarded as another promising technology for the effective removal of CO_2 and H_2S from natural gas, especially to remove ppm level contents. Here, the acid gases interact with the surface of highly porous solid material having high surface area. The gases can adsorb physically (physisorbents) as well as chemically (chemisorbents) on the surface of adsorbents. Physisorbents can remarkably reduce capital and operating cost as these can reversibly adsorb and desorb CO_2 at near ambient conditions. Activated carbons are commonly used adsorbents for gases and vapors. The major advantages of using adsorbents are operation simplicity and the opportunity of concurrent acid removal as well as gas dehydration. Surface modifications of these adsorbents with polymers and organic molecules is generally carried out in order to enhance their performance.

In this chapter, various polymer incorporated adsorption systems have been reviewed for effective separation of acid gases from natural gas. Special emphasis has also been given to the effect of polymers with basic nitrogen functionalities on the efficiency and selectivity of the adsorbents for acid gases. Various observations in terms of advantages and disadvantages of using different polymers have also been made. Thus, an up-to-date analysis of developing concepts, experimental results as well as theoretical predictions of the polymer enhanced adsorbents for acid gas removal, with particular emphasis on CO_2, H_2S, SO_x and NO_x, are highlighted.

8.2 Polymer Enhanced Adsorbents for CO_2 Capture

For ideal CO_2 capturing technology, Khatri *et al.* [10] suggested three key aspects as (1) high CO_2 adsorption capacity which should be greater than 1 $mmol.g^{-1}$ (2) energy necessity for regeneration should be less compared to aqueous amine process, (3) long term regeneration capacity in a power plant flue gas atmosphere. Several research studies have been performed on different types of adsorbents for the separation of CO_2, including amine-supported adsorbents [11], metal organic frameworks (MOF) [12], carbon based adsorbents [13] and zeolites [14]. Carbon based adsorbents offer plentiful advantages such as low regeneration energy, fast adsorption kinetics, high adsorption capacity and low cost initial sources [15]. Nano-casting is considered as one of the dominant methods for the preparation of mesoporous carbon materials, where the mesopores contribute remarkably to the adsorption process.

For CO_2 adsorption, the presence of basic nitrogen functionalities is very important, thus, requiring to maximize the number of nitrogen functional groups in the surface modification of the adsorbents. As a result, amine functionalized adsorbents have gained considerable research interest because of their high efficiency as well as selectivity for CO_2 removal from gas mixtures. Generally, amine based CO_2 removal functionalities are added into porous supports by employing two approaches. First method is based on changing the carbon matrix surface chemistry by consolidating hetero-atoms like nitrogen to increase the characteristic adsorbate-adsorbent interactions. On the other hand, the second method employs the modification of the porous substrate surface chemistry by impregnation with liquid organic polymers, such as polyethyleneimine (PEI) [16,17]. Many research studies have reported the use of PEIs for enhancing the CO_2 adsorption capacity [16-19]. These polymers have higher CO_2 adsorption capacity due to the high number of amine groups present in the polymer structure. The primary as well as secondary amino groups present in PEI undergo reaction with CO_2 to generate carbamates. These carbamates react further to form bicarbonates in the presence of water, thereby, enhancing the CO_2 adsorption capacity. As expected, long polymer chains with amine groups grafted on each repeat unit exhibit higher CO_2 adsorption capacity than the short polymer chains. This is due to the presence of large number of CO_2 anchoring sites available to form carbamate group. However, these should meet some requirements for maximizing their ability as a polymeric adsorbent and enhancing the CO_2 removal capacity.

Post combustion CO_2 removal by using adsorbents consisting of basic nitrogen functional groups has also proved to be very efficient, as the addition of nitrogen groups to the carbon structure enhances the specific interaction between the CO_2 molecules and carbon surface [20]. In nano-casting approach, an inorganic removable template such as silica is employed to prepare porous adsorbent polymers having high surface area. Here, high temperature activation is not required for preparing adsorbents, and nitrogen is included inside the polymer matrix. On dissolution of inorganic template, the polymer remains as a mirror image and inherits the template porosity.

8.2.1 Carbon Adsorbents

Carbon based materials are effective for gas sorption as well as storage due to abundance, tunable surface area, strong pore structure, chemical stability and easy preparation at industrial scale. Different types of carbon based materials have been studied for CO_2 removal such as activated carbons [21], carbon molecular sieves [22], carbon nanotubes [23] and graphene [24]. Activated carbons are suitable for CO_2 capture because of advantages like high adsorption capacity at high pressure, low cost and are easy regeneration. However, the use of naturally occurring precursor for the activated carbon preparation limits the strength, purity and physical form of end products. This limitation can be overcome by the application of polymeric precursors. Here, the purity and reproducibility can be controlled, and the physical form of end products can be tailored during. The surface properties of activated carbons can be optimized by controlling the synthetic process of polymers, thereby, increasing the CO_2 adsorption capacity. Recently, preparation of activated carbons from synthetic materials has acquired immense research attention. Recently, few studies have reported increased adsorption capacity of activated carbon for H_2S [25] and CO_2 [26] by synthesizing adsorbents using nitrogen containing compounds. Carbon adsorbents consisting of nitrogen can be generated using nitrogen accommodating compounds like melamine, aniline, acetonitrile, etc., along with templates such as zeoliteY, SBA-45 and MCM-41 [27]. The cost of such systems is dependent on the cost of the precursor as well as the carbon yield during the activation and carbonization stages. Phenolic resin (PF), prepared from the phenol and formaldehyde, is one of the most commonly used types of precursors. Phenolic resin based activated carbons provide benefits such as good physical strength and very low level of impurities. In one such study, Martin *et al.* [28] employed different types of phenolic resins as the precursor materials, and ultimate CO_2 adsorption capacities till 10.8 wt% were obtained. It was observed that the narrow micropores were extremely active towards CO_2 separation at atmospheric pressure. It was concluded that for the pre-combustion CO_2 removal, the PF resin derived activated carbons exhibited higher potential as adsorbents. In another work by Martin *et al.* [29], the pre-combustion CO_2 adsorption capacity of two PF resin based activated carbons (Resol PFNA and Novolac PFCLA) were analyzed under dynamic and static conditions. Both adsorbents exhibited similar CO_2 adsorption under equilibrium at 298 K and 15 bar. For a mixture of 60% H_2 and 40% CO_2, at the aforementioned pressure and temperature conditions, PFNA exhibited higher CO_2/H_2 selectivity than PFCLA, due to the narrower average pore width. At 298 K, the breakthrough tests were performed at a total pressure of 15 bar in a fixed bed column and the CO_2 adsorption capacity was observed to be 6.5 mmol.g^{-1} and 6.4 mmol.g^{-1} for PFCLA and PFNA, respectively. However, at 313K, a decrease in the CO_2 adsorption capacity was observed (5.3 mmol.g^{-1} for PFCLA, 5.8 mmol.g^{-1} for PFNA). Further, both activated carbon adsorbents displayed excellent cyclability as well as regenerability over successive pressure swing adsorption (PSA) cycles. Overall, it was concluded that the adsorbents exhibited higher potential for the application of pre-combustion CO_2 adsorption by PSA process.

Long *et al.* [30] generated activated carbons through physical activation process from phenol-melamine-formaldehyde gels. The activated carbons had a specific area of about 525–685 m^2.g^{-1}. In a similar study by Ru-Ling *et al.* [31] for the CO_2 adsorption, highly porous activated carbon was prepared from melamine modified phenol formaldehyde (MPF) resin by steam activation at different ranges of activation temperature (700 °C - 950 °C) at atmospheric pressure. The modification of PF resins helped to provide high nitrogen content for activated carbon. The vessel containing MPF resin beads was placed in an oven and the temperature was increased at a rate of 5 °C.min^{-1} until the activation temperatures were attained (700 °C, 750 °C, 800 °C, 850 °C, 900 °C and 950 °C). Subsequently, activation was carried out using deionized water. It was observed that on enhancing the activation temperature from 700 °C to 950 °C, the specific surface area developed from 382 m^2.g^{-1} to 1439 m^2.g^{-1}. Further, the elemental and FTIR analysis confirmed on increasing the activation temperature, nitrogen atom content present in the MPF resin decreased. The CO_2 adsorbed amount on MPF850 at 1 atm and 0 °C was observed to be 6.71 mol.kg^{-1}. The adsorption capacity of MPF 850 remained unaltered even after five adsorption/desorption cycles at atmospheric pressure.

Pevida *et al.* [32] generated a range of highly porous melamine-formaldehyde adsorbents, without any high temperature activation for CO_2 removal. Adsorbents were prepared by template synthesis of MF resin and carbonization at varying temperatures, using silica as template material. The adsorbents were characterized in terms of their elemental composition, textural properties as well as surface chemistry. The materials contained 42 wt% nitrogen and had a surface area of 880 $m^2.g^{-1}$. The synthesis method was observed to be very efficient as the adsorbents exhibited CO_2 capture ability of upto 2.25 $mmol.g^{-1}$ at 25 °C, thereby exceeding the adsorption capacity of several commercial activated carbons. The adsorption capacity was observed to be highest at 25°C and eventually decreased with increasing the temperature. The behavior was different from the performance of PEI based adsorbents, where the adsorption capacity remained constant until 90°C [33]. This was due to the difference in chemistry of the nitrogen present in PEI based adsorbents and MF resin. In another study, Pevida *et al.* [34] also developed nitrogen enriched activated carbon adsorbents by silica templating technique using melamine formaldehyde resin. Four different silica templated samples were generated, and the carbonization was carried out by heating the samples up to 400 °C, 500 °C, 600 °C and 700 °C for 1 h. It was observed that both texture as well as chemistry of the adsorbents played a remarkable role in determining the CO_2 removal capacity. It was confirmed that associating a convenient textural advancement with an agreeable chemistry resulted in an improvement in CO_2 adsorbing capacity of the adsorbents. The 600 °C carbonized adsorbents had the greatest adsorption capacity of 2.25 $mmol.g^{-1}$ under pure CO_2 atmosphere at a temperature of 25 °C. The observed value was remarkably higher when compared to the previous values attained for MF polymer based sorbents without silica template (ultimate CO_2 removal capacity of 1.02 $mmol.g^{-1}$ at 25 °C [20]). The addition of silica template enhanced the textural advancement of MF based adsorbents which enhanced the CO_2 adsorption.

Hao *et al.* [35] pyrolyzed copolymer of formaldehyde, lysine and resorcinol for the preparation of nitrogen doped porous carbon monolith. The CO_2 adsorption capacity of the generated material was observed to be 3.13 $mmol.g^{-1}$ for 100% CO_2 at 1 atm and 25 °C. Drage *et al.* [36] prepared nitrogen enriched carbons from UF and MF resins, which were polymerized in the presence of K_2CO_3, the chemical activation agent. In this system, activation was performed over a range of temperatures. Further, carbons generated by chemical activation were observed to be more efficient for the CO_2 removal, when compared to the carbons prepared by CO_2 physical activation. The adsorbents, generated by the chemical activation of UF resin at 500°C, were able to capture 8 wt%

CO_2 at a temperature of 25 °C. This exhibited the usefulness of surface chemistry of melamine functionalized carbon for the preparation of effective CO_2 adsorbents. In another study, Goel *et al.* [37] prepared a series of mesoporous carbon adsorbents with high nitrogen content, through nano-casting method. The mesoporous silica was employed as the template while melamine formaldehyde resin served as the precursor. A varying carbonization temperature range of 400-700 °C was used for the preparation of the adsorbents. The highest CO_2 adsorption capacity of about 0.83 mmol.g^{-1} was observed at 12.5% CO_2 (rest nitrogen) atmosphere at a temperature of 30 °C. Exothermic process was observed to take place and the adsorption capacity decreased with increasing temperature.

Wang *et al.* [38] reported the development of functional adsorbents by using commercially available carbon materials, such as activated carbons and carbon blacks, to prepare carbon-based "molecular basket" sorbent (CB-MBS) by modifying the surface with CO_2-philic PEI. Molecular basket sorbent (MBS) are solid amine materials which consists of mesoporous or nanoporous materials along with functional polymers. CO_2 sorption capacity of 154 mg-CO_2/g-sorb was observed for carbon black based MBS with 65 wt% PEI. The carbon black-supported PEI adsorbents were observed to have high sorption capacity due to high pore volume and large pore size. Figure 8.1 also exhibits CO_2 sorp-

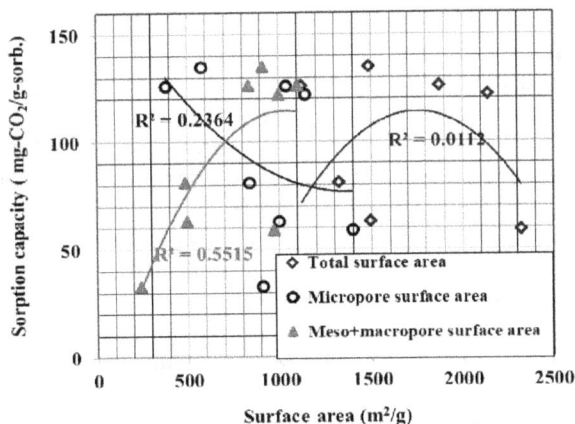

Figure 8.1 CO_2 sorption capacity of carbon black based MBS as a function of surface area. The capacity was measured by TGA at 75 °C and ambient pressure under a pure CO_2 flow at a flow rate of 100 mL/min. Reproduced from Reference 38 with permission from American Chemical Society.

tion capacity of carbon black based MBS as a function of surface area. In an earlier study to gain enhanced adsorption capacity of the PEI modified adsorbents, Ma *et al.* [39,40] also used molecular basket system using silica support as basket, which helped in retaining large quantity of PEI. Figure 8.2 also exhibits sorption breakthrough curves for CO_2 and H_2S in both single-stage and two-stage sorption processes using the as-generated MBS, when a model gas containing 0.40 v% H_2S, 2.40 v% CO_2, and 20 v% H_2 in N_2 was employed. Another study by the same researchers studied the CO_2 capture capacity of a series of molecular basket sorbents polyethylenimine-SBA-15 (mesoporous silica) containing different amount of PEI [41]. The PEI-SBA-15 sorbents were generated using the wet-impregnation techniques. It was observed that the CO_2 sorption by PEI-SBA-15 occurred through chemisorption even under high pressure, while in the system with SBA-15 alone, physisorption occurred at high pressure. Further, it was also confirmed that the best loading for CO_2 adsorption ability depended on the sorption temperature as well as on the pore structure of the supports. The CO_2 dissipation in the PEI bulk played a remarkable role in determining the CO_2 sorption ability as well as adsorption/desorption rate at lower temperature. At 75 °C, the highest CO_2 removal ability of 154.0 mg/g was obtained over PEI-60/SBA-15 sorbents. However, the highest ratio of CO_2 adsorption to the possible amine sites was exhibited by PEI-30/SBA-15, which was due to the improved dispersion inside the pores with lesser CO_2 adsorption barricade in the PEI-30/SBA-15 adsorbent.

Jitong *et al.* [42] reported CO_2 removal by PEI loaded mesoporous carbon sorbents. The sorbents were observed to combine the advantages of impregnated PEI as well as excellent porous characteristics of mesoporous carbons. It was observed that the PEI loading of 65 wt% was ideal due to less mass transfer resistance as well as the highest utilization ratio of amine compound. As the loading was further increased, the benefit of mesoporous structure and large pore volume of the support material was reduced. The carbon based sorbents offered better CO_2 sorption ability, as compared to mesoporous silica supports. Overall, it was concluded that the PEI loaded carbon sorbents provided benefits like high CO_2 removal capacity, good regeneration ability, low regeneration temperature at 100°C and high utilization ratio of the amine groups.

Carbon nanotubes (CNT) possess good ability to separate CO_2 from natural gas because of their high thermal stability, chemical stability and unique physico-chemical properties. CO_2 adsorption in carbonaceous materials like multi-walled carbon nanotubes (MWCNT) occurs near the carbon surface solid because of the physical forces that carbon atoms exert on the CO_2 molecules.

Ngoy *et al.* [43] reported about increasing the CO_2 adsorption behavior by using ethylenediamine for grafting onto polysuccinimide (PSI) to prepare polyaspartamide (PAA), which was covalently linked to MWCNT. An increase in CO_2 adsorption ability was observed due to the fact that primary amine reacted with CO_2 by chemisorption process to form carbamate group. The PAA modified MWNTs exhibited higher CO_2 adsorption capacity of 70 gCO_2/kg as compared to 46.17 gCO_2/kg for PAA, 26.90 gCO_2/kg PSI, and 15.20 gCO_2/kg MWNTs alone.

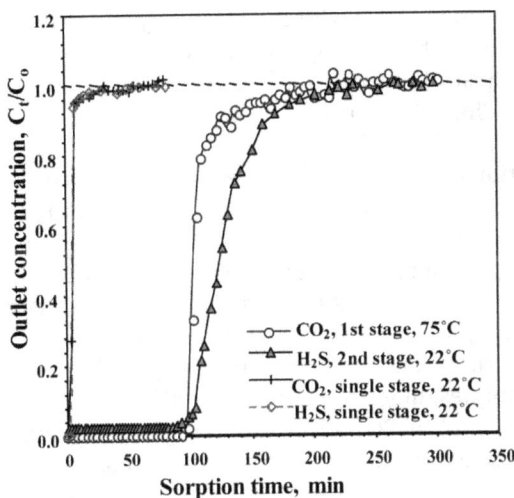

Figure 8.2 Sorption breakthrough curves for CO_2 and H_2S in both single-stage and two-stage sorption processes using the as-generated MBS, when a model gas containing 0.40 v% H_2S, 2.40 v% CO_2, and 20 v% H_2 in N_2 was employed. Reproduced from Reference 39 with permission from American Chemical Society.

Nowadays, graphene, a novel two dimensional carbon nanomaterial, has acquired immense importance as a solid adsorbent for CO_2 removal from flue gases [44]. This is because of its exclusive molecular structure and several interesting properties like large surface area, excellent chemical stability, tunable porosity, excellent thermal conductivity and good mechanical strength. As control on surface functionalization, doping, enhancement of the specific surface area as well as pore structure are of substantial importance to increase the gas adsorption capacity, thus, graphene offers tremendous opportunities for preparing tailor made carbon based adsorbents [45]. As discussed before, amine functionalized adsorbents have enhanced CO_2 adsorption capacity, which increases even further when nitrogen is effectively added to the support. Hence, the amino group surface density has a remarkable role in the CO_2 adsorption capacity. As a result, polyaniline (PANI), which is an abundant source of nitrogen containing groups, was used with graphene to produce graphene-polyaniline nanocomposites with good CO_2 adsorption capacity [46].

In addition to graphene, its oxidized derivative, graphene oxide (GO) has also received significant research attention for development as a CO_2 adsorbent. The incorporation of GO with different polymers may lead to the preparation of CO_2 adsorbents with significantly superior performance. The structural as well as physical properties of GO-PEI porous materials was examined by Sui *et al.* [47]. The authors used a facile approach for the generation of GO-PEI materials having three dimensional (3D) interconnected networks under moderate conditions. The materials possessed lower density along with high adsorption capacity for CO_2. The generated material was regarded as a unique adsorbent as it exhibited combined properties like excellent porous structure together with high amine density for increasing the CO_2 adsorption capacity. The samples had an extensive surface area (472 $m^2.g^{-1}$) as well as total pore volume (1.3 $cm^3.g^{-1}$). Figure 8.3 compares the carbon dioxide adsorption isotherms of GO, hydrothermal reduced graphene (HTG), and GO-PEI porous material (GEPM) samples, along with the adsorption capacity. The results confirmed enhanced level of CO_2 adsorption of the GEPM materials due to the presence of basic sites and large specific surface area.

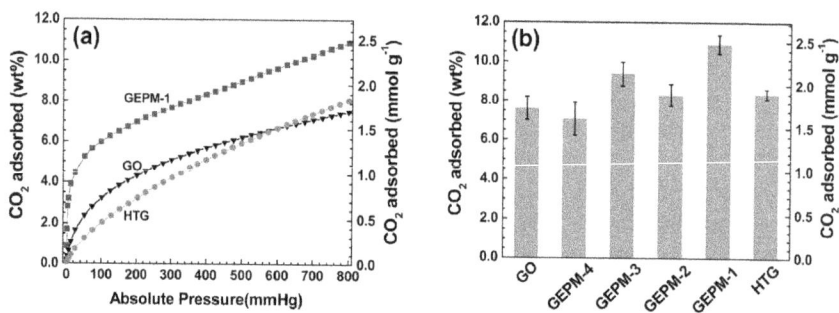

Figure 8.3 Comparison of the (a) carbon dioxide adsorption isotherms of GO, hydrothermal reduced graphene (HTG), and GO-PEI porous material (GEPM) samples and (b) comparison of the capacity of these materials. Reproduced from Reference 47 with permission from American Chemical Society.

Preparation of porous adsorbents for CO_2 capture from renewable sources has also been reported. Out of the different amino groups enriched natural polymers, chitosan has been utilized for various applications like waste water treatment, heterogeneous catalysis. This is because of its various advantages like presence of beneficial functional groups (-NH_2 and -OH) and environmental compatibility. Valechha *et al.* [48] reported a series of chitosan based adsorbents for CO_2 capture. The authors studies the interaction of free available amine with carbon dioxide molecules, along with studying the effect of crosslinking and functionalization. It was observed that chitosan glutaraldehyde exhibited superior adsorption than chitosan due to the formation of imine functionality.

In general, the mechanical as well as physical properties of chitosan are not as high as many synthetic polymers. Hence, nano-fillers like graphene, GO, CNT are generally incorporated in the polymer matrix in order to achieve these benefits. Alhwaige *et al.* [49] reported chitosan-GO hybrid aerogels as an interesting adsorbent for CO_2 capture. Aerogels, which are continuous pore materials with large surface area and ultra-low density, have been recently investigated for CO_2 removal from natural gas. Organic aerogels are pyrolyzed in an inert atmosphere for generating porous carbon materials with large surface area [50]. Examples include melamine-formaldehyde, resorcinol-formaldehyde, chitosan carbon gel, etc. [51]. In the studies by Alhwaige *et al.*, chitosan-GO aerogels were generated with different compositions using freeze drying method and the CO_2 adsorption was examined at various operating conditions. It was observed that the addition of 20 wt% GO led to the doubling of the quantity of CO_2 adsorbed on chitosan-GO hybrid aerogels as compared to chitosan alone. In addition, the adsorbents were observed to have good stability, easy regeneration and cost effectiveness over multiple cycles.

PPy is a basic CO_2 sorbent with good thermal stability and can be prepared at large scales. However, high cost and lower surface area limit its use for CO_2 removal. The limitations can be overcome by interfacing with graphene, which specifically leads to increased surface area. Chandra *et al.* [52] prepared N-doped porous carbons by chemical activation of graphene-polypyrrole (PPy) composites. Specifically, PPy functionalized graphene was prepared by chemical polymerization of pyrrole in GO utilizing ammonium persulfate and subsequent consecutive reduction employing hydrazine. The CO_2 removal capacity was observed to be 4.3 $mmol.g^{-1}$ for the samples activated at a higher temperature of 500-600 °C, due to the enhanced adsorbent-adsorbate interaction on the micro-porous surface. It was confirmed that the graphene-PPy composites had high adsorption capacity as well as selectivity for CO_2 removal, along with a relatively inexpensive and easy synthesis process for the purpose of industrial scaling. Similar to the aforementioned work, a series of S- or N- doped graphene based adsorbents using various kinds of polymer-graphene nanocomposites (r-GO /polythiophene [53], r-GO/polyaniline[54], GO/ polyindole [55]) have been prepared by chemical activation at 400-800 °C. The materials were observed to have large pore volume, deep micro-porosity and high surface area. These doped samples exhibited higher CO_2 adsorption under ambient conditions (>4 mmol.g^{-1}), which was higher than several other solid adsorbents like activated carbon, MOFs and amine functionalized silica. In addition, the materials also exhibited excellent selectivity for CO_2 over N_2 (post combustion CO_2 removal), CH_4 and H_2 (pre-combustion CO_2 removal) and could also be regenerated without any difficulty for repeated use.

In another study by Mishra and Ramaprabhu [56], polyaniline-graphene nanocomposite was confirmed to be a promising adsorbent for CO_2 removal. An ultimate adsorption capacity of 75 mmol.g^{-1}, 47 mmol.g^{-1} and 31 mmol.g^{-1} was observed at 11 bar and temperatures of 25, 50, and 100 °C, respectively. The adsorption capacities were significantly higher as compared to pure graphene and the adsorbents exhibited higher degree of recyclability. A decrease in the adsorption capacity was observed with increase in temperature, which was opined to be due to the enhancement in the kinetic energy of gas molecules at higher temperature. Yang *et al.* [57] prepared a graphene based organic-inorganic ternary (PEI-G-silica) CO_2 solid adsorbent by nano-casting method. A unique and effective structure was generated by strong collaborative influence of each component. For instance, graphene offered large surface area and high thermal conductivity, while flexible morphology and uniform porosity was provided by mesoporous silica. With the presence of PEI, large CO_2 adsorption capacity was attained. Thus, the ternary adsorbent offered excellent CO_2 removal capacity and long term cycling stability.

Recently, Tsoufis *et al.* [58] intercalated diaminobutane poly(propylene imine) dendrimers into the interlayer spacing between GO sheets without employing any cross linking agent. The CO_2 adsorption capacity of resulting hybrids was observed to exhibit faster kinetics and higher values under wet circumstances, as compared to dry conditions. Thus, it was clear that the presence of water had an important role on the adsorption capacity of the graphene material.

Overall, from the existing research studies, it is clear that the unique structure and interesting properties of polymer-graphene nanocomposites provide immense opportunities for designing adsorbents suited for industrial applications.

8.2.2 Metal Organic Frameworks (MOF) and Porous Polymer Networks (PPN)

MOF, also referred as porous co-ordination polymers (PCP), are considered as one of the most promising porous materials with high degree of crystallinity, very high surface area (1000–8000 m^2/g) and porosity (50-90% free volume)[59]. In these physical adsorbents, the metal clusters are bonded strongly by organic linkers. This combination creates wide variety of porous materials with exclusive properties like chemical and structural tenability [60]. These materials have been reported to have significant CO_2 uptake ability and CO_2/N_2 selectivity at room temperature [61]. However, majority of MOFs have trouble in meeting the strict industrial requirements [62]. On the other hand, purely organic porous polymers, a class of adsorbents with proportionate pore size as well as surface areas, have good physico-chemical properties and stability due to covalent bonding in the network structure [63]. In spite of the fact that most of these materials are amorphous, their stability is highly useful in industrial applications. For porous material to be industrially acceptable, in addition to the high physico-chemical stability, high CO_2/N_2 selectivity as well as high CO_2-uptake ability is equally important. Both of these factors are influenced by the addition of CO_2-philic moieties in the structure. Lu *et al.* [64] prepared polyamine tethered PPN for CO_2 separation from flue gas. The addition of polyamine groups to PPN resulted in significant CO_2 adsorption behavior at low pressure and 295 K. The results confirmed that the CO_2 adsorption capacity was related to the amine loading rather than the surface area, and the results were consistent with the results obtained in another study by Dawson *et al.* [65]. In a study by Lin *et al.* [66], it was observed that the PEI incorporated MOF adsorbents were capable to achieve good selectivity of CO_2 over nitrogen up to 1200 at 50 °C and 770 at 25 °C. In another study, the adsorption ability of PEI impregnated resins for flue gas containing CO_2 concentration of 400 ppm and 15 vol % was observed to be 99.3 mg/g and 181 mg/g respectively at 25°C [67] (Figure 8.4).

Lu *et al.* [68] reported the preparation of a porous polymer network PPN-6 with permanent porosity and grafted the PPN-6 with lithium sulfonate and sulfonic acid. The resulting products, i. e., PPN-6-SO$_3$Li and PPN-6-SO$_3$H exhibited significant enhancement in CO_2 uptake ability and significantly large CO_2/N_2 adsorption selectivities at ambient conditions. At zero loading, the materials exhibited heat of adsorption of 30.4 kJ.mol^{-1} (PPN-6-SO$_3$H) and 35.7 kJ.mol^{-1} (PPN-6-SO$_3$Li), which were significantly higher than non-grafted PPN-6 (17 kJ.mol^{-1}). Figure 8.5 demonstrates the increases in isosteric heats of CO_2

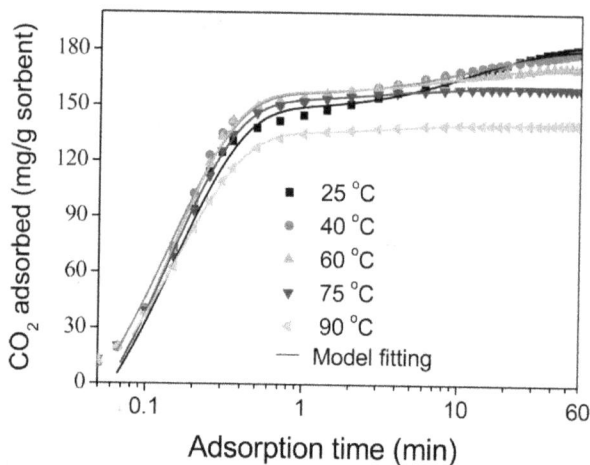

Figure 8.4 CO_2 adsorption (60 min) kinetics on PEI-impregnated resin (HP20/PEI-50) as a function of temperature. Reproduced from Reference 67 with permission from American Chemical Society.

adsorption and CO_2 uptake capacities of these materials.

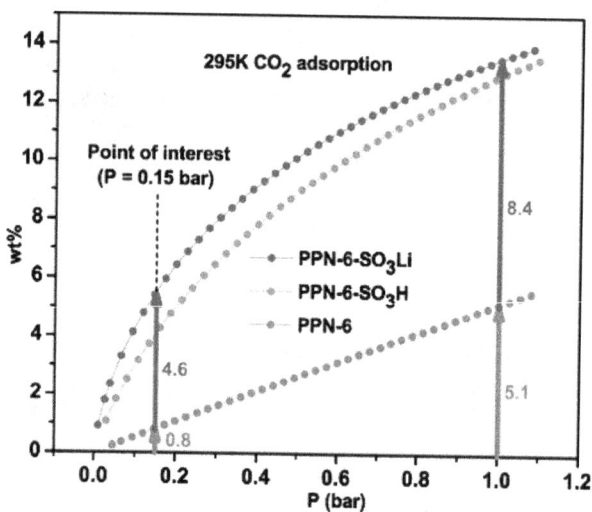

Figure 8.5 PPN grafted with sulfonic acid (PPN-6-SO_3H) and its lithium salt (PPN-6-SO_3Li). Reproduced from Reference 68 with permission from American Chemical Society.

8.2.3 Silica Based Adsorbents

Amine modified mesoporous silica has also been widely investigated for CO_2 capture. The unmodified silica materials have large pore size that can be tuned by proper selection of organic surfactants which act as template during the preparation, along with the surface hydroxyl groups, which can be chemically modified. For CO_2 capture, solid adsorbents are generated by the surface grafting of primary or secondary amines onto silica. This grafting can be achieved chemically (covalent bonding) as well as physically by wet impregnation. Amines are physically consolidated into silica materials by immersing the silica in a solution of a polymeric amine like polyethyleneimine. Thus, weak van der Waals forces generate between the amine and pore surface. Silica nano-particles that are modified by PEI are relatively easy to prepare, inexpensive and regenerable adsorbent with good capacity to capture CO_2 [69]. The main disadvantage of using amine grafted silica materials for the CO_2 removal is the hydrolysis of siloxane linkages which is promoted by either amine [70] or acidic conditions, thereby, decreasing the performance. Further, the ability of silica materials with physically attached amine may decrease over time due to the leaching process or redistribution within the material at the time of usage. Recently, several studies have reported PEI loaded mesoporous silica materials like SBA-12 [71], SBA-15 [72], SBA-16 [73], MCM-48 [74] and KIT-6 [75], and have observed high CO_2 capacity of these materials in the range of 1.5 to 3.5 mol.kg^{-1}. In these studies, the mesoporous silica materials were generally prepared using synthetic polymers as the template and tetra-ethyl-ortho-silicate as the silica source. In one of these studies comparing the performance of different silica materials, Son et al. [72] studied the CO_2 adsorption performance of different mesoporous silicas supported by PEI adsorbents. The authors observed that the SBA-15 supported PEI adsorbent exhibited a higher CO_2 capacity when compared to MCM-41 and MCM-48 supported PEI adsorbents. In another study, Witoon [76] studied the influence of PEI content and bimodal porous silica supports on CO_2 adsorption capacity at various adsorption temperatures. Rice husk was employed as the silica source. At lower PEI content (10-20 wt%), the CO_2 adsorption capacity of PEI loaded bimodal porous silica was slightly lower than PEI loaded unimodal silica. Recently, hyper-branched aminosilicas generated by in-situ aziridine polymerization on porous solids, were also used as adsorbents for the CO_2 removal from air [77]. Goeppert et al. [18] prepared adsorbents based on fumed silica (FS) with PEI impregnation. Samples with PEI loading of 33 wt % (FS-PEI-33) and 50 wt% (FS-PEI-50) were generated. The materials had superior adsorption capacity for the removal of CO_2 from air. The adsorbents were also able to adsorb CO_2 reversibly in repeated cycles under mild conditions. At lower loadings, the polymeric amine was observed to be better dispersed on the support surface. However, at higher loadings, a significant portion of the amino groups present in the PEI was not accessible because of the poor dispersion on the surface of the support. Further, the adsorbents could perform well under both humid as well as dry conditions. Figure 8.6 illustrates the results attained on FS-PEI-50 and FS-PEI-33 for the CO_2 adsorption under humid and dry conditions. These findings indicated that the generated materials were superior to zeolites, which lose CO_2 adsorption capacity under humid conditions. Thus, FS-PEI adsorbents can be used for CO_2 separation from gas streams in submarines as well as other closed environments.

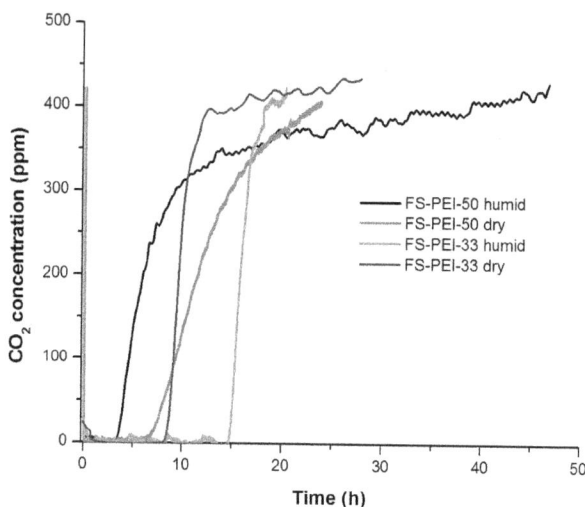

Figure 8.6 Adsorption of CO_2 from the air at 25 °C on FS-PEI-50 and FS-PEI-33 under humid and conditions. Reproduced from Reference 18 with permission from American Chemical Society.

Lively *et al.* [78,79] prepared a new type of CO_2 removal adsorbent system based on hollow fiber based solid adsorbent that operated in rapid temperature swing adsorption (TSA) mode. This prevented many deficiencies related to the cyclic sorption processes. The solid adsorbent particles were embedded in a polymeric porous hollow fiber matrix, which provided several advantages as compared to polymer fibers which are traditionally used for membrane applications. In the hollow fiber system, high volume of adsorbent materials could be incorporated, which allowed rapid mass transfer to the adsorbent due to large voids. Rezaei *et al.* [80] prepared silica/amine/polymer hollow fiber adsorbents using post spinning infusion method, as shown in Figure 8.7. The fibers exhibited good CO_2 removal capacity from the simulated flue gas. Thus, the post spinning infusion technique provided a new platform for preparing silica/amine/polymer hollow fibers for flue gas CO_2 capture applications.

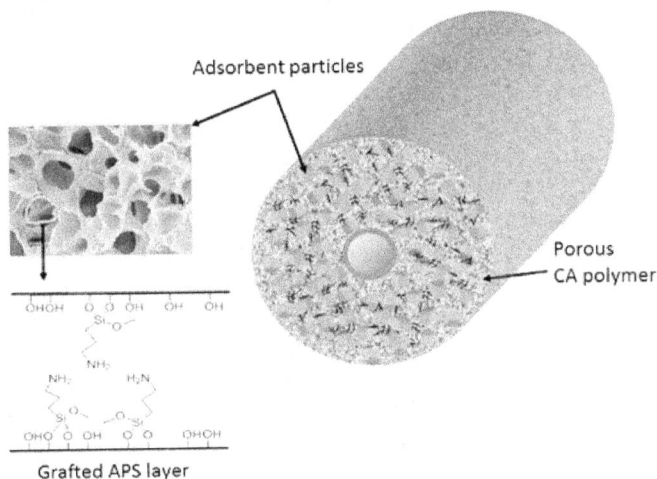

Adsorbent particles

Porous
CA polymer

Grafted APS layer

Figure 8.7 Generation of silica/amine/polymer hollow fiber adsorbents using post spinning infusion method. Reproduced from Reference 80 with permission from American Chemical Society.

8.3 Polymer Enhanced Adsorbents for H₂S Capture

H_2S is an immensely toxic gas and causes damage to the pipelines, along with deterioration of natural gas quality, thus, requiring its removal to ppm levels. As mentioned earlier, the amount of H_2S in natural gas can be minimized by reaction with amines. The reaction is highly exothermic and regardless of the amine structure, H_2S reacts with primary, secondary as well as tertiary amines. Several types of adsorbents like activated carbon [81], MOF [82], metal oxides [83] and graphite [84] have also been used for the effective H_2S removal. Other research studies have also reported different solid materials for

H₂S separation at high temperature (>300 °C) from a series of industrial gas streams [85,86]. As mentioned earlier for CO2 removal, Ma *et al.* prepared molecular basket regenerable adsorbent (PEI-loaded MCM-41), which could adsorb CO₂ as well as H₂S from a model gas. Using the same molecular basket sorbent concept, Wang *et al.* [87] prepared sorbent by loading a PEI polymer on the mesoporous molecular sieve to remove H₂S by an acid-base reaction. The results indicated that the material could separate H₂S from gas streams to a minimum of 2 ppmv at ambient conditions. The amount of PEI on the molecular sieve was observed to have a strong effect on the sorption performance. The mesoporous molecular sieve with three dimensional channel structure as well as large pore size helped to enhance the kinetic ability of the PEI sorbent. The generated sorbents exhibited a saturation capacity of 3.02 mmol H₂S/g-sorb and breakthrough capacity of 0.79 mmol H₂S/g-sorb at 22 °C, gas hourly space velocity (GHSV) of 674 h⁻¹ with a gas consisting of 4000 ppmv of H₂S. Further, it could also be regenerated readily under mild conditions (75-100 °C temperature range). Thus, considering the remarkable sorption performance, stability and regenerability of PEI/SBA-15 sorbent, it represented one of mist effective sorbents for H₂S separation from gas streams in more energy efficient and environment friendly operation. Chen *et al.* [88] generated PEI loaded porous silica monolith having high stability and high H₂S breakthrough capacity of about 1.27 mmol of H₂S/g sorbent at a temperature of 22 °C. The adsoprbent could be regenerated at 75 °C. Further, it was also observed that H₂S adsorption performance was influenced by the amount of PEI added to the adsorbent as well as the interactions between the H₂S and amine functional group present in PEI. Figure 8.8 also shows the sorption results of the sorbents as a function of PEI molecular weight. It was observed that the sorbent with M_w of 600 exhibited the best sorption performance. Increasing the molecular weight of PEI led to reduction in the sorption due to steric hindrance and low fluidity, which hindered the internal diffusion of H₂S, thus, reducing the utilization efficiency of amine groups. PEI with lowest molecular weight also did not exhibit high sorption probably due to its composition as it was a mixture of linear monomers and branch polymers. In addition, interaction of H₂S with the polymer chains of different molecular weight contributed to different performance of these PEI polymers.

In another study by Wang *et al.* [89], it was observed that the nano-porous composite adsorbent PEI/SBA-15 had high adsorption capacity for H₂S and the presence of moisture had a promoting effect on the H₂S separation from gas streams. Recently Jaiboon *et al.* [90] prepared PEI modified high porosity adsorbent silica xerogels for low temperature H₂S separation. Using fixed bed

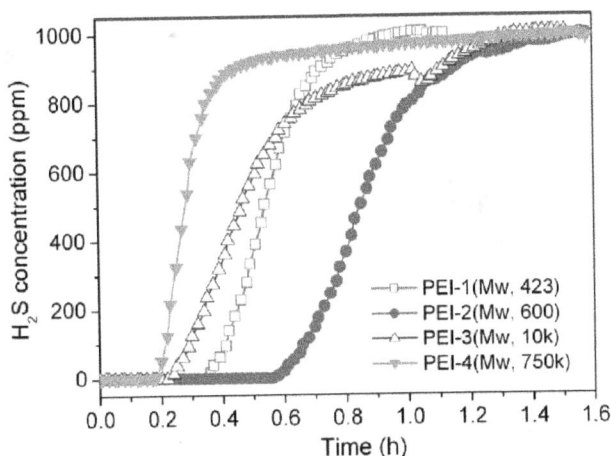

Figure 8.8 Breakthrough curves of H_2S for porous silica sorbents loaded with 65 wt % PEI as a function of PEI molecular weight. Reproduced from Reference 88 with permission from American Chemical Society.

system, the H_2S adsorption as well as desorption capacities were characterized with respect to operating temperature, amine type, loading quantity and gas flow rate. On enhancing the amine loading level and reducing the temperature, the H_2S removal ability could be increased. At 30 °C, the xerogel containing 50 wt% PEI (fumed silica -PEI800-50) exhibited the highest saturation capacity and breakthrough time. Out of the different types, the optimal sorbent (fumed silica-PEI800-50) was observed to regenerate easily for minimum ten successive adsorption-desorption cycles, at a mild temperature, without any decrease in the regeneration ability or adsorption capacity.

8.4 Polymer Enhanced Adsorbents for SO_x and NO_x Capture

Adsorption of weak acidic gases like SO_2, CO_2, NO and NO_2 onto amine containing polymeric adsorbents has been examined [91,92]. The authors analyzed the influence of amine structure on the adsorption of acid gases as well as thermal reversibility of adsorption-desorption processes. It was observed that the thermal reversibility of gas removal process reduced in the order $CO_2 >$ $SO_x> NO_x$. Poor CO_2 adsorption capacity was observed for the tertiary amine functionalized polymer. However, the adsorbent exhibited good affinity for the weak acidic gases. Further, it was also noticed that the SO_2 adsorption mechanism on amino-polymers was the same as CO_2 adsorption. Only few studies have dealt with the SO_x/NO_x activated degradation process at molecular level in supported amine adsorbents. The mechanism of reaction of amino-polymers with NO under ambient condition is as follows:

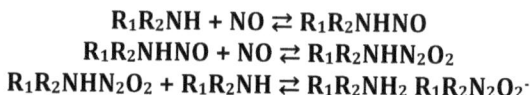

$$R_1R_2NH + NO \rightleftarrows R_1R_2NHNO$$
$$R_1R_2NHNO + NO \rightleftarrows R_1R_2NHN_2O_2$$
$$R_1R_2NHN_2O_2 + R_1R_2NH \rightleftarrows R_1R_2NH_2\ R_1R_2N_2O_2^-$$

Hallenbeck and Kitchin [93] conducted research on the polymeric CO_2 adsorbent functionalized using primary amines and the stability was examined in the presence of SO_2. There was a decrease in the CO_2 adsorption capacity of the polymer from 1.4 to 0.1 mmol/g on exposing it to 431 ppm SO_2 over nine cycles at 50 °C. Figure 8.9 demonstrates CO_2 capture capacity of resin in the

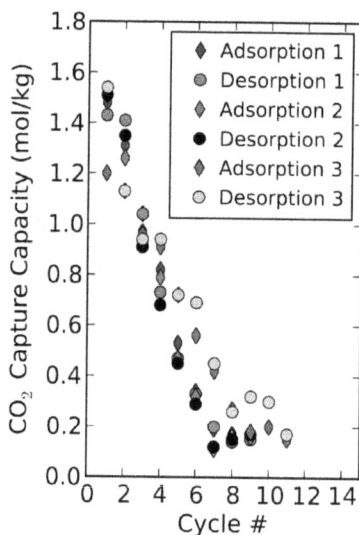

Figure 8.9 CO_2 capture capacity of resin in presence of SO_2. Reproduced from Reference 93 with permission from American Chemical Society.

the presence of SO_2. It was obvious that the CO_2 adsorption capacity of the adsorbent decreased in each cycle, till achieving a stage where the capacity was no longer available. More specifically, the activity decreased initially almost linearly, and stabilized afterwards at the minimum capacity of 0.2 mol/kg.

Analysis in mass spectrometer also revealed no presence of SO_2 for the first six cycles, indicating that all of the SO_2 was captured by the adsorbent. The authors also confirmed that the reduction in CO2 adsorption capacity was not attributed to the reduction in surface area or pore volume, but predominantly resulted from the interaction of SO_2 with amine groups in the adsorbent. In another recent study, Fan *et al.* [94] analyzed the cyclic stability of polymeric fibers functionalized with amine. Good stability was observed over 120 cycles for hollow fibers containing secondary amines with 25% capacity loss for CO_2, after exposing to 200 ppm SO_2 at 35°C.

8.5 Summary and Outlook

In summary, the chapter reviewed the performance of various polymers that are extensively employed for acid gas adsorption from the natural gas. The state of art in this direction was critically analyzed with a focus on the adsorbent material, mechanism of adsorption and challenges to be resolved.

To summarize, the new developments have resulted in the design and development of advanced materials with required chemical, physical and structural properties for effective acid gas removal when compared to the currently used technologies. Different types of synthetic polymers like polyethyleneimine, phenolic resin, melamine formaldehyde resin, urea formaldehyde resin, polyaniline, polypyrrole and natural polymers like chitosan have been reported to significantly contribute to the separation of CO_2 and H_2S from natural gas. Most of the polymer based adsorbents have been confirmed to have high efficiency towards the adsorption of acid gases. In addition, the concentration of amine groups available in the adsorbent as well as the interactions between the acid gases and amine groups are the two dominant factors that influence the amine efficiency and adsorption capacity. In addition, several other factors like the method of activation of carbons, length of polymer chains, degree of polymerization, pore structure and sorption temperature of the supports also play a vital role in determining the adsorption capacity. It was observed that the carbons prepared by chemical activation process were more efficient for the CO_2 separation than the carbons prepared by physical activation. Higher CO_2 adsorption ability was observed in long chain polymers due to the incorporation of more amine group in the adsorbent. Thus, a well developed and adequate porosity along with a favorable chemistry remarkably increased the acid gas adsorption capacity of the adsorbents. As a result, adsorbents need to be generated with sufficient basic sites to increase the affinity towards CO_2 and H_2S.

In spite of high adsorption capacity of the adsorbents, several economical and technical barriers hamper their large scale application for acid gas removal. The greatest obstacles for the usage of these adsorbents are absence of long term stability, limited range of convenient operating conditions, high preparation cost and inability for simultaneous multiple impurity gas removal. In addition to this, despite the fact that the new emerging materials like MOF exhibit good acid gas adsorption capacity, such adsorbents are unstable in wet conditions. Hence, more insights need to be gained for improving the stability under humid conditions. Significant research is still being carried out in order to investigate different classes of polymers so as to develop adsorbents with sufficient amount of nitrogen to enhance the adsorption as well as to overcome the aforementioned barriers. Despite several recent approaches, most of the potential remains unexplored and need to be translated into widespread use for the acid gas removal process through technology development.

References

1. Berg, S. V. (1998) Lessons in electricity market reform: regulatory processes and performance. *The Electricity Journal*, **11**, 13-20.
2. Rufford, T. E., Smart, S., Watson, G. C. Y., Graham, B. F., Boxall, J., Diniz da Costa, J. C., and May, E. F. (2012) The removal of CO_2 and N_2 from natural gas: a review of conventional and emerging process technologies. *Journal of Petroleum Science and Engineering*, **94**, 123-154.
3. Iliuta, M. C., Larachi, F., and Grandjean, B. (2004) Solubility of hydrogen sulfide in aqueous solutions of Fe(II) complexes of trans-1,2-cyclohexanediaminetetraacetic acid. *Fluid Phase Equilibria*, **218**(2), 305-313.
4. Engineering Data Book, Gas Processors Suppliers Association (GPSA), USA (2004).
5. Abdulrahman, R. K., and Sebastine, I. M. (2013) Natural gas sweetening process simulation and optimization: a case study of Khurmala field in Iraqi Kurdistan region. *Journal of Natural Gas Science and Engineering*, **14**, 116-120.
6. Dortmundt, D., and Doshi, K. (1999) Recent developments in CO_2 removal membrane technology, UOP, USA. Online: http://www.membrane-guide.com/download/CO2-removal-membranes.pdf
7. Rochelle, G. T. (2009) Amine scrubbing for CO_2 capture. *Science*, **325**, 1652-1654.
8. Xie, H.-B., Zhou, Y., Zhang, Y., and Johnson, J. K. (2011) Reaction mechanism of monoethanolamine with CO_2 in aqueous solution from molecular modeling. *Journal of Physical Chemistry A*, **114**, 11844-11852.
9. Herzog, H. J., and Drake, E. M. (1993) Greenhouse gas R&D programme, IEA/93/OE6.
10. Khatri, R. A., Chuang, S. S. C., Soong, Y., and Gray, M. (2006) Thermal and chemical stability of regenerable solid amine sorbent for CO_2 capture. *Energy and Fuels*, **20**, 1514-1520.

11. Gray, M. L., Champagne, K. J., Fauth, D., Baltrus, J. P., and Pennline, H. (2008) Performance of immobilized tertiary amine solid sorbents for the capture of carbon dioxide. *International Journal of Greenhouse Gas Control*, 2(1), 3-8.

12. Li, J. R., Sculley, J., and Zhou, H. C. (2012) Metal-organic frameworks for separations. *Chemical Reviews*, 112(2), 869-932.

13. Lu, C., Bai, H., Wu, B., Su, F., and Hwang, J. F. (2008) Comparative study of CO_2 capture by carbon nanotubes, activated carbons, and zeolites. *Energy and Fuels*, 22(5), 3050-3056.

14. Cavenati, S., Grande, C. A., and Rodrigues, A. E. (2004) Adsorption equilibrium of methane, carbon dioxide, and nitrogen on zeolite 13× at high pressures. *Journal of Chemical Engineering Data*, 49(4), 1095-1101.

15. Sevilla, M., Valle-Vigón, P., and Fuertes, A. B. (2011) N-doped polypyrrole-based porous carbons for CO_2 capture. *Advanced Functional Materials*, 21(14), 2781-2787.

16. Arenillas A., Smith K., Drage T. C., and Snape C. E. (2005) CO_2 capture using some fly ash-derived carbon materials. *Fuel*, 84, 2204-2210.

17. Wang, X. R., Li, H. Q., Liu, H. T., and Hou, X. J. (2011) As-synthesized mesoporous silica MSU-1 modified with tetraethylenepentamine for CO_2 adsorption. *Microporous & Mesoporous Materials*, 142(2-3), 564-569.

18. Goeppert, A., Czaun, M., May, R. B., Surya Prakash, G. K., Olah, G. A., and Narayanan, S. R. (2011) Carbon dioxide capture from the air using a polyamine based regenerable solid ddsorbent. *Journal of American Chemical Society*, 133, 20164-20167.

19. Satyapal, S., Filburn, T., Trela, J., and Strange, J. (2001) Performance and properties of a solid amine sorbent for carbon dioxide removal in space life support applications. *Energy and Fuels*, 15, 250-255.

20. Drage, T. C., Arenillas, A., Smith, A. M., Pevida, C., Piippo, S., and Snape, C. E. (2007) Preparation of carbon dioxide adsorbents from the chemical activation of urea–formaldehyde and melamine–formaldehyde resins. *Fuel*, 86, 22-31.

21. Heuchel, M., Davies, G. M., Buss, E., and Seaton, N. A. (1999) Adsorption of carbondioxide and methane and their mixtures on an activated carbon: simulation and experiment. *Langmuir*, 15, 8695-8705.

22. Okoye, I. P., Benham, M., and Thomas, K. M. (1997) Adsorption of gases and vapors on carbon molecular sieves. *Langmuir*, 13, 4054-4059.

23. Yim, W. L., Byl, O., Yates, J. T., and Johnson, J. K. (2004) Vibrational behavior of adsorbed CO_2 on single-walled carbon nanotubes. *Journal of Chemical Physics*, 120, 5377-5386.

24. Ghosh, A., Subrahmanyam, K. S., Krishna, K. S., Datta, S., Govindaraj, A., Pati, S. K., and Rao, C. N. R. (2008) Uptake of H_2 and CO_2 by graphene. *Journal of Physical Chemistry C*, 112, 15704-15707.

25. Adib, F., Bagreev, A., and Bandosz, T. J. (2000) Adsorption/oxidation of hydrogen sulfide on nitrogen-containing activated carbons. *Langmuir*, 16, 1980-1986.

26. Arenillas, A., Drage, T. C., Smith, K. M., and Snape, C. E. (2005) CO_2 removal potential of carbons prepared by co-pyrolysis of sugar and nitrogen containing compounds. *Journal of Analytical and Applied Pyrolysis*, 74, 298-306.

27. Vinu, A., Anandan, S., Anand, C., Srinivasu, P., Ariga, K., and Mori, T. (2008) Fabrication of partially graphitic three-dimensional nitrogen-doped mesoporous

carbon using polyaniline nanocomposite through nano-templating method. *Microporous Mesoporous Materials*, **109** (1-3), 398-404.

28. Martin, C. F., Plaza, M. G., Garcia, S., Pis, J. J., Rubiera, F., and Pevida, C. (2011) Microporous phenol-formaldehyde resin-based adsorbents for pre-combustion CO_2 capture. *Fuel*, **90**, 2064-2072.

29. Martin, C. F., Garcia, S., Beneroso, D., Pis, J. J., Rubiera, F., and Pevida, C. (2012) Pre-combustion CO_2 capture by means of phenol–formaldehyde resin-derived carbons: From equilibrium to dynamic conditions. *Separation and Purification Technology*, **98**, 531-538.

30. Long, D., Zhang, J., Yang, J., Hu, Z., Cheng, G., Liu, X., Zhang, R., Zhan, L., Qiao, W., and Ling, L. (2008) Chemical state of nitrogen in carbon aerogels issued from phenol-melamine-formaldehyde gels. *Carbon*, **46**, 1253-1269.

31. Tseng, R.-L., Wu, F.-C., and Juang, R.-S. (2015) Adsorption of CO_2 at atmospheric pressure on activated carbons prepared from melamine-modified phenol–formaldehyde resins. *Separation and Purification Technology*, **140**, 53-60.

32. Pevida, C., Snape, C. E., and Drage, T. C. (2009) Templated polymeric materials as adsorbents for the postcombustion capture of CO_2. *Energy Procedia*, **1**, 869-874.

33. Drage, T. C., Arenillas, A., Smith, K. M., and Snape, C. E. (2008) Thermal stability of polyethylenimine based carbon dioxide adsorbents and its influence on selection of regeneration strategies. *Microporous and Mesoporous Materials*, **116**(1), 504-512.

34. Pevida, C., Drage, T. C., and Snape, C. E. (2008) Silica-templated melamine-formaldehyde resin derived adsorbents for CO_2 capture. *Carbon*, **46**(11), 1464-1474.

35. Hao, G.-P., Li, W.-C., Qian, D., and Lu, A.-H. (2010) Rapid synthesis of nitrogen-doped porous carbon monolith for CO_2 capture. *Advanced Materials*, **22**(7), 853-857.

36. Drage, T. C., Arenillas, A., Smith, K. M., Pevida, C., Piippo, S., and Snape, C. E. (2007) Preparation of carbon dioxide adsorbents from the chemical activation of urea-formaldehyde and melamine formaldehyde resins. *Fuel*, **86**(1-2), 22-31.

37. Goel, G., Bhunia, H., and Bajpai, P. K. (2015) Mesoporous carbon adsorbents from melamine-formaldehyde resin using nano-casting technique for CO_2 adsorption. *Journal of Environmental Science*, **32**, 238-248.

38. Wang, D. X., Ma, X. L., Sentorun-Shalaby, C., and Song, C. S. (2012) Development of carbon-based "molecular basket" sorbent for CO_2 capture. *Industrial and Engineering Chemistry Research*, **51**(7), 3048-3057.

39. Ma, X. L., Wang, X. X., and Song, C. S. (2009) "Molecular basket" sorbents for separation of CO2 and H2S from various gas streams. *Journal of the American Chemical Society*, **131**(16), 5777-5783.

40. Wang, X. X., Schwartz, V., Clark, J. C., Ma, X. L., Overbury, S. H., Xu, X. C., and Song, C. S. (2009) Infrared study of CO_2 sorption over "molecular basket" sorbent consisting of polyethylenimine-modified mesoporous molecular sieve. *Journal of Physical Chemistry C*, **113**, 7260-7268.

41. Wang, X., Ma, X., Song, C., Locke, D. R., Siefert, S., Winans, R. E., Mollmer, J., Lange, M., Moller, A., and Glaser, R. (2013) Molecular basket sorbents polyethylenimine–SBA-15 for CO_2 capture from flue gas: Characterization and sorption properties. *Journal of Microporous and Mesoporous Materials*, **169**, 103-111.

42. Wang, J., Chen, H., Zhou, H., Liu, X., Qiao, W., Long, D., and Ling, L. (2013) Carbon dioxide capture using polyethylenimine-loaded mesoporous carbons. *Journal of*

Environmental Science (China), **25**(1), 124-32.
43. Ngoy, J. M., Wagner, N., Riboldi, L., and Bolland, O. (2014) A CO_2 capture technology using multi-walled carbon nanotubes with polyaspartamide surfactant. *Energy Procedia*, **63**, 2230-2248.
44. Najafabadi, A. T. (2015) Emerging applications of graphene and its derivatives in carbon capture and conversion: Current status and future prospects. *Renewable Sustainable Energy Reviews*, **41**, 1515-1545.
45. Gadipelli, S., and Guo, Z. X. (2015). Graphene-based materials: Synthesis and gas sorption, storage and separation. *Progress in Materials Science*, **69**, 1-60.
46. Kemp, K. C., Chandra, V., Saleh, M., and Kim, K. S. (2013) Reversible CO_2 adsorption by an activated nitrogen doped graphene/polyaniline material. *Nanotechnology*, **24**, 235703.
47. Sui, Z.-Y., Cui, Y., Zhu, J.-H., and Han, B.-H. (2013) Preparation of three-dimensional graphene oxide-polyethylenimine porous materials as dye and gas adsorbents. *ACS Applied Materials and Interfaces*, **5**(18), 9172-9179.
48. Valechha, A., Thote, J., Labhsetwar, N., and Rayalu, S. (2012) Biopolymer based adsorbents for the post combustion CO_2 capture. *International Journal of Knowledge Engineering*, **3**(1), 103-106.
49. Alhwaige, A. A., Agag, T., Ishida, H., and Qutubuddin, S. (2013) Biobased chitosan hybrid aerogels with superior adsorption: Role of graphene oxide in CO_2 capture. *RSC Advances*, **3**, 16011-16020.
50. Tao, Y., Endo, M., and Kaneko, K. (2008) A review of synthesis and nanopore structures of organic polymer aerogels and carbon aerogels. *Recent Patents on Chemical Engineering*, **1**, 192-200.
51. Sevilla, M., and Fuertes, A. B. (2011) Sustainable porous carbons with a superior performance for CO_2 capture. *Energy and Environmental Science*, **4**, 1765-1771.
52. Chandra, V., Yu, S. U., Kim, S. H., Yoon, Y. S., Kim, D. Y., Kwon, A. H., Meyyappan, M., and Kim, K. S. (2012) Highly selective CO_2 capture on N-doped carbon produced by chemical activation of poly- pyrrole functionalized graphene sheets. *Chemical Communications*, **48**, 735-737.
53. Seema, H., Kemp, K. C., Le, N. K., Park, S.-W., Chandra, V., Lee, J. W., and Kim, K. S. (2014) Highly selective CO_2 capture by S-doped microporous carbon materials. *Carbon*, **66**, 320-326.
54. Kumar, N. A., Choi, H.-J., Shin, Y. R., Chang, D. W., Dai, L., and Baek, J.-B. (2012) Polyaniline-grafted reduced graphene oxide for efficient electrochemical supercapacitors. *ACS Nano*, **6**(2), 1715-1723.
55. Saleh, M., Chandra, V., Kemp, K. C., and Kim, K. S., (2013) Synthesis of N-doped microporous carbon via chemical activation of polyindole-modified graphene oxide sheets for selective carbon dioxide adsorption. *Nanotechnology*, **24**, 255702.
56. Mishra, A. K., and Ramaprabhu, S. (2012) Nanostructured polyaniline decorated graphene sheets for reversible CO_2 capture. *Journal of Materials Chemistry*, **22**, 3708-3712.
57. Yang, S., Zhan, L., Xu, X., Wang, Y., Ling, L., and Feng, X. (2013) Graphene-based porous silica sheets impregnated with polyethyleneimine for superior CO_2 capture. *Advanced Materials*, **25**, 2130-2134.

58. Tsoufis, T., Katsaros, F., Sideratou, Z., Romanos, G., Ivashenko, O., Rudolf, P., Kooi, B. J., Papageorgiou, S., and Karakassides, M. A. (2014) Tailor-made graphite oxide–DAB poly(propylene imine) dendrimer intercalated hybrids and their potential for efficient CO_2 adsorption. *Chemical Communications*, **50**, 10967-10970.

59. Furukawa, H., Muller, U., and Yaghi, O. M. (2015) "Heterogeneity within order" in metal-organic frameworks. *Angewandte Chemie International Edition*, **54**, 3417-3430.

60. Zhou, H.-C., and Kitagawa, S. (2014) Metal-organic frameworks (MOFs). *Chemical Society Reviews*, **43**, 5415-5418.

61. Mason, J. A., Sumida, K., Herm, Z. R., Krishna, R., and Long, J. R. (2011) Evaluating metal-organic frameworks for post-combustion carbon dioxide capture via temperature swing adsorption. *Energy and Environmental Science*, **4**, 3030-3040.

62. D'Alessandro, D. M., Smit, B., Long, J. R. (2010) Carbon dioxide capture: prospects for new materials. *Angewandte Chemie International Edition*, **49**, 6058-6082.

63. Rabbani, M. G., and El-Kaderi, H. M. (2011) Template-free synthesis of a highly porous benzimidazole-linked polymer for CO_2 capture and H_2 storage. *Chemistry of Materials*, **23**, 1650-1653.

64. Lu, W., Sculley, J. P., Yuan, D., Krishna, R., Wei, Z., and Zhou, H.-C. (2012) Polyamine-tethered porous polymer networks for carbon dioxide capture from flue gas. *Angewandte Chemie International Edition*, **51**, 7480-7484.

65. Dawson, R., Stockel, E., Holst, J. R., Adams, D. J., and Cooper, A. I. (2011) Microporous organic polymers for carbon dioxide capture. *Energy and Environmental Science*, **4**, 4239-4245.

66. Lin, Y., Yan, Q., Kong, C., and Chen, L. (2013) Polyethyleneimine incorporated metal-organic frame works adsorbent for highly selective CO_2 capture. *Scientific Reports*, **3**, 1859.

67. Chen, Z., Deng, S., Wei, H., Wang, B., Huang, J., and Yu, G. (2013) Polyethylenimine-impregnated resin for high CO_2 adsorption: an efficient adsorbent for CO_2 capture from simulated flue gas and ambient air. *ACS Applied Materials and Interfaces*, **5**, 6937-6945.

68. Lu, W., Yuan, D., Sculley, J., Zhao, D., Krishna, R., and Zhou, H.-C. (2011) Sulfonate-grafted porous polymer networks for preferential CO_2 adsorption at low pressure. *Journal of the American Chemical Society*, **133**, 18126-18129.

69. Meth, S., Goeppert, A., Prakash, G. K. S, and Olah, G. A. (2012) Silica nanoparticles as supports for regenerable CO_2 sorbents. *Energy and Fuels*, **26**, 3082-3090.

70. Smith, E. A., and Chen, W. (2008) How to prevent the loss of surface functionality derived from aminosilanes. *Langmuir*, **24**, 12405-12409.

71. Zelenak, V., Badanicova, M., Halamova, D., Cejka, J., Zukal, A., Murafa, N., Goerigk, G. (2008) Amine-modified ordered mesoporous silica: Effect of pore size on carbon dioxide capture. *Chemical Engineering Journal*, **144**, 336-342.

72. Son, W. J., Choi, J. S., and Ahn, W. S. (2008) Adsorptive removal of carbon dioxide using polyethyleneimine-loaded mesoporous silica materials. *Microporous and Mesoporous Materials*, **113**, 31-40.

73. Wei, J., Shi, J., Pan, H., Zhao, W., Ye, Q., and Shi, Y. (2008) Adsorption of carbon dioxide on organically functionalized SBA-16. *Microporous and Mesoporous Materials*, **116**, 394-399.

74. Jang, H. T., Park, Y. K., Ko, Y. S., Lee, J. Y., and Margandan, B. (2009) Highly siliceous MCM-48 from rice husk ash for CO2 adsorption. *International Journal of Greenhouse Gas Control*, **3**, 545-549.

75. Liu, Y., Shi, J., Chen, J., Ye, Q., Pan, H., Shao, Z. H., and Shi, Y. (2010) Dynamic performance of CO_2 adsorption with tetraethylenepentamine loaded KIT-6. *Microporous and Mesoporous Materials*, **134**, 16-21.

76. Witoon, T. (2012) Polyethyleneimine-loaded bimodal porous silica as low-cost and high-capacity sorbent for CO_2 capture. *Materials Chemistry and Physics*, **137**, 235-245.

77. Nhut, J. M., Vieira, R., Pesant, L., Tessonnier, J. P., Keller, N., Ehret, G., Cuong, P. H., and Ledoux, M. J. (2002) Synthesis and catalytic uses of carbon and silicon carbide nanostructures. *Catalysis Today*, **76**, 11-32.

78. Lively, R. P., Leta, D. P., DeRites, B. A., Chance, R. R., and Koros, W. J. (2011) Hollow fiber adsorbents for CO_2 capture: Kinetic sorption performance. *Chemical Engineering Journal*, **171**, 801-810.

79. Lively, R. P., Chance, R. R., Mysona, J. A., Babu, V. P., Deckman, H. W., Leta, D. P., Thomann, H., and Koros, W. J. (2012) CO2 sorption and desorption performance of thermally cycled hollow fiber sorbents. *International Journal of Greenhouse Gas Control*, **10**, 285-294.

80. Rezaei, F., Lively, R. P., Labreche, Y., Chen, G., Fan, Y., Koros, W. J., and Jones, C. W. (2013) Aminosilane-grafted polymer/silica hollow fiber adsorbents for CO2 capture from flue gas. *ACS Applied Materials and Interfaces*, **5**, 3921-3931.

81. Choi, S., Drese, J. H., Chance, R. R., Eisenberger, P. M., and Jones, C. W. (2011) US Patent Application 2011/0179948A1.

82. Britt, D., Tranchemontagne, D., and Yaghi, O. M. (2008) Metal-organic frameworks with high capacity and selectivity for harmful gases. *PNAS*, **105**, 11623-11627.

83. Elseviers, W. F., and Vereist, H. (1999) Transition metal oxides for hot gas desulfurization. *Fuel*, **78**, 601-612.

84. Huang, P. H. (2014) Molecular dynamics investigation of separation of hydrogen sulfide from acidic gas mixtures inside metal-doped graphite micropores. *Physical Chemistry and Chemical Physics*, **17**, 22686-22698.

85. Novochinskii, I. I., Song, C. S., Ma, X. L., Liu, X., Shore, L., Lampert, J., and Farrauto, R. J. (2004) Low-temperature H2S removal from steam-containing gas mixtures with ZnO for fuel cell application. 1. ZnO particles and extrudates. *Energy and Fuels*, **18**(2), 576-583.

86. Karayilan, D., Dogu, T., Yasyerli, S., and Dogu, G. (2005) Mn-Cu and Mn-Cu-V mixed oxide regenerable sorbents for hot gas desulfurization. *Industrial and Engineering Chemistry Research*, **44**, 5221-5226.

87. Wang, X., Ma, X., Xu, X., Sun, L., Song, C. (2008) Mesoporous-molecular-sieve-supported polymer sorbents for removing H2S from hydrogen gas streams. *Topics in Catalysis*, **49**, 108-117.

88. Chen, Q., Fan, F., Long, D., Liu, X., Liang, X., Qiao, W., and Ling, L. (2010) Poly(ethyleneimine)-loaded silica monolith with a hierarchical pore structure for H2S adsorptive removal. *Industrial and Engineering Chemistry Research*, **49**, 11408-11414.

89. Wang, X., Ma, X., Sun, L., and Song, C. (2007) A nanoporous polymeric sorbent for deep removal of H_2S from gas mixtures for hydrogen purification. *Green Chemistry*, **9**, 695-702.

90. Jaiboon, V., Yoosuk, B., and Prasassarakich, P. (2014) Amine modified silica xerogel for H_2S removal at low temperature. *Fuel Process Technology*, **128**, 276-282.

91. Diaf, A., Garcia, J. L., and Beckman, E. J. (1994) Thermally-reversible polymeric sorbents for acid gases: CO_2, SO_2, and NO_x. *Journal of Applied Polymer Science*, **53**, 857-875.

92. Diaf, A., and Beckman, E. J. (1995) Thermally reversible polymeric sorbents for acid gases, IV. Affinity tuning for the selective dry sorption of NO_x. *Reactive Polymers*, **25**, 89-96.

93. Hallenbeck, A. P., and Kitchin, J. R. (2013) Effects of O_2 and SO_2 on the capture capacity of a primary-amine based polymeric CO_2 sorbent. *Industrial and Engineering Chemistry Research*, **52**, 10788-10794.

94. Fan, Y., Rezaei, F., Labreche, Y., Lively, R. P., Koros, W. J., Jones, C. W. (2015) Stability of amine-based hollow fiber CO_2 adsorbents in the presence of NO and SO_2. *Fuel*, **160**, 153-164.

9

Polymeric Pipeline Coatings for Oil and Gas Industry

9.1 Introduction

9.1.1 Pipeline Coatings

In oil and gas industry, about 10% of the overall production cost of oil and gas products is due to the pipelines and their maintenance [1]. In this respect, oil and gas industries continuously strive to diminish overall transmission pipelines cost. Pipeline coatings are immensely important for the maintenance of the pipelines and many research studies have focused on the development of functional coatings for the pipelines. Both external as well as internal pipeline coatings have been recommended to achieve protection of pipelines, thereby, ensuing a longer service life. More specifically, the internal protection is possible by the usage of liners, which can be used for applications demanding the corrosion protection as well as the rehabilitation of corroded surfaces. The usage of polymeric liners reduces the inner surface irregularities and roughness, thereby, creating substantial smoother surfaces. This lessens the maintenance cost of pipelines by offering improved flow efficiency of oil and gas through it [2,3]. In addition, the external protection of pipelines can be achieved through the usage of pipeline coverings, which provides sizeable anti-corrosive and wear resistance properties [4,5]. The selection of most effective coating is an important parameter that determines the durability of the lodged pipelines in the industries. In this respect, polymer based pipeline coatings have gained significant interest in oil, gas and petrochemical industries owing to their superior characteristics compared to other materials. These materials offer combination of scratch, wear and mechanical damage resistance along with good anti-corrosive properties to the pipelines, thus, creating enhanced life span. Among the various polymer categories useful for such coatings, epoxies, polyurethane (PU), polypropylene (PP), polyethylene (PE), copolymers of ethylene (e.g.: EBA and EVA), fluoropolymers, etc. have gained more usage industrially due to their benefits in terms of both performance and cost [6].

Anish M. Varghese and Vikas Mittal*, The Petroleum Institute (part of Khalifa University of Science and Technology), Abu Dhabi, UAE
*Current address: Bletchington, Wellington County, Australia

9.1.2 Requirements for Long Lasting Performance of Coatings

The final performance of polymeric coatings depends greatly on the things happening during manufacture, application, transportation, installation and field operation stages of their lifetime. The freshly applied polymeric coatings successfully isolate the pipelines from the adverse environment conditions, thereby, preventing the corrosion [7]. Over the span of usage, the coating materials can undergo changes in the properties, which can affect the overall efficiency and performances of pipelines. The predominant reasons for these kinds of changes include permeation of air and water molecules in the coating, loss of adhesion and cohesion, disbondment with passage/prevention of cathodic protection current, increase in cathodic protection current, etc. [8]. Due to these reasons, many studies are currently being carried out to gain more insights about under coating corrosion in order to enhance the adhesion strength between the pipelines and polymeric materials, thereby, reducing the maintenance cost. The under coating corrosion is induced only when both water and oxygen are present together and lead to metallic dissolution/anodic effect, thus, resulting in corrosion. The important factors which decide the corrosion rate in pipelines include temperature, pressure, wind velocity profiles and composition of carrying fluids [9]. As the replacement of coatings for some parts of pipelines, for example for underground portions, is not possible in short duration of time, thus, oil and gas industries need long life of coating materials. Thus, due to these reasons, coating materials should possess properties such as resistance towards water molecules and moisture, pressure variations, bacteria and mushrooms, capillary effect of water, temperature variations, solvents (especially oils and their derivatives) and mechanical damages [10]. This review describes the progress in the common polymer based coatings used to protect oil and gas pipelines.

9.2 Advancements in Polymeric Pipeline Coatings: A Brief Summary

The first successful industrial pipeline coating was coal tar, which was introduced about 85 years ago. Up to 1970, coal tar was the major coating material and is still in use in some specific areas because of its outstanding moisture resistance. Subsequently, the use of coal tar enamel (CTE) was reported, which was initially a plasticized mixture of coal tar pitch, coal and distillates. To improve the properties of CTE, inert fillers were added to the composition [11]. Afterwards, four component CTE based coatings were developed in order to overcome the drawbacks of coal tar. Coal tar melts in the hot environment and becomes stiff in cold conditions and furthermore caused health problems. The four component CTE system consisted of primer, CTE, inner and outer wraps of glass fiber. In this system, epoxy was largely used as the primer, thus, it is also considered as one of the first polymer based pipeline coatings systems. Epoxy provided good adhesion, UV resistance and thermal stability to the system. Later, asphalt enamels were also introduced as a replacement for coal tar. These enamels provided good corrosion resistance, however, studies revealed the presence of lower amount of carcinogenic contents in asphalt enamels [12].

In the meantime, polymeric tapes also got industrial attention for use as pipeline coatings. The polymeric tape is a composite material of a soft elastomer based adhesive inner layer and an outer monolithic polymer layer. Commonly, the outer layer includes polyolefins, polyvinylchloride, butyl rubber, etc. as these polymers provide good mechanical strength along with outstanding thermal, electrical and corrosion resistance to the coatings [13]. Following this, heat shrinkable tape coatings were introduced in 1980' s. Generally, these consist of a high shear strength hot melt adhesive of thermoplastic with a thick radiation crosslinked polyethylene based backing. These coatings need preheated pipeline surfaces to melt the adhesive layer and, thus, generate strong adhesion with the pipeline surfaces. Heat shrinkable structures were obtained through the heating of crosslinked polyethylene backing with the aid of propane torches from outside [14,15].

In the early 1960s, fusion bonded epoxy (FBE) was generated as an effective pipeline protective coating. Extensive developments concerning the performance improvements of FBE have been carried out since then. Generally, it is a 100% solid thermosetting epoxy powder, which offers strong adhesion to metallic surface through heat induced crosslinking [16,17]. Also, spray-applied liquid coatings have received attention as effective pipeline coating systems. Normally, these coatings are based on high build epoxy, polyurethane or a combination of two along with curing agents [18]. Following the developments of both epoxy and polyolefin based coatings, coating systems based on the combination of these two also generated industrial interest. Both two-layer and three-layer PE and PP coatings were developed. These coatings presented comprehensive characteristics of both epoxies and polyolefins [19]. Following the modifications of PE based coatings, high performance composite coatings (HPCC) were developed and applied for pipeline protection. It is a three component powder system consists of FBE primer, chemically modified polyethylene tie layer and polyethylene outer layer [13]. Biopolymer based coatings have also been explored for the purpose of generating anti-corrosion property. Figure 9.1 shows the electrochemical impedance spectroscopy (EIS) plots of polyvinylbutyral (PVB) and chitosan (Ch) coatings cross-linked with glutaraldehyde (Glu), developed for potential application as external pipeline coatings [20]. More recently, epoxy nanocomposite coatings, PU nanocomposite coatings and hybrid fluoropolymer resins have been introduced to enhance the performances of epoxy, PU and fluoropolymer based pipeline coating systems.

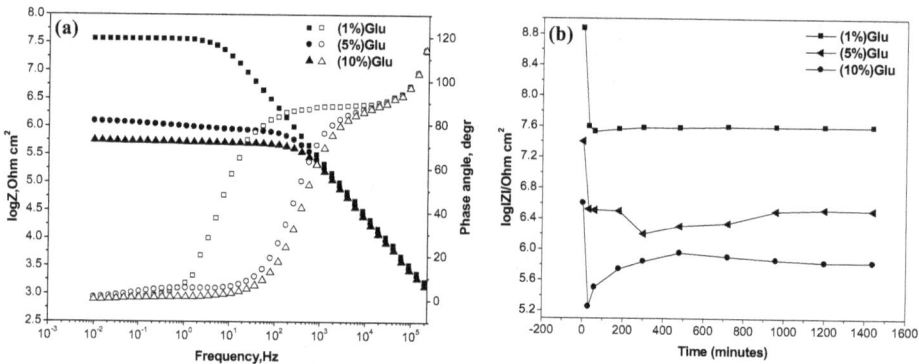

Figure 9.1 EIS plots of PVB_Ch/x%Glu_PVB coating having different percentages of glutaraldehyde; (a) Bode and phase plots obtained after 2 h immersion in 0.3M salt solution and (b) logZ at low frequency from the Bode plot vs. time of immersion. Reproduced from Reference 20 with permission from Springer.

9.3 Polymeric Pipeline Coatings

9.3.1 Polyolefin Based Pipeline Coatings

Polyolefin is a group of thermoplastic polymeric materials obtained from the polymerization of monomer olefin with a general formula of C_nH_{2n}. These classes of polyolefin based coatings include either PE or PP. These polymers possess excellent corrosion resistance and outstanding mechanical strength for pipeline protection. Furthermore, health and safety issues associated with these polymers are minimal as compared to other materials. However, their low degree of adhesion or bonding to metallic surfaces is a significant limitation [21,22]. In order to overcome this, significant developments have been achieved in the last decades in this area. Among the various advancements, two-layer and three-layer polyolefin based coatings gained more industrial interest.

Two-layer polyolefin coatings were introduced in 1960's. This coating system provides a combination of bonding and corrosion protection. This consists of an extruded polyolefin over coating and an inner layer of adhesive or sealants. The application of polyolefin coatings need the adhesive layer based on butyl or asphalt pipeline. Adequate flow of adhesive inner layer over pipelines upon heating is necessary to get uniform coating and which is then covered with polyolefin extrudates (side extruded or cross head extruded polyolefin) based on either high density PE (HDPE) or PP [15,23]. Three-layer polyolefin coatings are one of the most acceptable multilayer polymeric coating systems and have been in application since 1980's. These coatings are composed of an inner layer of either FBE or liquid epoxy, a tie layer of olefin copolymer adhesive and an outer layer of polyolefin. The inner layer of primer is applied with the aid of electrostatic spraying, followed by heating of pipeline surfaces to suitable temperature to obtain the uniform coating. Afterwards, the tie layer of adhesive is applied over primer surface using either side extrusion or spraying, followed by final coating of polyolefin outer layer [15,23].

PE Based Pipeline Coatings

PE is one of the most commonly used commodity thermoplastic polymeric material with a chemical formula of $(C_2H_4)_n$. Among the various grades of PE, high density PE (HDPE) and medium density PE (MDPE) have received more acceptance for pipeline coating applications. Both of these grades are prepared using Ziegler-Natta catalysts. PE based pipeline coatings provide outstanding performance such as good mechanical properties, superlative corrosion resistance, fine chemical stability and inexpensiveness. However, PE possesses low softening point, which usually reduces the applications overreaching the temperature of 80°C [24].

Due to its excellent properties, PE has been widely used in many kinds of pipeline coatings formulations and has undergone diverse modifications in order to suit the pipeline coating applications. The history of PE as a successful pipeline coating material has started through its usage as polymeric tapes. As mentioned earlier, polymeric tape is a coating material composed of an adhesive inner layer and a monolithic polymer outer layer. In PE tapes, outer covering layer of PE offers good mechanical properties as well as combination of corrosion, electrical and chemical resistance properties to the coatings. Normally, a soft elastomeric adhesive inner layer is used to enhance the adhesion of PE outer layer over pipelines, which serves as primer. Specifically, the anti-corrosive PE outer layer is wrapped throughout the adhesive coated pipelines [5]. As a continuation of PE tapes, PE based heat shrinkable tape coatings were developed. As also mentioned earlier, it has an inner lining of thermoplastic hot melt adhesive and a monolithic radiation crosslinked PE cover coating. In this system, the outer coverings are applied over the pipelines with hot melted adhesive inner layer. Heat shrinkable structure can be generated through the heating of radiation crosslinked PE from the outside with the aid of propane torches. Moreover, a three component system of PE has also been developed, which uses an additional epoxy primer [11].

As a result of continuous developments in PE adhesion over the pipeline surfaces, two-layer and three-layer PE coatings were introduced. More details of these are available in the earlier section of this review. Among these two coating systems, three-layer PE coatings have received more interest. Structurally, these are composed of epoxy inner lining, PE copolymer adhesive tie layer and PE surface layer. The performance of such a system results from the combined contribution from individual layers. Epoxy lining provides cathodic disbonding resistance and improved cohesion to the system. Due to good moisture, oxygen and chemical resistance as well as excellent mechanical properties of PE, it acts as an outstanding protective barrier. Generally, epoxy primers are applied by electrostatic spraying whereas middle and PE surface coatings are employed on the pipelines using extrusion [25,26].

Kamimura *et al.* [27] studied the cathodic disbonding mechanism of three layer PE pipe coating consisting of liquid epoxy primer, maleic acid anhydride modified PE adhesive layer and PE protective layer. The impact of coating thickness, dissolved oxygen, sodium chloride (NaCl) concentration and cathodic potential on the cathodic disbonding of PE in NaCl solution under an elevated temperature of 65 °C was investigated. Moreover, the migration routes of cations, water and oxygen molecules to the coatings were studied. From the results of the disbonding test over 14 days, it was observed that the disbonding radius (r) of coatings decreased with increase in PE thickness from 0.9 to 8.5 mm. The migration of oxygen and water molecules through PE contributed to disbanding, however, the increase in thickness of PE was observed to shield the migration of oxygen and water molecules through it and thereby resulted in a decrease in r value. The excellent chemical resistance of PE prevented the sodium migration for all coating thicknesses. At the same time, the migration of sodium through the interface of coatings and steel was observed. Furthermore, the authors reported strong disbonding inhibiting effect of coatings at elevated temperature with strong evolution of hydrogen along with a cathodic potential of -1500 mV$_{SCE}$ [27]. In another study, Guermazi *et al.* [28] reported the hygrothermal ageing consequences of HDPE pipeline coatings on physico-chemical, mechanical and tribological properties. The authors submerged the HDPE pipeline coatings in distilled water as well as synthetic sea water (saline solution) at an elevated temperature of 70 °C to accelerate the ageing process. The existence of a low diffusion process in both the solvents was observed and a comparatively higher diffusion of water molecules was found in distilled water. A decrease in glass transition temperature (T_g) of HDPE coatings was observed with increase in immersion duration, which resulted from the increase in flexibility of amorphous molecular part of the coatings through the plasticizing effect of water molecules. The decreased tensile strength and elastic modulus values of coatings after hygrothermal ageing revealed the deterioration of mechanical properties. Furthermore, the authors reported that decrease in wear resistance for aged coating samples, which was attributed to the plasticizing effect of the penetrated solvent molecules in the amorphous part [28]. In a similar study, Guermazi *et al.* [6] investigated the effect of hygrothermal ageing time and temperature on the structural as well as mechanical properties of HDPE pipeline coatings. The samples were immersed in synthetic sea water at varying temperatures of 23, 70 and 90 °C for different immersion periods up to several months. An increase in solvent diffusion in to the coatings was observed with increase in temperature. Also, deterioration in mechanical properties such as tensile modulus, tensile strength and stress at 500% of strain was observed with enhancement in ageing parameters such as temperature and time. Furthermore, the authors displayed that the degradation behavior of aged samples through the structural variations using Fourier transform infrared (FTIR) spectroscopy. FTIR spectroscopic analysis demonstrated a decrease in peak intensities as well as vanishing of some functional groups in the coating materials after hygrothermal ageing [6].

Tribological behavior of pipeline coatings is an important parameter that determines the application in oil and gas industries. Guermazi *et al.* [29] investigated the friction and wear responses of both unaged and hygrothermally aged HDPE coating samples under varying applied loads and testing times at room temperature in a pin-on-disk tribometer. The applied load and test duration had strong impact on the friction coefficients and wear resistance of coatings. Increasing tendency of friction coefficients and wear volume were observed with increase in load. Also, the authors reported the independency of friction coefficient and the dependency of wear volume on the accelerated ageing. The remarkable deterioration of wear resistance was noticed for aged samples, particularly for samples aged for long durations. Moreover, the authors proposed a wear mechanism of coatings and confirmed the existence of direct relationship between wear volume and energy dissipation using an energetic quantitative approach [29]. In another study, the response of PE coatings towards the scratch damage was investigated [30]. For this purpose, unaged and aged coating samples were analyzed under varying scratch parameters of sliding velocity, angle of attack and applied normal load at room temperature. A decrease in friction coefficient with increasing sliding velocity and an increase in friction coefficient with increasing attack angle of the indenter were reported. The variation of applied load had no significant effect on the friction coefficient. The existence of combined permanent viscous as well as temporary elastic deformations during the application of scratch load was reported because of the viscoelastic behavior of PE. Furthermore, variation in wear volume of the coatings with increase in attack angle of indenter and applied normal load was noticed. The degradation behavior of samples was examined using hygrothermally aged coatings and remarkable decline in the wear resistance of samples was observed [30].

In another study, Samimi *et al.* [9] studied the causes of corrosion in three layer PE coated steel pipelines. The authors confirmed the influence of initial adhesiveness and contact environment on the performance of coatings. Moreover, the authors demonstrated the impact of moisture as well as cathodic disbonding resistance of coatings on the durability [9]. In another study, the occurrence of disbonding at the steel/FBE interface of the three-layer PE pipeline coating was investigated [31]. The authors reported the existence of under cured FBE layer when the applied temperature for adhesive-FBE bonding was low. In addition, a weak bonding between the FBE and adhesive was also observed when the temperature was high because of the fully crosslinked structure of FBE.

As mentioned earlier, high performance composite coating (HPCC) is a multi-component powder system composed of FBE primer, tie layer of chemically modified PE copolymer adhesive and outer layer of MDPE. In order to reduce the interlayer delamination and, thus, to create a single coating structure, a blend of adhesive and FBE is used as the tie layer for this system. As a result, a strong interlocked structure without well-defined interfaces is formed between the components. This results from the structural similarities of adhesive layer and PE and which helps to intermingle effectively with the structures containing FBE [13,32]. Howell *et al.* [33] evaluated physical, chemical and mechanical properties of coating membranes and coated steel pipes based on HPCC. The authors analyzed microstructure, impact strength, adhesion, water permeability, resistance towards cathodic disbondment and electrochemical impedance of coating samples. The presence of PE in the HPPC coatings was observed to result in a uniform coating structure with improved water and chemical diffusion resistance. Further, superior adhesion of HPCC to the steel pipelines was observed according to both ASTM and CSA standards. The authors also observed good impact energy of 10.2 J at low temperature conditions (0 °C). The results of impedance measurements underlined outstanding anti-corrosive characteristics of HPCC because of capillary behavior. Moreover, a little cathodic disbondment was reported for HPCC. In another study, Singh *et al.* [34] compared the properties of powder coated HPCC and conventional 3 layer PE coatings. Application of HPCC was beneficial for special needs such as coatings for raised welds and thickness adjustments. Excellent adhesion as well as good barrier properties for HPCC were observed. The authors also reported excellent corrosion resistant behavior and mechanical properties for the coating. In another work, Guan *et al.* [32] also observed outstanding cost effective performance of HPCC as compared with typical coating systems. The performance was confirmed through net lifelong performance-cost satisfaction investigations of miscellaneous coating systems [32].

In order to resolve the usual complications associated with the conventional three layer PE coatings such as weld tenting and low coating thickness on the outside weld parts, Lam *et al.* [35] developed a new system of polyethylene coatings. The authors introduced a side extruding HDPE above the graded structure PE coating (GSPE), which led to uniform coating over the external weld, thus, avoiding weld tenting. Moreover, the authors observed considerably low residual stresses on the coatings, which resulted in low temperature flexibility and good impact strength and, thus, improved adhesion of these coatings [35].

PP Based Pipeline Coatings

PP is a versatile semi-crystalline commodity thermoplastic polymer with a chemical formula of $(C_3H_6)_n$, prepared using Ziegler-Natta catalysts [36]. PP based pipeline coatings have been in use since 1980's and have gained attention because of their enhanced mechanical and thermal properties along with advantageous chemical resistance as compared to conventional PE and epoxy based systems [37]. Till now, PP has been used in diverse pipeline coatings include polymeric tapes, heat-shrinkable sleeves and two-layer and three-layer PP coatings.

Suzuki *et al.* [38] reported PP coated steel pipes suitable for oil and gas transmission at -30°C to 120°C temperature range. Structurally, the system contained steel pipe, modified polyolefin based adhesive inner layer and PP outer layer. The authors developed the PP based pipeline coatings to resolve the problems of PE based system and to obtain good coating performance even at higher temperatures. PP coatings exhibited good chemical stability and elevated softening temperature, however, inferior mechanical properties were observed at lower temperatures when compared to PE coatings. As a continuation of advancements in PP based pipeline coatings, Guidetti *et al.* [39] made use of three-layer PP coating system for oil and gas transmission pipelines. The three-layer coating system comprised of epoxy resin inner layer, modified PP copolymer based tie layer and outer covering of PP. Significant bonding of epoxy resin with oxides of metal surfaces as well as polar groups of adhesive tie layer was observed. Also, the existence of compatible structures was observed due to the structural resemblance of tie layer and PP outer layer. Thus, appreciable performance of the three-layer PP coatings was obtained due to the synergistic effect of epoxy resin and PP. Epoxy resin contributed adhesion, excellent cathodic disbonding resistance and good interfacial properties to the system, whereas PP developed outstanding physical and mechanical properties, high temperature stability and resistance towards chemicals, corrosion and water permeation. In another study, Moosavi *et al.* [40] examined the feasibility of three-layer PP pipeline coatings at elevated temperatures (more than 100°C) for long term performance. Total disbondment of coatings from the steel pipes as well as the serious cracking and tearing were observed. This was attributed to the low temperature resistance of FBE layer in the three-layer PP pipeline coatings. Further, the increased operating temperatures and heat applied maintenance activities were observed to reduce the life span of the coating. The authors observed decrease in mechanical properties on changing the FBE layer material to harder one. The authors recommended the use of FBE primer having a glass transition temperature (T_g) in the range of pipeline design temperature [40]. In another work, Moosavi *et al.* [41] investigated the failure history of the untimely failed three-layer PP gas pipeline coatings at high temperature. The authors observed the cracking of polypropylene top layer and the disbondment of the coatings from the surface of steel pipe. The cracking of PP outer layer was observed to result from the net effect of thermo-oxidative degradation and high residual stresses, whereas the disbondment of three-layer PP coating resulted from the adhesion loss as well as high residual stresses [41]. Overall, PP has emerged as a useful material for the pipeline coatings.

9.3.2 Epoxy Based Pipeline Coatings

Epoxies are high performance thermosetting polymeric material with wide range of applications especially for polymer based coatings. These materials are prepared by the condensation reaction of bisphenol-A (BPA) and epichlorohydrin (ECH), which creates molecules having two or more epoxide groups [42]. As compared with other polymeric materials, epoxy based coating materials provide excellent performance profile, which makes these coatings the materials of choice in many applications. Epoxy based coatings impart good adhesion, chemical resistance, outstanding cathodic protection, stress cracking resistance and excellent resistance towards microorganisms [16]. Thus, epoxy based coating systems have been developed in various forms and applications. These materials find use in applications such as liquid epoxy coatings, fusion bonded epoxy (FBE) coatings, base primer coatings for many conventional coating systems and so on. Epoxy based coatings are well accepted for pipeline protection through both external and internal coating applications. The excellent properties such as suitable coating thickness in a single step, 100% solid nature, corrosion resistance, easy pipeline cleaning and improved fluid flow efficiencies of epoxy based coatings make them advantageous for inner liners [3,43].

Coatings based on liquid epoxies comprise of two-part system of epoxy resin and curing agent, where the curing agent normally used is either polyamine or polyamide. Also, liquid epoxies can be used in combination with polyurethane as a spray applied coating system [18]. Epoxies also find applications as base coat primer for many coating systems including coal tar enamels, two-layer and three-layer polyolefin coating systems, HPCC and so on because of their excellent adhesion characteristics to steel surfaces. Another significant application area of epoxies is FBE, available as single layer and dual coat FBE. Generally, FBE is a heat curable one part powder form of thermosetting epoxy resin and exhibits good adhesion, fine surface finish and outstanding resistance towards abrasion, chemicals and soil stress. Single layer FBE coatings have been used since 1960's and are a single layer monolithic structure of FBE. These are applied over the well cleaned and preheated pipelines using electrostatic spraying. Dual coat FBE coatings are advanced form of single layer FBE coatings with versatile performance profile and have been in use since 1990's. Structurally, dual coat FBE comprise of top and base FBE coatings. This multilayer structure of coating system imparts many advantages to the pipelines such as resistance to mechanical damages caused by impact load, abrasion, gouge, friction with organic and inorganic surfaces, etc., along with ultraviolet or weathering resistance as well as high temperature performance [43].

The friction factor of the flowing gas through the pipelines greatly depends on the nature of internal coating. Yang *et al.* [3] conducted an aerodynamic evaluation of the internal epoxy coatings of natural gas pipelines to study this phenomenon. The authors compared the Colebrook-White equation based numerical friction factors with the data obtained from both field tests and model experiments. The authors observed consistency in both numerical and experimental readings, which proved the usefulness of numerical model for the analysis of epoxy internal coated gas pipelines. The application of internal epoxy coatings for the gas pipelines was observed to lower the frictional pressure drop, thereby, reducing the payback time [3]. Wei *et al.* [44] reported the influence of flow conditions on the performance degradation of FBE coatings in corrosive environments. For this purpose, three miscellaneous FBE coatings in 3% NaCl solution at 60 °C were studied under flowing and steady conditions and the degradation behavior was analyzed using electrochemical impedance spectroscopy (EIS). The authors observed strong effect of subjective conditions and flowing environment on the performance deterioration of protective coatings. At flowing conditions, the ions present in the corrosive environments diffused easily through the coatings when compared to water molecules.

In many recent studies, different advanced epoxy based coatings systems have also been reported, which confirm high potential of these systems for use in pipeline coatings [45,46]. For instance, Luo and Mather [45] reported an advanced system based on shape memory assisted self-healing coatings. This was achieved by distribution of electrospun thermoplastic polycaprolactone in a shape memory epoxy matrix. As shown in Figure 9.2, the damage to the coatings could be healed automatically on the application of heat. In another study, Augustyniak *et al.* [46] reported smart epoxy coatings here early detection of steel corrosion could be gauged through fluorescence. It was achieved through fluorescence indicator added in the epoxy coatings systems, which formed complex with the ferric ions generated during corrosion process. The indicator was observed to become fluorescent in and around the areas where the corrosion was initiated, however, this was achieved before any observable damage to the metal occurred, thus, indicating the advanced nature of detection. Figure 9.3 also shows the performance of the indicator in various test conditions. The fluorescence effect was clearly visible in the images, which resulted in early detection of any corrosion process using the advanced coating system. In another study, Wei *et al.* [47] reported the effect of carbon black (CB) on the performance improvement of FBE coatings under corrosive environments. The FBE coatings with varying CB loadings (0.5-4 wt %) were immersed in a 3% NaCl solution at room temperature. The degradation behavior of CB filled FBE pipeline coatings was analyzed using EIS, thermo-gravimetry, differential scanning calorimetry and visual inspections. FBE coating with CB concentration above the percolation value exhibited peculiar electrochemical behavior. T_g value was observed to increase significantly after crossing the percolation value. Further, the addition of CB in to FBE improved its barrier properties through the CB network formation, which correspondingly contributed to the enhancement of the coating's performance.

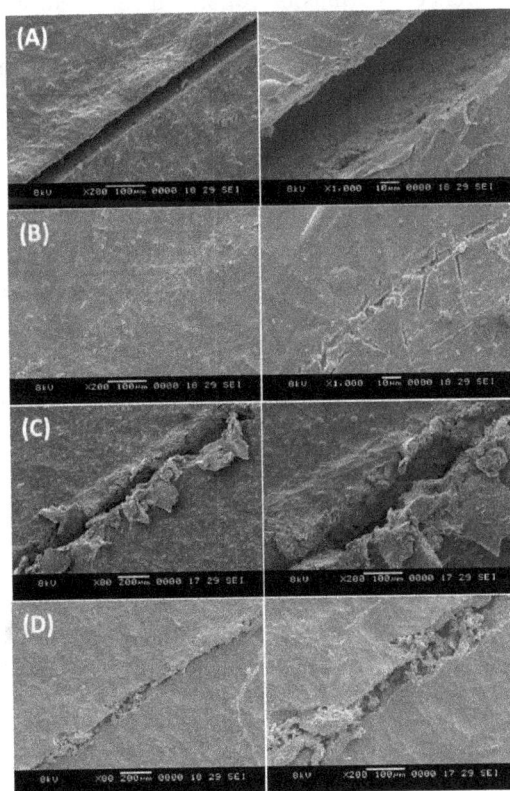

Figure 9.2 SEM micrographs of (A) cracked coating, (B) crack coating after self-healing, (C) scribed coating, and (D) scribed coating after self-healing. Reproduced from Reference 45 with permission from American Chemical Society.

Goertzen *et al.* [48] investigated the creep behavior of the coating system based on carbon fiber-epoxy composite to understand the long-term deflection and failure behavior. Creep tests were carried out at both room temperature and elevated temperatures with the help of respective tensile and flexural creep analysis. The authors observed consistent creep curves in both experiments. Tensile creep tests revealed outstanding creep rupture resistance of composite coatings and the coating material had the capability to withstand a load of 77% of the ultimate tensile strength (UTS) for up to 1600 h. Elevated temperature flexural creep analysis was performed on a dynamic mechanical analyzer (DMA) with a temperature profile of 30 to 75 °C. The authors predicted the creep levels of coatings extrapolated to the 50 years life span and observed the stress value of 84% of UTS at 30 °C and 42% of UTS at 50 °C. Further, the modulus reduction of 18% at 30 °C and 58% at 50 °C after 50 years of composite coating life was predicted. Another study by Alamilla *et al.* [49] reported the failure inspection as well as the mechanical behavior of FBE coated oil pipelines. The authors followed five different methodologies to examine the failure behavior including visual inspection, environmental analysis, interfacial characterization, mechanical and metallurgical analysis. It was revealed that the existence of both iron oxide and iron sulfide accelerated corrosion mechanisms on the FBE coated oil pipelines. Also, the authors underlined that the leakage caused ductile type failure on the oil pipelines, which was in good agreement with the previous reported works.

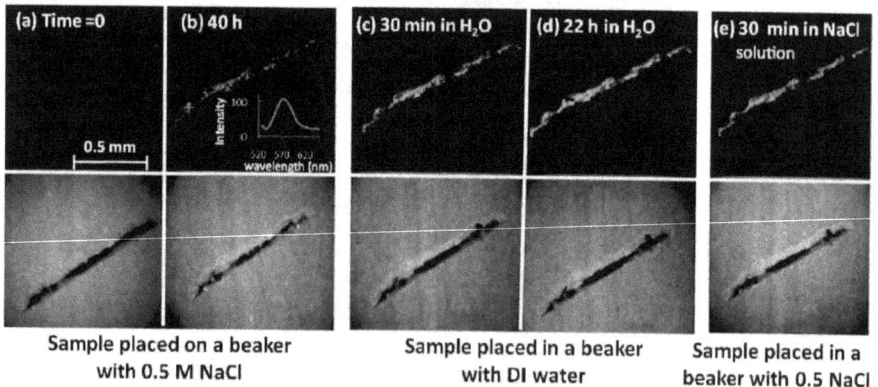

Figure 9.3 Scribed area on the coated sample as a function of different exposure time to different corrosive environments. Top row: fluorescent images and bottom row: digital camera images. Reproduced from Reference 46 with permission from American Chemical Society.

Zhou *et al.* [50] developed high T_g FBE coatings to protect the oil/gas pipe-lines at inflated service temperatures. The higher T_g of final coatings was ob-tained through the incorporation of brominated epoxy resin in the composi-tion of thermoset epoxy powder coatings. In the coating system, the T_g en-hanced up to 159 °C, which was about 39% greater than that of general FBE. Enhanced cathodic disbondment resistance and water soak adhesion were observed when the coatings were analyzed for 28 days at 95 °C. The system exhibited enhanced UV resistance, flexibility and better impact properties as compared to general FBE. In another study, Moon *et al.* [51] also compared the performance of high T_g FBE and general FBE. The experimental results confirmed the outstanding performance of high T_g FBE at elevated tempera-tures. To protect and strengthen the pipelines, Duell *et al.* [52] also introduced a fiber reinforced polymer (FRP) system based on carbon fiber and epoxies. The authors used an epoxy putty to eliminate the surface roughness of defect-ed surfaces and carbon fiber-epoxy composites as the protecting covering. Various defect geometries on the pipeline surfaces were detected using finite element analysis. Both modeling and field tests were applied to confirm the effectiveness of carbon fiber-epoxy composite based coatings. Furthermore, a little effect of varying defect geometries on the failure pressure was observed. In another study, Yuan *et al.* [53] reported self-healing polymer coatings based on epoxy/mercaptan as healant. In this system, epoxy and mercaptan hardener were encapsulated individually and were subsequently embedded in epoxy. The material exhibited significantly enhanced self-healing perfor-mance (Figure 9.4) even at much lower capsule content. For instance, 43.5% healing efficiency was observed with 1 wt % capsules and 104.5% healing ef-ficiency was reported with 5 wt % capsules. The healing occurred at or below room temperature, thus, further confirming the potential of the developed system to be effective for pipeline protection. Much better balance between strength and toughness of the healed system could be achieved.

In another study, Bakhshandeh *et al.* [54] reported the development of anti-corrosive organic-inorganic hybrid coatings based on epoxy-silica nanocom-posites. The coatings were fabricated using silane functionalized diglycidyl ether of bisphenol A epoxy resin and pre-hydrolyzed tetraethoxysilane (TE-OS) with the aid of 3-aminopropyl triethoxysilane (APTES) as coupling agent. The extent of compatibility between organic and inorganic phases was varied using different amounts of APTES. It was noticed that the smaller silica do-mains were generated for the coating system with epoxide to amine ratio of 4:1. Hybrid composites with 12.5 wt% of TEOS content exhibited optimal ad-hesion strength as well as micro-hardness. The enhanced barrier performance

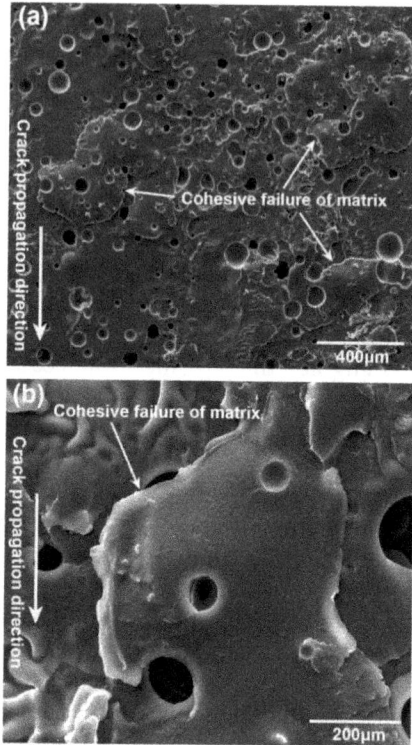

Figure 9.4 SEM images of the fracture surface of a healed specimen, containing 10 wt % epoxy-loaded capsules and 10 wt % hardener-loaded capsules. Reproduced from Reference 53 with permission from American Chemical Society.

and the resultant corrosion resistance of coatings were attributed to the creation of silica intermediate layer at the coating-substrate interface. In another study reporting the advancement of epoxy based coating systems, Weng *et al.* [55] generated advanced anti-corrosion coatings by mimicking fresh plant leaves, thorough the combination of superhydrophobicity and redox catalytic capability. The authors generated superhydrophobic elecroactive epoxy (SEE) coating on steel using nano-casting method from the surface structure of Xanthosoma sagittifolium leaves. Due to such structured coating, the anti-corrosion performance was observed to enhance significantly as compared to the smooth coating on the steel substrate. Figure 9.5 exhibits the atomic force microscopy (AFM) images of the SEE surface topography.

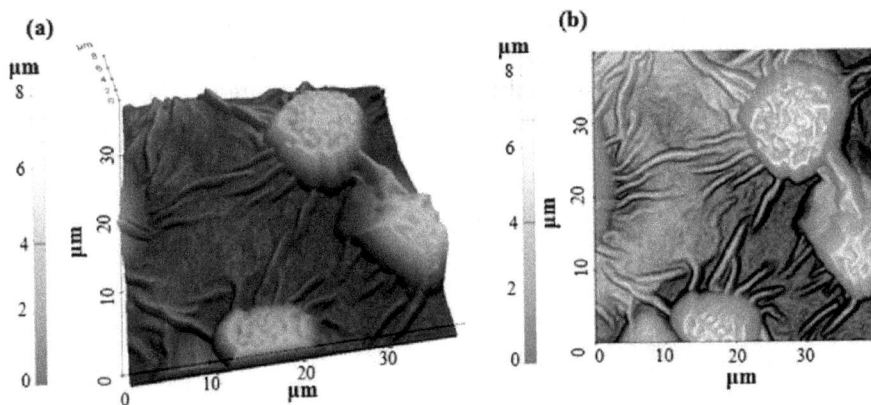

Figure 9.5 AFM images of the SEE topography; (a) 2-dimensional and (b) 3-dimensional. Reproduced from Reference 55 with permission from American Chemical Society.

9.3.3 Polyurethane Based Pipeline Coatings

Polyurethanes (PUs) are class of polymeric materials first introduced in 1937 and are available in thermoplastic, thermoset and elastomeric forms. These materials find wide variety of commercial and technical applications because of their attractive property profile. Commonly, these materials are prepared by the polyaddition reaction between polyisocyanates and macro-polyols. The final structure of PU is driven by the nature of isocyanates and polyols, along with the extent of crosslinking [56-58]. PU pipeline coatings offer enhanced corrosion, abrasion, scratch, environmental degradation and tear propagation resistances along with better adhesion, low temperature impact strength, adherence to health and safety stipulations and rapid curing speed. Due to the crosslinked structure in the thermoset PU coatings, further improvement in tensile strength, abrasion resistance and chemical stability of the coatings is achieved [22]. Also, rigid PU foams are widely used for the thermal insulation of pipelines in the oil and gas industries owing to their enhanced insulating properties.

Ghosal *et al.* [59] reported soya polyurethane based silica hybrid composite coatings for anti-corrosion performance. In-situ generation of silica in the polymer led to enhanced thermal, physico-mechanical and corrosion resistance of the hybrid coatings. Figure 9.6 shows the potentiodynamic polarization (PDP) studies of the composites in comparison with pure polymer and pure substrate. The corrosion protection ability of the coatings enhanced with increasing the amount of silica in the composites. The enhanced corrosion protection efficiency of the composite coatings was attributed to the blocking effect and good adhesion of coating with the substrate surface. In another study,

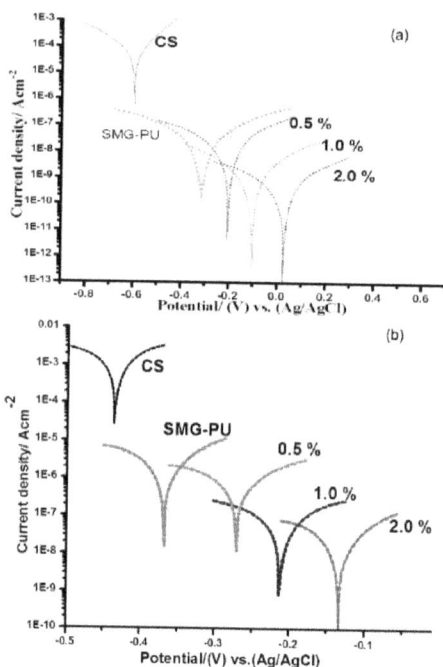

Figure 9.6 PDP curves of (a) CS, (b) SMG-PU, and composites with (c) 0.5%, (d) 1% and (e) 2% silica content in 3.5 wt % NaCl medium (SMG: soy oil monoglyceride, CS: carbon steel). Reproduced from Reference 59 with permission from American Chemical Society.

Chattopadhyay *et al.* [60] studied the effect of chain extender on the phase mixing as well as coating performance of polyurethane ureas. It was concluded that the bulky diol chain extenders helped to enhance phase mixing. Sufone (SUL) based chain extender was also observed to enhance mechanical properties of the polymer, along with surface segregation. Figure 9.7 also demonstrates the G' plots for SUL as a function of temperature at various angular frequencies. Another study by Samimi *et al.* [61] reported the properties and the application of 100% solid PU based pipeline coatings. The authors reported outstanding performance such as good adhesion, excellent resistance to corrosion, stroke, chemicals and frication, satisfactory flexibility and high temperature resistance. Furthermore, non-toxicity along with easy handling and fast curing of the coatings was also reported. Guan *et al.* [62] also investigated the performance of 100% solid rigid PU coatings used for the welded pipeline joints and rehabilitation of oil and gas pipelines. These advanced coatings were of castable or sprayable type and exhibited much faster curing. The authors also confirmed the superior performance of the coatings in comparison with other commercially available coating systems based on PU and other matrices. Furthermore, the coatings exhibited adherence to health and safety regulations as well as easy handling due to absence of volatile organic compounds (VOC), cold temperature curing, low curing time and balanced viscosity.

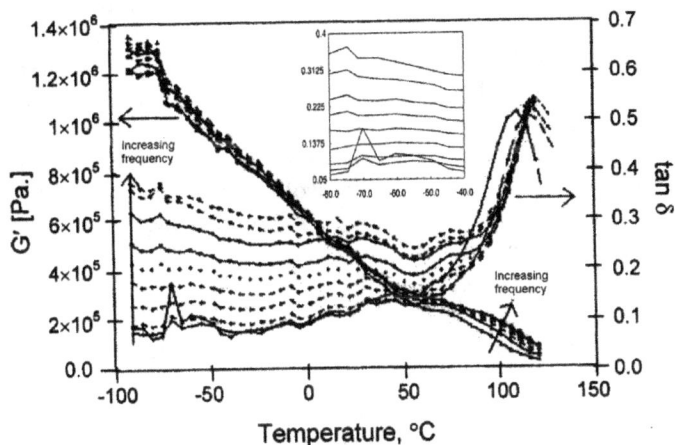

Figure 9.7 Variation of G' for SUL as a function of temperature at various angular frequencies (1, 5, 10, 20, 30, 40, and 80 rad/s). Reproduced from Reference 60 with permission from American Chemical Society.

Kong *et al.* [63] developed high-solid PU coating system using vegetable oil derived polyols. The generation of polyols possessing substantial functionality and low viscosity from 5 different vegetable oils such as canola oil (two grades), sunflower oil, flax oil and camalina oil was reported, followed by treatment with petrochemical derived diisocyanates to develop high-solid PU coatings exhibiting excellent mechanical and thermo-mechanical properties. In comparison, flax oil based PU had higher T_g and crosslinking density, low solvent swelling ratio, improved hydrophobic nature, excellent tensile properties, hardness and abrasion resistance over other vegetable oil based PUs [63]. Hygrothermal ageing of thermoplastic PU coatings was examined by Boubakri *et al.* [64] using PU coating samples immersed in water at 70 °C for 6 months. The authors evaluated water absorption as well as thermal, mechanical and tribological properties of the coatings. The coating samples exhibited high rate of Fickian water diffusion. A decrease in T_g value was observed owing to the plasticizing effect of coatings resulting from the absorption of water molecules. Ageing of PU coating samples was confirmed using FTIR results. Also, the ageing led to the reduction in the mechanical and tribological properties.

In another study, Rassoul *et al.* [65] investigated the cathodic protection of thermally insulated PU coated pipelines. The effect of specific gravity of PU and electrolytic (NaCl) concentration on the cathodic protection of coatings was studied. Samples with four distinct densities (35, 68, 86 and 113 kg/m³) in four different electrolytic concentrations (0, 1.5, 3.5 and 5% NaCl) were examined at room and higher temperatures. PU coatings with density of 68 kg/m³ in 3.5 and 5% NaCl concentrations were observed to have lower resistivity towards cathodic protection owing to inadequate PU resistivity and elevated electrolytic conductivity. Also, a current density of 30.9 mA/m² was observed to be suitable for the cathodic protection of PU with a density of 68 kg/m³ in 3.5% NaCl solution after 1608 h of complete saturation. An adequate shielding potential of coatings to protect pipelines up to primary saturation of 240 h was observed. Furthermore, the coatings presented stabilized cathodic protection up to 60 °C. A study by Sousa *et al.* [66] evaluated corrosion under thermal insulation of rigid PU foams on the pipelines and the compatibility of the foams with anti-corrosive base coatings. For this purpose, the authors developed aqueous extracts of PU foams according to ASTM C871 and analyzed using chemical, electrochemical and mass loss characterizations. Substantial impact of temperature increase on pH, conductivity and contents of phosphate, chloride and fluoride was observed using chemical analysis. Also, the influence of other components in the PU foams besides flame retardant on under coating corrosion was revealed through the halides generation studies. Remarkable effect of chloride content in the PU foams on the corrosion was observed, which can be resolved only through proper control of chloride during PU foam preparation. Furthermore, the compatibility tests using two anti-corrosive coatings indicated that FBE was the best choice for new pipelines and a traditional maintenance coating (BAR RUST) for old pipeline repairs.

Motamedi *et al.* [67] reported the corrosion protection capability and adhesion strength of PU coatings on mild steel surfaces in the presence of 1 M sulfamic acid cleaning solution containing surfactants such as dodecyltrimethylammonium bromide (DTAB) and its counterpart 12-4-12. Enhanced adhesion strength and corrosion resistance behavior of PU coatings in both DTAB and 12-4-12 were detected. Acid cleaning solution containing DTAB presented slightly better performances when compared to 12-4-12. Low molecular weight of DTAB was advantageous to enhance the adsorption rate and, thereby, to enrich the performance. Akbarian *et al.* [68] evaluated the corrosion protection capability of nanoparticulate silver added PU coatings in 3.5% NaCl solution. The effect of silver nanoparticles on the corrosion performance of water based PU (WPU) and high solid PU (HPU) was analyzed using EIS, SEM and FTIR spectroscopy. The incorporation of silver nanoparticles had no remarkable effect on the corrosion protection of HPU, whereas caused coating deterioration in the case of WPU. Lesser amount of carbonyl groups and their decomposition under moisture environment attributed to the deterioration of silver nanoparticles impregnated WPU. Furthermore, silver nanoparticles offered antibacterial properties to the PU coating. Khun *et al.* [69] studied the cathodic delamination of PU composite coating containing multiwalled carbon nanotube (MWCNT) from the steel surfaces with the aid of scanning Kelvin probe (SKP). Increase in anti-corrosion properties as well as considerable decrease in cathodic delamination for PU/MWCNT composite coatings was observed with increase in MWCNT content from 0 to 0.5 wt%. Also, the composite coatings exhibited enhanced cathodic delamination resistance due to the increased barrier resistance against oxygen and water molecules resulting from uniform dispersion of highly dense MWCNT structures in PU matrix. In another study, the individual effect of graphene oxide (GO), mildy reduced graphene oxide (RGO) and functionalized graphene (FG) on the corrosion resistance properties of waterborne PU (WPU) was reported by Li *et al.* [70]. The corrosion behavior of composite coatings in 3.5% NaCl solutions was analyzed using EIS and salt spray tests. Enhanced corrosion resistance for all PU/graphene composites was observed. Good dispersion of GO and RGO in the PU matrix was observed, whereas a submicron sized aggregation was reported in the case of FG. Thus, the fine dispersion of GO and RGO offered outstanding corrosion barrier performances to PU over FG. PU composites containing 0.2 wt% RGO were reported to have optimal anti-corrosion properties and constant impedance modulus of 10^9 Ω at 0.1 Hz for up to 235 h [70].

9.3.4 Fluoropolymer Based Pipeline Coatings

Fluoropolymers are the fluorocarbon based polymers containing stronger carbon-fluorine bond in their chain structure. The history of this important class of polymeric materials dates back to 1938 with the accidental invention of polytetrafluoroethylene (PTFE or Teflon) by DuPont. Fluoropolymers have been utilized for a wide variety of coating applications because of their excellent performance characteristics. The unique properties of these polymers include high temperature stability, resistance to corrosion, abrasion and chemicals, high electrical resistivity, low surface energy and refractive index. However, fluoropolymers generally have poor solubility in conventional coating solvents and poor adhesion to metals and other substrates. To enhance the usage of fluoropolymers as coating materials, many developments have been reported to overcome the drawbacks. Generally, fluoropolymer coatings are used for the purposes of corrosion resistance as well as fouling-release in the oil/gas pipelines. Examples of fluoropolymers used for coating purposes comprise polytetrafluororoethylene (PTFE), tetrafluoroethylene/hexafluoroethylene copolymers (FEP), TFE/perfluoroalkyl vinyl ether copolymers (FEP), polyvinylidene fluoride (PVDF), fluoroethylene vinyl ether (FEVE), etc.

In a recent study, Lee *et al.* [71] reported transparent superhydrophobic and translucent superamphiphobic coatings via spraying silica–fluoropolymer hybrid nanoparticles (SFNs). Figure 9.8 demonstrates the fabrication process of the fluoropolymer modified silica nanoparticles. No specific modification of the substrate was required and the coatings could be applied to a variety of substrates. The modified silica nanoparticles exhibited strong potential of application in a wide variety of advanced coatings systems. In another

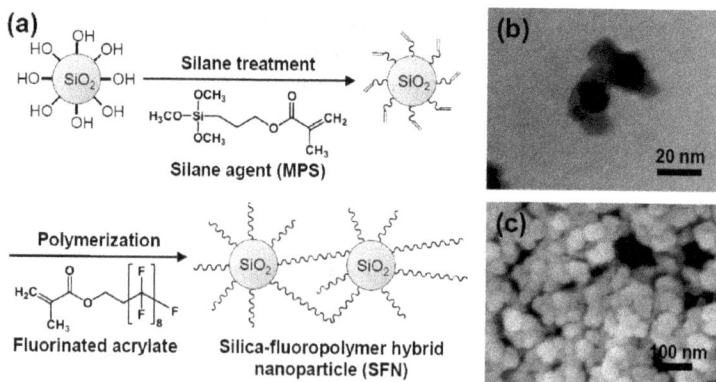

Figure 9.8 (a) Schematic of the fabrication process of fluoropolymer modified silica nanoparticles. TEM (b) and SEM (c) images of the hybrid nanoparticles (SFNs). Reproduced from Reference 71 with permission from American Chemical Society.

study reporting the enhancement of fluoropolymers for coatings applications, Gudipati *et al.* [72] developed crosslinked hyperbranched fluoropolymer (HBFP)-poly(ethylene glycol) (PEG) composite coatings (Figure 9.9). The coatings inhibited protein adsorption and marine organism settlement.

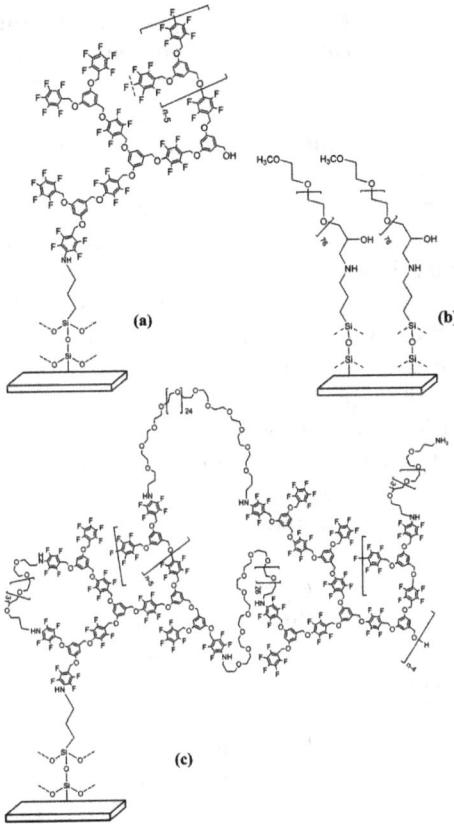

Figure 9.9 Representation of covalent attachment of (a) HBFP, (b) PEG, and (c) HBFP-PEG cross-linked networks to the substrate. Reproduced from Reference 72 with permission from American Chemical Society.

Darden *et al.* [73] discussed the modifications in the long-life fluoropolymer based coating (such as FEVE) to adjust with the environmental regulations. In this respect, new FEVE coating systems were developed, which included FEVE solid resins, FEVE water dispersions, FEVE water emulsions in blended resin system, silanol functional FEVE resins and fluorourethanes. Significant improvement in weathering resistance was observed for coatings based on FEVE blend. Fluorourethanes were also attractive for industrial use because of favourable life cycle. McKeen *et al.* [74] reported a sequence of new fluoropolymer based multilayered coating systems with outstanding adhesive strength for the inside surfaces of the oil and gas pipelines. Structurally, these coatings composed of blend of fluoropolymer and binder resins such as polyamide-imide, polyether sulfone or polyphenylene sulfide as primer, permeation barrier additives as midcoat and fluoropolymers as topcoat. The coatings were analyzed under various conditions of high temperature and pressure in sweet, sour and hydrochloric acid environments. The properties of the coatings, especially adhesion strength, remained unchanged after exposure to variety of laboratory testing environments. Another study by McKeen *et al.* [75] generated a fluoropolymer based internal coating system for the protection of downhole production tubulars in the oil and gas wells. The coatings were examined using corrosion autoclave tests under salt and sweet environments at 8000 psi pressure and 325 °F temperature for 24 h and followed by a rapid decompression from 8000 psi to 1500 psi within 34 sec. In addition, the coatings were analyzed for abrasion resistance and deposition reduction. It was concluded from the study that the fluoropolymer based coatings were favored due to corrosion resistance at high temperatures and pressures as well as fouling-release against organic and inorganic depositions include asphaltene, paraffin and $BaSO_4$ scales [75].

Bayram *et al.* [76] studied the performance of perfluoroalkoxy (PFA) fluoropolymer and a hybrid epoxy-fluoropolymer resin for the corrosion protection of the interior surfaces of the oil and gas pipelines. The morphological, adhesion, hardness and corrosion resistance behavior of the coatings towards chlorine, hydrochloric acid and salt fog environments was investigated. PFA coatings were a three-layer system composed of pure PFA as top layer, blend of perfluoroethylene propylene (FEP) (59 wt%) and two non-fluoropolymer binders polyamide-imide (PAI) (5 wt%) and polyether sulfone (PES) (36 wt%) as primer and a blend of PFA and FEP as midcoat. In the case of PFA coatings, good adhesion of coatings with substrates was reported. Also, good interlayer adhesion and chemical resistance was observed under corrosive environments. PFA coatings along with primer and midcoat were confirmed to be suitable for corrosion resistance. Hybrid epoxy-fluoropolymer resin was a blend of diglycidyl ether of bisphenol A (DGEBA) and 10 wt% low molecular weight PTFE as single layer. The authors observed good adhesion strength of these hybrid coatings to the substrates, which was attributed to the incorporation of epoxy in the coating formulations. Further, the usage of PTFE provided high service temperature and corrosion resistance to the coatings.

Table 9.1 presents advantages, limitations and application of PE, PP, epoxies, PU and fluoropolymer based pipeline coating systems reported in various literature studies.

9.4 Conclusions

In this chapter, polymer coatings used for the pipeline protection in oil and gas industries have been reviewed, progressing from epoxy primer of four components coal tar enamel (CTE) system in the early stages to the next generation polymer nanocomposite based coatings under development today. Polymer based coatings provide effective solutions for external as well as internal protection of pipelines from corrosion, fouling and mechanical damage. Of the various polymeric materials, polyethylene, polypropylene, epoxies, polyurethanes and fluoropolymers have attractive property profiles. Polyolefin based coatings have excellent mechanical properties, corrosion, moisture and chemical resistance, healthy and safety adherence and low cost. Performance characteristics of both polyethylene and polypropylene based coatings have gained remarkable developments over the last decades so as to achieve suitability for application in challenging environments. Protective coatings based on epoxies either as stand-alone liquid/powder or as part of other coating systems also exhibit good performance profiles, including good adhesion strength to metallic surfaces, cathodic protection and resistance to chemicals, stress cracking and biological environments. Similarly, PU pipeline coatings have good adhesion, low temperature impact strength, adherence to health and safety regulations, fast curing speed, thermal insulation and resistance to corrosion, abrasion, scratch and tear propagation. Fluoropolymer based coating systems have also acquired wide acceptance and usefulness for pipeline protection. These polymers offer outstanding corrosion resistance at elevated temperatures and pressures, along with good fouling release owing to their unique features such as high temperature stability, low surface energy and refractive index, resistant to corrosion, abrasion and chemicals, high electrical resistivity, etc. Thus, polymer materials enhance the life span of oil and gas pipelines through either corrosion control or rehabilitation of corroded parts as protective coating materials.

Table 9.1 Comparison of commonly used polymeric pipeline coatings, specifically their application

Polymer	Coating systems	Advantages	Limitations
Polyethylene	Polymeric tapes; Heat shrinkable tape coatings; 2-layer and 3-layer polyethylene coatings; High performance composite coatings (HPCC)	Good mechanical properties; Corrosion resistance; Chemical stability; Electrical resistance; Water and moisture resistance; Inexpensive	Inferior adhesion to metallic surfaces; Limited temperature range
Polypropylene	Polymeric tapes; Heat shrinkable sleeves; 2-layer and 3-layer polypropylene coatings	Good mechanical properties; Corrosion resistance; High temperature stability; Electrical resistance; Water and moisture resistance; Good chemical resistance; Inexpensive	Inferior adhesion to metallic surfaces
Epoxies	Base coat primer; Fusion bonded epoxy (single and dual layer); Liquid coatings; Epoxy nanocomposites	Good adhesion; UV stability; Chemical resistance; Cathodic protection; Stress cracking resistance; Resistance to biological environments; Improved fluid flow characteristics	Low curing speed
Polyurethanes	Liquid coatings; Thermoplastic PU; 100% solid PU; Rigid PU foams; PU nanocomposites	Good corrosion resistance; Chemical resistance; Tear propagation resistance; Environmental degradation resistance; Abrasion and scratch resistance; Excellent thermal resistance, Low curing time	Low moisture resistance; Complexity in application
Fluoropolymers	Solid resins; Water dispersions; Emulsions in blended resin system; Silanol functionalized system; Fluorourethanes; Multilayered coating system; Hybrid fluoro-resins	Corrosion resistance; High temperature stability; Good abrasion and chemical resistance; Low surface energy and refractive index	Inferior adhesion to metallic surfaces; Poor solubility in conventional coating solvents

References

1. Banach, J. L. (1987) Pipeline Coatings - Evaluation, Repair, and Impact on Corrosion Protection Design and Cost. *Corrosion '87*, USA, CONF-870314.
2. Rueda, F., Otegui, J. L., and Frontini, P. (2012) Numerical tool to model collapse of polymeric liners in pipelines. *Engineering Failure Analysis,* **20**, 25-34.
3. Yang, X.-H., Zhu, W.-L., Lin, Z., and Huo, J.-J. (2005) Aerodynamic evaluation of an internal epoxy coating in nature gas pipeline. *Progress in Organic Coatings,* **54**, 73-77.
4. McGill, J. C., McGill, J. M., and Key, B. L. (1999) Pipeline Coating, US Patent 5984581 A.
5. Samour, C. M., Jackson, E. G., Thomas, S. J., and Davidson, L. E. (1980) Coated Pipe and Process for Making Same, US Patent 4213486 A.
6. Guermazi, N., Elleuch, K., and Ayedi, H. (2009) The effect of time and aging temperature on structural and mechanical properties of pipeline coating. *Materials & Design,* **30**, 2006-2010.
7. Papavinasam, S., Attard, M., Balducci, B., and Revie, R. W. (2009) Testing coatings for pipeline: new laboratory methodologies to simulate field operating conditions of external pipeline coatings. *Journal of Protective Coatings & Linings,* 32-51.
8. Papavinasam, S., Attard, M., and Revie, R. W. (2006) External polymeric pipeline coating failure modes. *Materials Performance,* **45**, 28-30.
9. Samimi, A., and Zarinabadi, S. (2011) An analysis of polyethylene coating corrosion in oil and gas pipelines. *Journal of American Science,* **7**, 1032-1036.
10. Samimi, A., Dokhani, S., Neshat, N., Almasinia, B., and Setoudeh, M. (2012) The application and new mechanism of universal produce the 3-layer polyethylene coating. *International Journal of Advanced Scientific and Technical Research,* **2**, 465-473.
11. Romano, M., Dabiri, M., and Kehr, A. (2005) The ins and outs of pipeline coatings: Coatings used to protect oil and gas pipelines. *Journal of Protective Coatings & Linings,* **22**, 40-47.
12. Sloan, R. N. (2001) Pipeline coatings. In: *Control of Pipeline Corrosion,* Peabody, A. W., and Bianchetti, R. L. (eds.), 2nd edition, NACE International, USA, pp. 7-20.
13. Niu, L., and Cheng, Y. (2008) Development of innovative coating technology for pipeline operation crossing the permafrost terrain. *Construction and Building Materials,* **22**, 417-422.
14. Jack, T. R., Wilmott, M. J., Sutherby, R. L., and Worthingham, R. G. (1996) External corrosion of line pipe - A summary of research activities. *Materials Performance,* **35**, 18-24.
15. Roche, M., and Melot, D. (2012) Recent experience with pipeline coating failures. In: *Protecting and Maintaining Transmission Pipe,* Technology Publishing Company, pp. 57-66. Online: http://www.paintsquare.com/store/assets/JPCL_transpipe_ebook.pdf#page=62 (accessed 18th February 2017).
16. Kehr, J. A. (2012) How fusion-bonded epoxies protect pipeline: single-and double-layer systems. In: *Protecting and Maintaining Transmission Pipe,* Technology

Polymers in Oil and Gas Industry

Publishing Company, pp. 13-22. Online: http://www.paintsquare.com/store/assets/JPCL_transpipe_ebook.pdf#page=62 (assessed 18th February 2017).

17. Enos, D., Kehr, J., and Guilbert, C. A high-performance, damage-tolerant, fusion-bonded epoxy coating. Online: alankehr-anti-corrosion.com (assessed 21st February 2017).

18. Kehr, J., Hislop, R., Anzalone, P., and Kataev, A. (2012) Liquid coatings for girthwelds and joints: proven corrosion protection for pipelines. In: *Protecting and Maintaining Transmission Pipe*, Technology Publishing Company, pp. 23-32. Online: http://www.paintsquare.com/store/assets/JPCL_transpipe_ebook.pdf#page=62 (assessed 18th February 2017).

19. Goldie, B. (2010) Developments in pipeline protection reviewed. *Journal of Protective Coatings & Linings*, 30-32.

20. Luckachan, G. E., and Mittal, V. (2015) Anti-corrosion behavior of layer by layer coatings of cross-linked chitosan and poly(vinyl butyral) on carbon steel. *Cellulose*, **22**, 3275-3290.

21. Soares, J. B., and McKenna, T. F. (2013) *Polyolefin Reaction Engineering*, John Wiley & Sons, USA.

22. Samimi, A., and Zarinabadi, S. (2012) Application of polyurethane as coating in oil and gas pipelines. *International Journal of Science and Investigations*, **1**, 43-45.

23. Roche, M. G. (2004) An experience in offshore pipeline coatings. *Corrosion 2004*, USA. Online: https://www.onepetro.org/conference-paper/NACE-04018 (assessed 28th February 2017).

24. Stafford, T. (1998) *Plastics in Pressure Pipes, Report 102*, Rapra Technology Limited, UK.

25. Soucek, M. D. (2012) The application and new mechanism of universal produce the 3-layer polyethylene coating," *International Journal of Chemistry*, **1**, 94-104.

26. Chang, B., Sue, H.-J., Wong, D., Kehr, A., Pham, H., Siegmund, A., Snider, W., Jiang, H., Browning, B., and Mallozzi, M. (2008) Integrity of 3LPE Pipeline Coatings – Residual Stresses and Adhesion Degradation. *7th International Pipeline Conference*, Canada, pp. 75-86.

27. Kamimura, T., and Kishikawa, H. (1998) Mechanism of cathodic disbonding of three-layer polyethylene-coated steel pipe. *Corrosion*, **54**, 979-987.

28. Guermazi, N., Elleuch, K., Ayedi, H., and Kapsa, P. (2008) Aging effect on thermal, mechanical and tribological behaviour of polymeric coatings used for pipeline application. *Journal of Materials Processing Technology*, **203**, 404-410.

29. Guermazi, N., Elleuch, K., Ayedi, H., Fridrici, V., and Kapsa, P. (2009) Tribological behaviour of pipe coating in dry sliding contact with steel. *Materials & Design*, **30**, 3094-3104.

30. Guermazi, N., Elleuch, K., Ayedi, H., Zahouani, H., and Kapsa, P. (2008) Susceptibility to scratch damage of high density polyethylene coating. *Materials Science and Engineering A*, **492**, 400-406.

31. Samimi, A. (2012) Study an analysis and suggest new mechanism of 3 layer polyethylene coating corrosion cooling water pipeline in oil refinery in Iran. *International Journal of Innovation and Applied Studies*, **1**, 216-225.

32. Guan, S. W., Gritis, N., Jackson, A., and Singh, P. (2005) Advanced Onshore and Offshore Pipeline Coating Technologies. *2005 China International Oil & GasPipeline Technology (Integrity) Conference & Expo*, China. Online: http://citeseerx.ist.psu.edu/viewdoc/download?doi=10.1.1.131.4370&rep=rep1&ty pe=pdf (assessed 21st February 2017).

33. Howell, G., and Cheng, Y. (2007) Characterization of high performance composite coating for the northern pipeline application. *Progress in Organic Coatings*, **60**, 148-152.

34. Singh, P. J., and Cox, J. J. (2000) Development of a Cost-effective Powder Coated Multi-component Coating for Pipelines. *Corrosion 2000*, USA. Online: https://www.onepetro.org/conference-paper/NACE-00762 (assessed 1st March 2017).

35. Lam, C., Wong, D. T., Steele, R., and Edmondson, S. (2007) A New Approach to High Performance Polyolefin Coatings. *Corrosion 2007*, USA. Online: http://w.brederoshaw.com/non_html/techpapers/BrederoShaw_TP_G_04.pdf (asessed 21st February 2017).

36. Karger-Kocsis, J. (2012) *Polypropylene Structure, Blends and Composites: Volume 3, Composites*, Springer Science & Business Media, USA.

37. Fairhurst, D., and Willis, D. (997) Polypropylene coating systems for pipelines operating at elevated temperatures. *Journal of Protective Coatings & Linings*, **14**, 64.

38. Suzuki, K., Ishida, M., Ohtsuki, F., Inuizawa, Y., Hinenoya, S., Tanaka, M., and Shindou, Y. (1986) Polypropylene Coated Steel Pipe, US Patent 4606953 A.

39. Guidetti, G., Rigosi, G., and Marzola, R. (1996) The use of polypropylene in pipeline coatings. *Progress in Organic Coatings*, **27**, 79-85.

40. Moosavi, A. N., Al-Mutawwa, S. O., Balboul, S., and Saady, M. R. (2006) Hidden Problems with Three Layer Polypropylene Pipeline Coatings. *Corrosion 2006*, USA. Online: https://www.onepetro.org/conference-paper/NACE-06057 (assessed 25th February 2017).

41. Moosavi, A. N., Chang, B. T., and Morsi, K. (2010) Failure Analysis Of Three Layer Polypropylene Pipeline Coatings. *Corrosion 2010*, USA. Online: https://www.onepetro.org/conference-paper/NACE-10002 (assessed 19th February 2017).

42. Pascault, J.-P., and Williams, R. J. (2009) *Epoxy Polymers*, John Wiley & Sons, USA.

43. Kehr, J. A., and Enos, D. G. (2000) FBE, A Foundation for Pipeline Corrosion Coatings. *Corrosion 2000*, USA. Online: https://www.onepetro.org/conference-paper/NACE-00757 (assessed 25th February 2017).

44. Wei, Y., Zhang, L., and Ke, W. (2006) Comparison of the degradation behaviour of fusion-bonded epoxy powder coating systems under flowing and static immersion. *Corrosion Science*, **48**, 1449-1461.

45. Luo, X., and Mather, P. T. (2013) Shape memory assisted self-healing coating. *ACS Macro Letters*, **2**, 152-156.

46. Augustyniak, A., Tsavalas, J., and Ming, W. (2009) Early detection of steel corrosion via "turn on" fluorescence in smart epoxy coatings. *ACS Applied Materials and Interfaces*, **1**, 2618-2623.

47. Wei, Y., Zhang, L., and Ke, W. (2007) Evaluation of corrosion protection of carbon black filled fusion-bonded epoxy coatings on mild steel during exposure to a quiescent 3% NaCl solution. *Corrosion Science*, **49**, 287-302.

48. Goertzen, W. K., and Kessler, M. (2006) Creep behavior of carbon fiber/epoxy matrix composites. *Materials Science and Engineering A*, **421**, 217-225.

49. Alamilla, J., Sosa, E., Sanchez-Magana, C., Andrade-Valencia, R., and Contreras, A. (2013) Failure analysis and mechanical performance of an oil pipeline. *Materials & Design*, **50**, 766-773.

50. Zhou, W., Jeffers, T. E., and Decker, O. H. (2007) Properties of a novel high T_g FBE coating for high temperature service. *Materials Performance*, **46**, 36-40.

51. Moon, B., Paek, B., Lee, J., Wang, H., and Lee, N. (2011) New Development of a High T_g FBE Coating. *Corrosion 2011*, USA. Online: https://www.onepetro.org/conference-paper/NACE-11032 (assessed 23rd February 2017).

52. Duell, J., Wilson, J., and Kessler, M. (2008) Analysis of a carbon composite overwrap pipeline repair system. *International Journal of Pressure Vessels and Piping*, **85**, 782-788.

53. Yuan, Y. C., Rong, M. Z., Zhang, M. Q., Chen, J., Yang, G. C., and Li, X. M. (2008) Self-healing polymeric materials using epoxy/mercaptan as the healant. *Mcromolecules*, **41**, 5197-5202.

54. Bakhshandeh, E., Jannesari, A., Ranjbar, Z., Sobhani, S., and Saeb, M. R. (2014) Anti-corrosion hybrid coatings based on epoxy–silica nano-composites: Toward relationship between the morphology and EIS data. *Progress in Organic Coatings*, **77**, 1169-1183.

55. Weng, C.-J., Chang, C.-H., Peng, C.-W., Chen, S.-W., Yeh, J.-M., Hsu, C.-L., and Wei, Y. (2011) Advanced anticorrosive coatings prepared from the mimicked xanthosoma sagittifolium-leaf-like electroactive epoxy with synergistic effects of superhydrophobicity and redox catalytic capability. *Chemistry of Materials*, **3**, 2075-2083.

56. Rosu, D., Rosu, L., and Cascaval, C. N. (2009) IR-change and yellowing of polyurethane as a result of UV irradiation. *Polymer Degradation and Stability*, **94**, 591-596.

57. Engels, H. W., Pirkl, H. G., Albers, R., Albach, R. W., Krause, J., Hoffmann, A., Casselmann, H., and Dormish, J. (2013) Polyurethanes: Versatile materials and sustainable problem solvers for today's challenges. *Angewandte Chemie International Edition*, **52**, 9422-9441.

58. Guan, S. W. (2003) 100% Solids Rigid Polyurethane Coatings Technology and Its Application on Pipeline Corrosion Protection. *Pipeline Engineering and Construction International Conference 2003*. Online: http://ascelibrary.org/doi/pdf/10.1061/40690%282003%2928#sthash.5BKyd9zK.dpuf (assessed 1st March 2017).

59. Ghosal, A., Rahman, O. U., and Ahmad, S. (2015) High-performance soya polyurethane networked silica hybrid nanocomposite coatings. *Industrial and Engineering Chemistry Research*, **54**, 12770-12787.

60. Chattopadhyay, D. K., Sreedhar, B., and Raju, K. V. S. N. (2005) Effect of chain

extender on phase mixing and coating properties of polyurethane ureas. *Industrial and Engineering Chemistry Research,* **44,** 1772-1779.

61. Samimi, A. (2012) Use of polyurethane coating to prevent corrosion in oil and gas pipelines transfer. *International Journal of Innovation and Applied Studies,* **1,** 186-193.
62. Guan, S. W. (2003) Advanced 100% Solids Rigid Polyurethane Coatings Technology for Pipeline Field Joints and Rehabilitation. *Corrosion 2003,* USA. Online: http://www.penderlo.com/doc/PUTech.pdf (assessed 2nd March 2017).
63. Kong, X., Liu, G., Qi, H., and Curtis, J. M. (2013) Preparation and characterization of high-solid polyurethane coating systems based on vegetable oil derived polyols. *Progress in Organic Coatings,* **76,** 1151-1160.
64. Boubakri, A., Elleuch, K., Guermazi, N., and Ayedi, H. (2009) Investigations on hygrothermal aging of thermoplastic polyurethane material. *Materials & Design,* **30,** 3958-3965.
65. Rassoul, E.-S. A., Abdel-Samad, A., and El-Naqier, R. (2009) On the cathodic protection of thermally insulated pipelines. *Engineering Failure Analysis,* **16,** 2047-2053.
66. De Sousa, F., Da Mota, R., Quintela, J., Vieira, M., Margarit, I., and Mattos, O. (2007) Characterization of corrosive agents in polyurethane foams for thermal insulation of pipelines. *Electrochimica Acta,* **52,** 7780-7785.
67. Motamedi, M., Tehrani-Bagha, A., and Mahdavian, M. (2014) The effect of cationic surfactants in acid cleaning solutions on protective performance and adhesion strength of the subsequent polyurethane coating. *Progress in Organic Coatings,* **77,** 712-718.
68. Akbarian, M., Olya, M., Mahdavian, M., and Ataeefard, M. (2014) Effects of nanoparticulate silver on the corrosion protection performance of polyurethane coatings on mild steel in sodium chloride solution. *Progress in Organic Coatings,* **77,** 1233-1240.
69. Khun, N., and Frankel, G. (2016) Cathodic delamination of polyurethane/multiwalled carbon nanotube composite coatings from steel substrates. *Progress in Organic Coatings,* **99,** 55-60.
70. Li, J., Cui, J., Yang, J., Li, Y., Qiu, H., and Yang, J. (2016) Reinforcement of graphene and its derivatives on the anticorrosive properties of waterborne polyurethane coatings. *Composites Science and Technology,* **129,** 30-37.
71. Lee, S. G., Ham, D. S., Lee, D. Y., Bong, H., and Cho, K. (2013) Transparent superhydrophobic/translucent superamphiphobic coatings based on silica–fluoropolymer hybrid nanoparticles. *Langmuir,* **9,** 15051-15057.
72. Gudipati, C. S., Finlay, J. A., Callow, J. A., Callow, M. E., and Wooley, K. E. (2005) The antifouling and fouling-release perfomance of hyperbranched fluoropolymer (HBFP)–poly(ethylene glycol) (PEG) composite coatings evaluated by adsorption of biomacromolecules and the green fouling alga Ulva. *Langmuir,* **1,** 3044-3053.
73. Darden, W., and Parker, B. (2006) Advances in Fluoropolymer Resins for Long-life Coatings. *PACE,* USA. Online: http://lumiflonusa.com/wp-content/uploads/2010_PACE_Paper.pdf (assessed 5th March 2017).
74. Albert, R., McKeen, L. W., and Hofmans, J. (2008) Corrosion Testing of Fluorocoatings for Oil/Gas Production Tubing. *Corrosion 2008,* USA. Online:

https://www.onepetro.org/conference-paper/NACE-08029 (assessed 26th February 2017).

75. Hofmans, J., Nelissen, J., Tixhon, J.-M., and McKeen, L. W. (2012) Engineered Internal Downhole Coating Solutions for Corrosion & Deposition Control in Downhole Production Tubulars. *Abu Dhabi International Petroleum Conference and Exhibition,* UAE. Online: https://www.onepetro.org/conference-paper/SPE-161204-MS (assessed 2nd March 2017).

76. Bayram, T. C., Orbey, N., Adhikari, R. Y., and Tuominen, M. (2015) FP-based formulations as protective coatings in oil/gas pipelines. *Progress in Organic Coatings,* 88, 54-63.

10

Biopolymer Coatings

10.1 Introduction

Metallic materials and structures exposed to aggressive environments during application undergo deterioration and premature failure because of the processes of corrosion. Application of polymeric coatings is the standard procedure widely employed to control the corrosive destruction of the metal structures [1-3]. However, many of the conventional coating systems lead to several environment and health related issues such as emission of volatile organic solvents and difficulty in recycling or waste disposal of resins at the end of their usage. Thus, to overcome these issues, there is an increasing demand for new "green or bio" coating technologies based on waterborne or high solid content coating formulations for the corrosion protection of metallic structures [4-6]. In addition to overcoming the environmental issues, the trend for biopolymer coatings has also evolved from the realization that the fossil resources used for the conventional organic coatings are inherently finite. Thus, it is vital to develop effective biopolymer coating systems by using film forming polymers originated from bio-sources. This approach will not only cut down the growing prices of raw materials, but would also generate environmentally friendly coatings, and solve the waste disposal problems of resin materials. Cellulose, chitosan, starch, polylactide, vegetable oils, etc., are some of the bio-polymers extensively developed for the coating applications due to their inherent properties of biodegradability and low toxicity [7,8]. In addition to the corrosion resistant coatings and paints, biopolymers can also be formulated as corrosion resistant pigments, alloys and composites. These macromolecules are rich in functional groups which provide multiple adsorption sites suitable to form strong complexes with metal ions and, thus, protect them from corrosion [9-11]. Their performance can be further improved by suitable chemical modifications or by the use of reinforcing materials like nanoparticles. This chapter presents an overview on the development of various biopolymer based coatings and their ability in protecting metal substrates in different corrosive environments in different process industries.

Gisha E. Luckachan and Vikas Mittal, The Petroleum Institute (part of Khalifa University of Science and Technology), Abu Dhabi, UAE*
Current address: Bletchington, Wellington County, Australia

10.2 Cellulose

Cellulose, one of the most abundant biopolymers functionalized naturally with hydroxyl and amino groups, is commonly utilized to inhibit metal corrosion in acid media. Derivatives of cellulose such as carboxymethyl cellulose [12,13], hydroxypropyl cellulose [14], ethyl hydroxyethyl cellulose [15], hydroxypropyl methyl cellulose (HPMC) [16] and aminated hydroxyethyl cellulose [17] have been widely used as corrosion inhibitors. A mixed type of inhibition (i.e. inhibiting both anodic and cathodic processes of corrosion) was also reported for cellulose derivatives, especially for hydroxyethyl cellulose (HEC) in acid solutions [17]. Inhibition occurred by the coordinate bonding of the polar functional groups of the HEC polymer with the metal surface and the surface coverage by the bulky cyclic ring structures of cellulose adsorbed on it. Thus, HEC adsorption on the metal surface occurs through multiple OH functional groups and the aromatic ring structures. An enhanced inhibitive effect of HEC was reported by Arukulam *et al.* [17] by the incorporation of an inhibitor containing halide additive (KI). The iodide ions stabilized the cellulose adsorption on the metal surface, which occurred by the coulombic interaction between HEC^+ and I^- ions. Firstly, the iodide ions chemisorbed strongly on the metal surface, and HEC^+ adsorbed subsequently, which led to higher extent of surface coverage and, hence, greater corrosion inhibition. Potentiodynamic polarization measurements exhibited a mixed type corrosion inhibition using HEC+KI combination. The anodic and cathodic branches of Tafel plots were observed to shift towards lower values of corrosion current density, as compared to the plots corresponding to HEC alone, thus, confirming superior performance.

In another study, Mobin *et al.* [18] employed the mixing of a minimal concentration of surfactants, triton X 100 (TX), cetyl pyridinium chloride (CPC) and sodium dodecyl sulfate (SDS) with hydroxyethyl cellulose to enhance the inhibitive effect on A1020 carbon steel corrosion in acid solutions. The concentration of surfactants was kept very low to maintain the eco-friendly nature of HEC. The enhanced inhibition of HEC in the presence of the surfactant occurred by a protective film formation on the metal surface by the interaction between the HEC and the surfactant. In acid solution, Cl^- ions present in the electrolyte solution adsorbed on the metal surface to which HEC molecules, protonated in acid solution, attached by electrostatic interaction. Surfactant was then attached to HEC which led to a large surface coverage, thus, providing a better corrosion protection than pure HEC molecules. Each surfactant interacted differently with HEC in acid solution. In 1M HCl, SDS and TX were suggested to interact electrostatically with HEC, whereas CPC binding to the HEC in 1M HCl was opined occur through hydrophobic group i.e. an attractive lipophilic interaction between the surfactant tail and polymer chain. Moreover, the polymer backbone exhibited an expansion by the repletion of charged groups on ionic surfactant bound to HEC. This expansion helped the polymer molecules to occupy more surface area on carbon steel surface and, thus, protect it from aggressive medium.

HPMC has also been reported to exhibit a mixed-type corrosion inhibition in 0.5M H_2SO_4 and 1M HCl solutions and the extent of inhibition was influenced by the inhibitor concentration, temperature of the media and time of exposure [19,20]. The inhibitive effect of HPMC was by the formation of protective layer on the metal surface through the hydroxypropyl end and the bulky aromatic ring of the molecule. It was reported that in the absence of corrosion, a physical blocking of active sites and in the presence of corrosion, a geometric blocking of active sites on the mild steel occurred due to HPMC molecules. The compound carboxy methyl cellulose was also reported as good inhibitor for mild steel corrosion in aqueous environments [13,21,22]. Another derivative of cellulose, carboxymethyl cellulose-1-hydroxyethane-1,1- diphosphonic acid, was investigated by Rajendran *et al.* [23] for the inhibitive behavior of carbon steel corrosion. An inhibition efficiency of 83.34% was observed for sodium carboxymethyl cellulose (Na-CMC) at a concentration of 5 mg/L at 20 °C [24]. Similar to other cellulose derivatives, Na-CMC also provided protection by the adsorption of inhibitor molecules at metal/solution interface [25]. The calculations of thermodynamic parameters indicated that a chemical adsorption of carboxyl group of Na-CMC occurred on the metal surface following a Langmuir's adsorption isotherm. Thus, a large number of cellulose derivatives have been confirmed as effective corrosion inhibitors, which underlines the usefulness of bio-based polymers for advanced applications.

Yang *et al.* [26] reported transparent cellulose films prepared from aqueous alkali (NaOH or LiOH)/urea (AU) solutions. The films exhibited high oxygen barrier properties in comparison with conventional cellophane films. As shown in Figure 10.1, the cellulose films also exhibited significantly lower water vapor permeability as compared to cellophane over the entire range of relative humidity, thus, further confirming the potential of these cellulose coatings to be useful alternatives to the conventional coating materials. In another study, Qi *et al.* [27] fabricated composite coatings through layer-by-layer assembly of cellulose and chitin nanofibrils. The layer by layer coatings on PET films exhibited high transparency, surface hydrophilicity, flexibility and nanoporous structures. With further methodological developments, the authors suggested the potential use of developed coatings systems in a wide variety of applications.

Figure 10.2 shows the AFM height images of the surfaces of (a) 20- and (b) 21-layered films. Cellulose acetate (CA) films combined with nanoparticles have also been reported as an effective treatment for corrosion protection of metals [28-30]. Tamborim *et al.* [31] studied the corrosion inhibitive effect of cellulose acetate coating doped with amoxicillin on AA2024-T3 aluminum alloy in 0.05 M NaCl solution. An effective protection was reported for CA films doped with 2000 ppm of amoxicillin, which occurred due to its ability to form different complexes with aluminium ions [32].

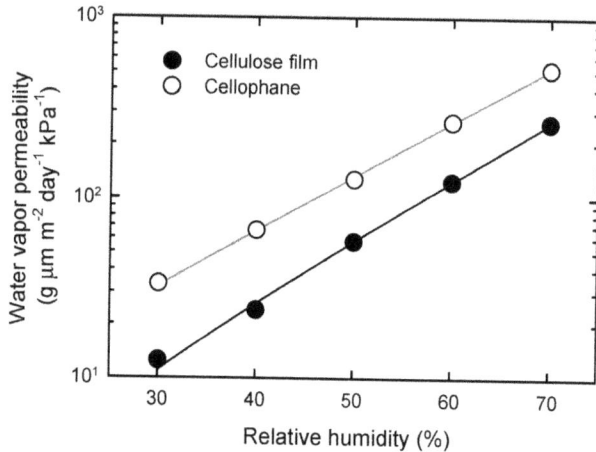

Figure 10.1 Effect of relative humidity on water vapor permeability of AU cellulose and cellophane. Reproduced from Reference 26 with permission from American Chemical Society.

Yabuki *et al.* [33,34] used cellulose nanofiber incorporated with corrosion inhibitor (calcium nitrite) to prepare self-healing polymer coatings. The coatings were scratched for polarization measurements and a higher resistance was observed for nanofiber/corrosion inhibitor coating as compared to the coating containing only corrosion inhibitor. The method represented an ideal methodology for the fabrication of biopolymer based self-healing coatings because of the network structure of cellulose nanofiber and distribution of the healing agent uniformly over the coating. Nanofibers provided a pathway for the release of healing agent through the coating. It was confirmed in the cross-section of the polymer scratch where empty holes of sizes equivalent to the diameter of cellulose nanofibers were observed after the corrosion test. Since cellulose nanofibers have several OH functional groups on their backbone chain, a pH

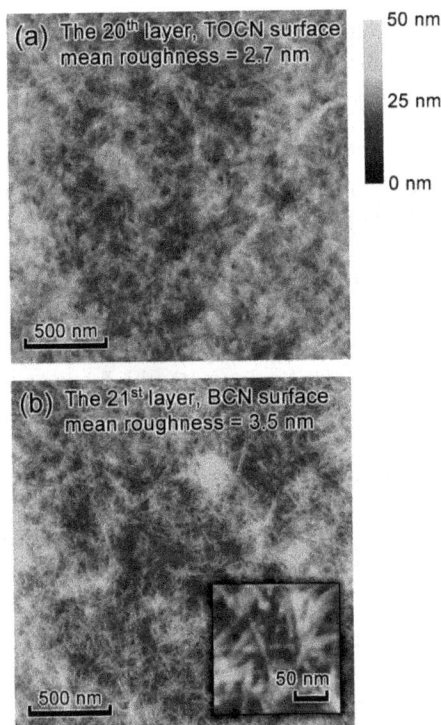

Figure 10.2 AFM height images of the surface of (a) 20- and (b) 21-layered films; TOCN: cellulose nanofibrils and BCN: chitin nanofibrils. Reproduced from Reference 27 with permission from American Chemical Society.

a pH controlled adsorption and desorption from the nanofiber chain is possible in the suitable environments. The authors also studied the behavior of nanofibers using oleic acid (OA) as corrosion inhibitor and sodium hydroxide as pH adjuster [35]. The polarization resistance of the scratched surfaces of nanofiber-oleic acid coatings were two orders of magnitude higher than the plain coating in a pH 11.4 solution. The protection occurred by the formation of a stable film on the metal surface by the interaction of oleic acid released through the nanofiber pathways with metal ions. The scratch exposed the metal surface towards the electrolyte that activated the anodic processes forming Fe^{2+} ions. The cathodic process occurring near the contact between the coating and the substrate formed OH^- ions which diffused through the coating, and resulted in an increase of local pH and promoted the release of oleic acid from the surface of nanofibers. The triggering action for such a self-healing process was the cathodic reaction on the metal surface.

10.3 Chitosan

Chitosan (CS), a derivative of polysaccharide chitin, is second to cellulose in natural abundance [36,37]. Its production is estimated around one billion tons per year. CS is a linear biopolymer of poly(d-glucosamine) connected to poly(N-acetyl-d-glucosamine) through a beta 1→4 linkage. The amine functional groups create positive charge density on the chitosan chain in slightly acidic medium [38]. Such a cationic property together with its biocompatibility make CS widely useful in biomedical applications. Its film forming and gelation properties enable it to be used as a barrier coating for the corrosion protection of metals as well. In addition, lone pairs of electrons on the O and N atoms of hydroxyl and amine functional groups of chitosan can interact ionically with the metal surface, so it is considered as a natural corrosion inhibitor [39,40,]. El-Haddad [41] reported 95% efficiency in copper corrosion inhibition by chitosan in acid medium (0.5 M HCl). Similar to cellulose, chitosan also acted as a mixed-type inhibitor which reduced both anodic and cathodic processes of copper corrosion. It was observed from the potentiodynamic measurements that E_{corr} shifted to cathodic side and the corrosion current density decreased with increase in chitosan concentration. The inhibition mechanism was driven by the electrostatic interaction of metal ions with Cl⁻ and positively charged chitosan in the acid medium. Copper surface, positively charged in acid solution, got adsorbed by Cl⁻ ions from the electrolyte, thus, keeping the metal surface more negatively charged in the solution. The positively charged chitosan molecules subsequently attach to the Cl⁻ ions on the metal forming a thin barrier layer [42]. In addition to the ionic adsorption, the distribution of electrons between the free heteroatoms of chitosan chain with vacant d orbitals of iron results in a chemical adsorption of cationic and neutral molecules on the metal surface. In addition to the pure chitosan, coatings of chitosan derivatives like acetyl thiourea chitosan and naturally available carboxymethyl chitosan also exhibited an efficient mild steel corrosion inhibition in acid solution [43,44]. Mild steel corrosion inhibition in sodium chloride solution was reported by Mohamed and Fekry using chitosan-crotonaldehyde Schiff's base [45]. The barrier property of chitosan coatings along with physical and mechanical properties could be enhanced by crosslinking reactions. Glutaraldehyde [46], sulfuric acid [47], epoxy [48] and dialdehyde starch [49] are the commonly used crosslinking agents for chitosan. Luckachan and Mittal [50] reported anti-corrosive chitosan coatings for mild steel protection using gluatraldehyde as crosslinking agent along with a hydrophobic poly(vinylbutyral) (PVB) over-coating. Figure 10.3 shows the

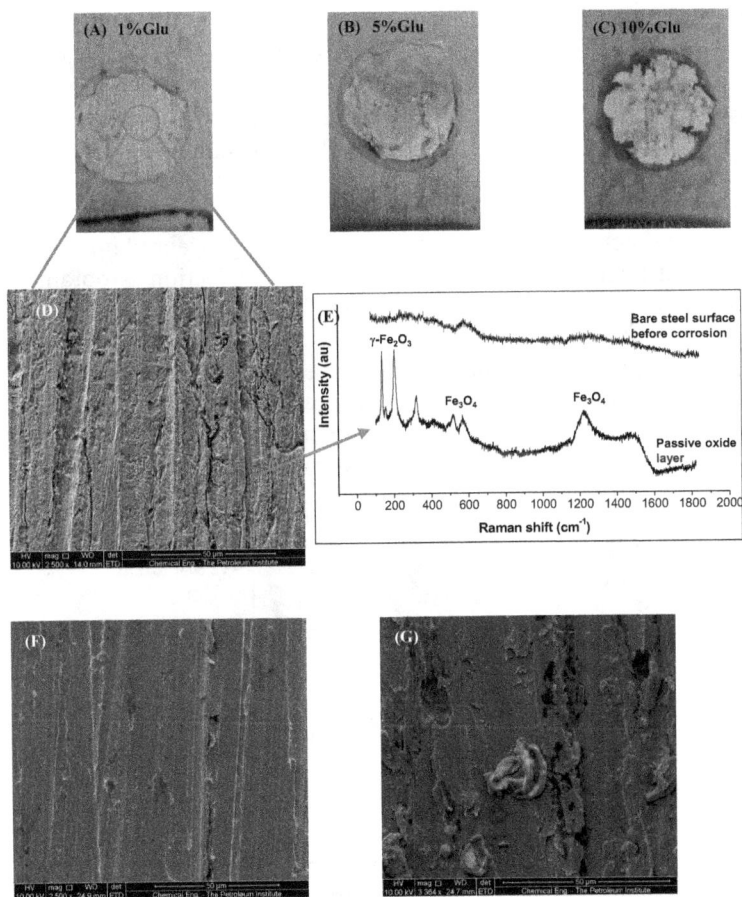

Figure 10.3 Optical photographs of PVB_Ch/1%Glu_PVB coated steel surface containing (A) 1%Glu, (B) 5%Glu and (C) 10%Glu after 24 h EIS in 0.3 M salt solution and film removal. Diameter of the circle was 1 cm². (D) SEM image and (E) Raman spectra of the area marked with red circle on optical photograph of steel surface taken after PVB_Ch/1%Glu_PVB coating removal. SEM images of bare steel (F) before and (G) after 24 h EIS measurement in 0.3 M salt solution for comparison of the corrosion performance with the composite coating. Reproduced from Reference 50 with permission from Springer.

optical photographs of steel surfaces coated with PVB_Ch/x%Glu_PVB layers containing different amount of glutaraldehyde after 24 h EIS measurent and film removal. Metal surface beneath the coating was observed to be uniformly covered by a thin film, which appeared in the SEM image as uniform less-porous thick precipitate. Raman spectra of the area exhibited characteristic bands of Fe_3O_4 and γ-Fe_2O_3 oxides. Hoeevr, the SEM image of the bare steel after corrosion analysis exhibited that corroison products were agglomerated. It indicated that chitosan coating prevented the agglomeration of passive iron oxide layers on the metal surface and, thus, helped to cover the surface uniformly. The chitosan stabilized iron oxide layer separated the corosive agents from reaching the metal surface and, thus, prevented further corrosion of the metal beneath the oxide layer. Such a protection is obtained by the chelating effect of chitosan with iron ions. These findings indicated that if the structure of chitosan was suitably modified, the obtained coating formulations represent high potential of large scale use in process industries.

Another approach to enhance the properties of chitosan is the addition of inorganic particles, which significantly enhances the oxygen barrier and adhesion strength. In a recent study, Luckachan and Mittal [51] (Figure 10.4) gener-

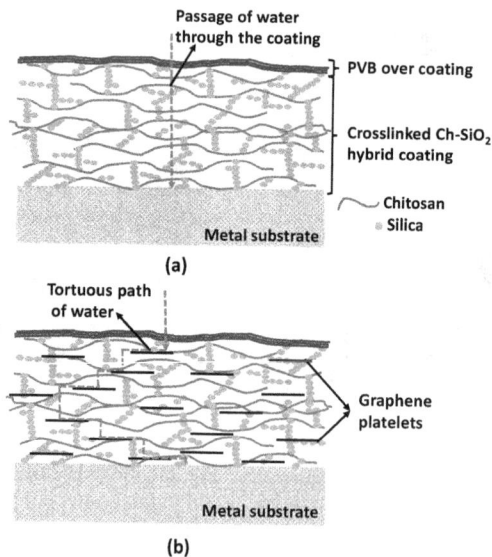

Figure 10.4 Schematic representation of corrosion protection provided by Ch-SiO₂ hybrid coatings in the absence (a) and presence (b) of graphene platelets. Reproduced from Reference 51.

ated stable chitosan-silica based protective coatings on carbon steel substrates [51]. Effect of chitosan content, cross-linking with silica, poly(vinyl butyral) (PVB) over-coat as well as graphene reinforcement on the anti-corrosion performance of the coatings in an aggressive environment was studied. An excellent corrosion resistance was obtained by the application of a PVB over-coating on to Ch-SiO$_2$ hybrid coating, even though the overall coating thickness was still very low (5-6 µm). Uniformly distributed graphene platelets contributed to the significantly enhanced corrosion resistance by acting as water barrier, which was reflected in the lower coating capacitance and I_{corr}. The schematic representation of the corrosion protection in the absence and presence of graphene platelets is presented in Figure 10.4.

Chitosan modified with β–cyclodextrin was also reported by Liu *et al.* [52] as a green inhibitor to carbon steel corrosion in 0.5 M hydrochloric acid solutions. The inhibition efficiency was observed to increase with inhibitor concentration (Figure 10.5), with the maximum inhibition efficiency of 96.02%.

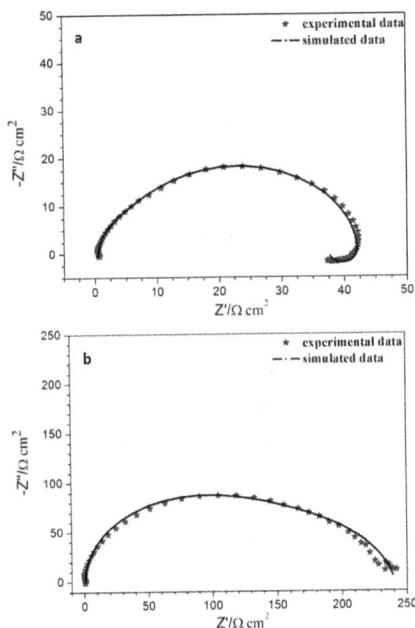

Figure 10.5 Simulated and experimental impedance spectra in 0.5 M HCl solution: (a) blank solution; (b) in the presence of 80 ppm inhibitor. Reproduced from Reference 52 with permission of American Chemical Society.

Inhibition process of β-cyclodextrin modified natural chitosan followed the Langmuir adsorption isotherm, where metal surface was covered and protected by the physical and chemical adsorption of chitosan.

In another study, Hoagland and Parris [53] casted chitosan/lactic acid films on pectin films using either glycerol or lactic acid as plasticizer. The storage modulus of the chitosan/pectin laminated films was observed to be significantly higher than the pure chitosan films (Figure 10.6).

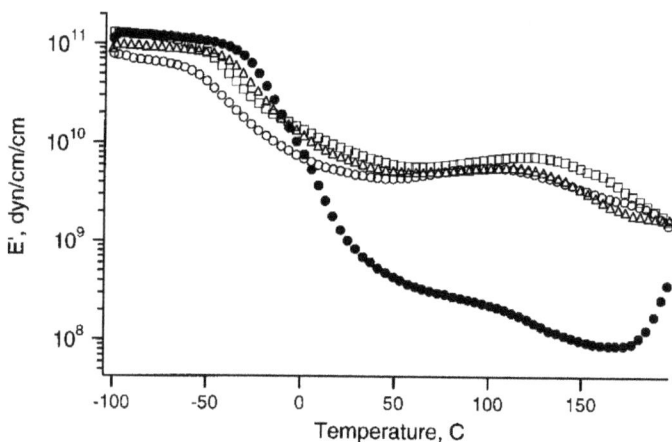

Figure 10.6 Storage modulus for pectin (P)/lactic acid (LA) film (P:LA 2:1, open circles), for chitosan (C)/lactic acid film (C:LA 2:2, solid circles), and for laminates (P:LA:C:LA 2:1:1:1, squares, and P:LA:C:LA 1:0.5:1:1, triangles). Reproduced from Reference 53 with permission from American Chemical Society.

The affinity of chitosan towards water is an undesirable property for its application as the corrosion mitigating coating films. Chitosan changes to a hydrogel by absorbing moisture from the environment [54,55]. It causes the easy degradation of the film and, thus, leads to the complete failure of chitosan as a protective barrier coating. Several works have reported the modification of chitosan structure using organic or bio-organic materials to reduce its water affinity. Sugama and Milian-Jimenez [56] reported the corn-starch derived dextrin modified chitosan for the protection of aluminium. Coating formulations were prepared by mixing chitosan dissolved in HCl with corn starch-derived dextrin (DEX) containing Ce nitrate as oxidizing agent. The mixture was dip-coated on aluminium surface, followed by heating to obtain a solid film. Under this condition, DEX oxidized to two fragmental products of 3,4-dihydroxybutanoic acid and glycolic acid. These fragments formed Ce-complexed carboxylate by reacting with Ce in the Ce nitrate. Ce-bridged carboxylate fragments had a strong affinity with the primary amine in chitosan chain and created a secondary amide linkage that kept the DEX grafted onto the chitosan chain. DEX grafting on CS resulted in the formation of uniform defect-free coatings with less susceptibility to moisture, along with reduced permeability to corrosive species through the coating. CS/DEX coating obtained from a formulation ratio of 70/30 withstood corrosion in salt-spray analysis for over 720 h. In addition, 70/30CS/DEX coating exhibited pore resistance an order of magnitude higher than that of single CS and DEX coatings. These results indicated lower permeability and higher barrier protection of DEX modified chitosan coatings on aluminium surface. Kumar and Buchheit [57] modified chitosan structure using epoxy functional silane and vanadate for the corrosion resistant coating for Al 2024 T3 alloy. Adhesive strength of chitosan coatings increased significantly by the addition of 1% GPTS ((3-glycidoxypropyl)-trimethoxysilane)-vanadate in the coating formulation to an extent which was comparable to the adhesive strength of pure epoxy coatings. A controlled release of vanadate from the chitosan coating was obtained by proper pH adjustment that provided a self-healing nature to these coatings for corrosion protection. The highest release of vanadate occurred in pH 10 solution and a negligible release was observed in pH 3 solution. The maximum adsorption of vanadate was in the pH 3-5 range because the amine functional groups of chitosan getting protonated in the acidic range. Hence, the adsorption and desorption of vanadate onto chitosan was observed to be reversible. CS-1%GPTS-vanadate coatings remained for 450 h in the salt spray chamber without any indication of corrosion initiation.

Another approach to solve the water affinity issue of chitosan is the modification with negatively charged polyacid electrolytes, since chitosan exists as a cationic polybase in acidic media. Sugama and Cook [58] attempted this approach by molecular modification of chitosan with poly(itaconic acid) (PIA). PIA modified CS coatings were prepared by dip-coating CS-PIA solution in HCl on aluminum substrate followed by heating at 200 °C for over 120 min. Under such high temperature conditions, the primary amine groups in CS interacted strongly with the carboxylic acid groups of poly(itaconic acid) and formed a secondary amide linkage which grafted the PI polymer onto the chitosan backbone and crosslinked between the chitosan chains. Also, PIA exhibited a direct affinity with aluminium surface by forming a –COO-Al linkage by the reaction of the COOH groups in PIA with OH groups of aluminium surface. Such an interfacial interaction of PIA with metal surface along with the crosslinked structure of chitosan enhanced the barrier protection by reducing moisture sensitivity and infiltration of corrosive species through the coating. CS/PIA coating formulation of 80/20 ratio possessed the aforementioned properties required for a corrosion protecting barrier coating and provided protection to aluminium metal for over 694 h in salt spray chamber.

Crosslinking chitosan with metal ions is a different approach used to enhance the barrier properties of chitosan coatings. The lone pairs of electrons on amine and hydroxyl functional chitosan can involve in coordination with the transition metal ions. Such an interaction occurs at neutral pH, as at low pH the protonated amine decreases the affinity of chitosan to chelate with metal ions [59]. This coordination mechanism was explained by so-called bridge and pendant models. In the bridge model, the metal ion is bound to the nitrogen atoms of either same or different chitosan chains. In the pendant model, the metallic ions are bound with the amino groups as a pendant [60-62]. Copper incorporated chitosan films were reported by Lundvall *et al.* [63], for providing protection to AA-2024-T3 aluminium alloy in corrosive environment. Chitosan dissolved in acetic acid solution was coated first on anodized surface of aluminium. The sample were dipped in copper salt of acetate for 24 h to incorporate copper ions in between the chitosan chains. An enhanced protection of chitosan-Cu coatings over pure chitosan coating was observed in borax solution, which was confirmed by a constant high open circuit potential. Changes in the salt concentration didn't change the performance of the coatings, but pH of the copper salt solution affected significantly the copper incorporation into the chitosan structure. An optimum pH of 4 favored coordination of copper ions onto chitosan amine groups and, thus, crosslinked the chitosan chains. Such a crosslinked structure enhanced the barrier protection of chitosan coatings by decreasing its permeability towards the moisture and corrosive species.

Recently several works have been reported on the use of chitosan as a reservoir of corrosion inhibitor for the controlled release at required areas of the coatings [64-66]. Carneiro *et al.* and Wang *et al.* [65,66] reported chitosan coatings loaded with organic corrosion inhibitor for the protection of aluminum alloys. Carneiro *et al.* [67] used chitosan solution loaded with 10 wt% of mercaptobenzotriazol (MBT) for the corrosion protection of AA2024 alloy. The main idea of testing CS-MBT was to use such thin biopolymer based coatings for a temporary corrosion protection and active protection of zones where barrier organic coatings cannot be applied. A localized nature of corrosion is usually observed on aluminum alloy 2024 in NaCl solutions within a neutral pH range. However, the local anodic and cathodic currents generated at the metal surface in the micrometer range can lead to considerable local pH changes, highlighting the importance of having a polymeric matrix responding to pH changes and an inhibitor which can suppress both anodic and cathodic corrosion processes. The release of MBT from CS is mainly based on the solubility of chitosan and MBT in different pH conditions. In acidic conditions, the solubility of chitosan allows partial dissolution of entrapped MBT, whereas in alkaline conditions, the solubility of MBT is the dominant parameter as chitosan is insoluble at this condition. In distilled water and NaCl solution, the range of pH is neutral and the combination of low solubility of CS and low solubility of MBT was accounted for a limited release of MBT. CS-MBT coated plates exhibited a clear and unchanged appearance after one week of immersion in 0.05 M sodium chloride solution, indicating the diminishing of corrosion processes by the inhibitor released from the chitosan matrix. Though incorporation of MBT into CS structure was beneficial in stopping the corrosion of aluminium alloy over a period longer than one week, it lacked the barrier nature because of its affinity towards moisture. Therefore, its use as a potential alternative to full coating systems is still not well established. The authors also attempted to solve this issue by functionalizing chitosan with hydrophobic entities like poly(ethylene-*alt*-maleic anhydride) (PEMA) and poly(maleic anhydride-*alt*-1-octadecene) (POMA) [65]. The aliphatic chains were grafted onto chitosan structure by reacting amine groups of CS with maleic anhydride functionality present in the aliphatic polymer chains. Hydrophobic nature of the coating was further enhanced by crosslinking with glutaraldehyde. It was observed by the water contact angle measurements that contact angle of chitosan increased from 69° ± 5° to 78° ± 6° by the loading of 10 wt% MBT. It was further increased to 137° ± 3° and 141° ± 6, by the grafting of PEMA and POMA respectively on to chitosan chains. Carneiro *et al.* [68] also incorporated Ce^{3+} ions into chitosan chains after chemical functionalization with 2,2,3,3-tetrafluoropropyl ether (GTFE). The coatings imparted an active corrosion protection by increasing the oxide layer resistance on the metal surface in presence of Ce^{3+} cations released from the chitosan matrix. However, the coatings didn't achieve the recommended barrier effect for high performance applications even after the chitosan modification. Therefore, such coatings are anticipated to be a better alternative to synthetic polymer coatings for short term applications and temporary protection. In another study, Hassannejad and Nouri [69] studied the self-healing behavior of chitosan-cerium ion (CE) nanocomposite coatings on AA5083-H321 prepared by incorporating CeO_2 in chitosan chain. The authors reported extensively increased corrosion resistance with increase of cerium ions concentration and maximum efficiency was attained at 5 $g.l^{-1}$ CeO_2 nanoparticles in the CS-CE formulation. The inhibitive efficiency of chitosan coating was attributed to the diffusion of cerium ions to the localized defects. As coating was placed in the corrosive solution, the cerium ion inhibited corrosion by creating thin protective layer of cerium oxide on the metal surface by changing the valency of Ce from +3 to +4. It indicated the self-healing nature of the coating, which was further confirmed from the high impedance values of CeO_2-CS-Ce (5 mM) coatings after a 120 h of exposure to the aggressive solution.

Chitosan is a suitable material for bio-corrosion protection because of its anti-bacterial property [70]. Bao *et al.* [71] studied this by using chitosan coatings on copper substrate to prevent the microbiologically influenced corrosion (MIC). Here, chitosan played a dual function, one as a carrier of corrosion inhibitor and second as a bacteriostatic matrix. In order to fabricate the chitosan coatings on copper substrate, a monolayer of alkane thiols was prepared firstly by immersing the pretreated copper surfaces in MUA (11-mercaptoundecanoic acid) solution of ethanol (5 mM). Chitosan-corrosion inhibitor (mercaptobenzotriazol (MBT)) formulation was then dip-coated onto the MUA modified copper surface. Corrosion analysis of these coatings showed a mixed type corrosion inhibition mechanism which decreased both cathodic and anodic current densities along with a positive shift of E_{corr} compared with the MUA and CS/MUA coatings. The authors reported a stable water insoluble product [Cu(I)-MBT]$_{ads}$ on the copper surface which was formed by the interaction of released MBT from chitosan coating with copper ions. Addition of inhibitor molecules directly to the corrosive environment is the generally used method for copper protection. The incorporation of inhibitor in the chitosan chains would be a beneficial alternative, as it minimizes the amount of inhibitor required for copper protection. In addition, corrosion protection performance of MBT-CS/MUA coating indicated effective anti-bacterial features as well. Copper coupons and MBT-CS/MUA coatings were exposed to sulphate reducing bacteria (SRB) culture media for 6 h. Corrosion was clearly observed in the FESEM images of the copper coupons with a rough and non-uniform surfaces along with the appearance of SRB bacteria growth. Such a settlement of SRB bacteria was not observed on the MBT-CS/MUA coated copper substrate indicating the protection of copper by the combined effect of MBT and chitosan.

Several studies have reported the chitosan nanocomposite coatings where the hydrophobicity and corrosion resistance of chitosan coatings were enhanced by the incorporation of suitable nano-fillers. Silica, graphene and graphene derivatives are the commonly employed nanofillers. Fayyad *et al.* [72] reported an enhanced hydrophobicity of chitosan coatings by incorporating graphene oxide (GO). A contact angle of 103.2° was obtained for CS/GO coatings compared with 88.4° of the pure chitosan coating. An oleic acid treatment of CS/GO films increased the contact angle further to 148°, which occurred due to the coverage of surface with bulky chains of oleic acid molecules. An excellent anti-corrosive nature occurred in CS/GO-OA coatings by the combined effect of GO and oleic acid moieties. GO reduced the permeability of water and oxygen through the coating by increasing tortuous path of diffusion and oleic acid reduced the hydrophilicity of the coating. Thus, the nano-filler reinforcement would be an effective method in the future studies to obtain biopolymer coatings for wide applications in the process industry.

10.4 Starch

Starch is a biodegradable polysaccharide composed of amylose (linear) and amylopectin (branched) molecules [73]. It is available at low cost in large amounts from agricultural crops. Since it is biodegradable starch, it is widely used in pharmaceutical, paper making, food preparation and textile applications. It is also considered as a natural corrosion inhibitor, however, its use as corrosion inhibitor is not well reported in the literature. Corn starch, potato, tapioca and wheat are commonly used refined starches. Rosliza *et al.* [74] examined the role of tapioca starch in improving the corrosion resistance of AA6061 alloy in seawater. Similar to other polysaccharides, tapioca starch also exhibited a mixed type corrosion inhibition mechanism with a little better control on the anodic reactions. A positive shift of corrosion potential was observed in the polarization curves along with a downward shift of anodic and cathodic branches of the curve with the addition of tapioca starch. The protection occurred by the formation of a thin film on the metal surface and, therefore, it was considered as a filming corrosion inhibitor which diminished the corrosion by creating a barrier between metal and the corrosive environment [75,76]. Effect of starch in preventing pitting corrosion of steel was studied by Abd El Haleem *et al.* [77] in aqueous environment. Bello *et al.* [78,79] used modified starch to protect carbon steel in alkaline conditions. Both activated starch (AS) and carboxymethylated starch (CMS) were analyzed for corrosion inhibition and the performance was observed to depend on the type and amount of end functional groups on the main chain of the polymer. CMS contained both carboxylate (-COO-) and alkoxy (-CO-) groups and AS contained alkoxy (-CO-) groups. AS had a strong affinity for ferrous ions which were attributed to a better anti-corrosion performance of AS over CMS. A thick inhibitive layer formed by the adsorption of AS on the metal surface was observed in the AFM images after 24 h of immersion showing the role of starch in providing steel corrosion protection. Sugama and DuVall [73] prepared a modified potato starch coating to inhibit the corrosion of aluminum substrates. Potato starch (PS) was modified by the opening of glycosidic rings with polyorganosiloxane (POS) grafting and then dip-coated on aluminum surface. A curing condition of 200 °C was employed to improve the hydrophobic characteristics of PS coating. The POS-grafted PS coating films deposited on the Al surface remained without any visible corrosion for over 288 h in salt spray analysis, confirming the barrier of the coating to moisture and corrosive species. The coating impedance also improved by a magnitude of two orders compared with the bare substrate. Though the above-mentioned polysaccharides have the ability to inhibit metal corrosion, they are generally limited in such applications because of their biodegradability and moisture sensitivity, unless the structure is suitably modified. As these are ecofriendly with low cost, thus, the coatings from these polymers have the potential to develop as suitable alternatives to the petroleum based polymer coatings, especially for short term applications.

10.5 Polylactide

Poly(lactic acid) or polylactide (PLA) is a biodegradable linear aliphatic thermoplastic polyester. It is widely produced from renewable resources like potato, corn and sugar beet, etc. It is mainly used for biomedical applications owing to its biodegradability, biocompatibility, mechanical properties and thermoplastic processability [80-82]. It has excellent oil resistance and gas impermeability and can also be processed easily. Apart from medical applications, it is used widely in the areas of packaging, textile, pharmaceutical uses and coatings [83-85]. Recently, PLA has attracted significant attention in the area of protective coatings of exterior surfaces of building materials because of its hydrophobic nature. It is useful in protecting buildings, especially historic monuments exposed to vigorous outdoor conditions. Svagan *et al.* [86] reported transparent films based on PLA and montmorillonite, as shown schematically in Figure 10.7. To achieve this, layers of montmorillonite and chitosan were generated on PLA films by a layer by layer addition method. The hybrid films exhibited superior properties as oxygen permeation of PLA was reduced respectively by 99 and 96%, at 20 and 50% RH, when 70 bilayers were deposited. Such systems represent potentially useful alternatives to achieve high performance coatings for industrial applications. Zeng *et al.* [87] studied the characteristics of a microarc oxidation (MAO) and poly(L-lactic acid) (PLLA) composite coating, which was fabricated on Mg–1Li–1Ca alloy by dip-coating and freeze-drying (Figure 10.8). MAO/PLLA composite coatings increased the corrosion resistance significantly that was attributed to an increase of E_{corr} from −1.66 V to −1.44 V and a decrease of I_{corr} by approx. 2 orders of magnitude.

Ocak *et al.* [88] reported the hydrophobicity on marble surfaces by the application of PLA coatings. It also diminished the formation of gypsum on marble surfaces when exposed to polluted environment. The authors noticed an increase in hydrophobicity of PLA coatings by the addition of nanoclay in the coating formulation. It enhanced the life time of the coating and the resistance of marble surfaces towards water and atmospheric pollutants [89]. A hydrophobic fluoro-functionalized PLA polymer suitable for stone protection was also prepared [90-92]. The fluorine atoms present in the PLA coating improved hydrophobicity as well as the resistance of coating to chemical and physical degradation agents. Pedna *et al.* [93] prepared highly hydrophobic coatings for buildings using SiO_2-fluorinated PLA bionanocomposites. The silica particles created roughness and the incorporated fluorine lowered the surface energy of the coating. The combined action of these two effects increased the hydrophobicity of the coating, which was confirmed from the water contact angle of about 140° on PLA coated building marble. To the best of knowledge, only a limited work on the use of PLA as corrosion resistant coating for metallic structures has been reported in the literature.

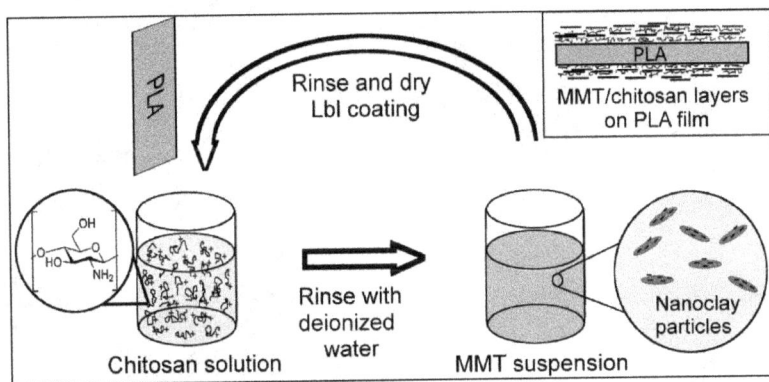

Figure 10.7 Schematic of layer by layer deposition process for the generation of multilayer MMT/chitosan structure on PLA films. Reproduced from Reference 86 with permission from American Chemical Society.

10.6 Vegetable oil based coatings

The economic as well as the environmental concerns of petroleum based materials leads to the necessity of finding out new generation of polymeric products from the raw materials like non-edible vegetable oils and their fatty acids. Vegetable oils (VO) are low cost, non-toxic, biodegradable and renewable materials available aplenty in nature and, thus, are ideal competitors to fossil derived raw materials [94]. The unique structure of VOs with unsaturation sites, esters, hydroxyls, epoxies and other functional groups enable them to undergo chemical modifications to form polymeric materials suitable for various applications especially as a main content in paints and coating formulations [95-101]. VO such as linseed, rubber seed, soybean, karanja oil, castor, cashew nut shell liquid, are being used for the preparation of alkyds [102], epoxies [103,104], polyols [105,106], polyurethanes [107,108], polyesteramides [109,110] and polyetheramides [111,112]. The polymeric formulations derived from vegetable oils reduce or completely avoid the use of volatile organic solvents (VOCs), thus,

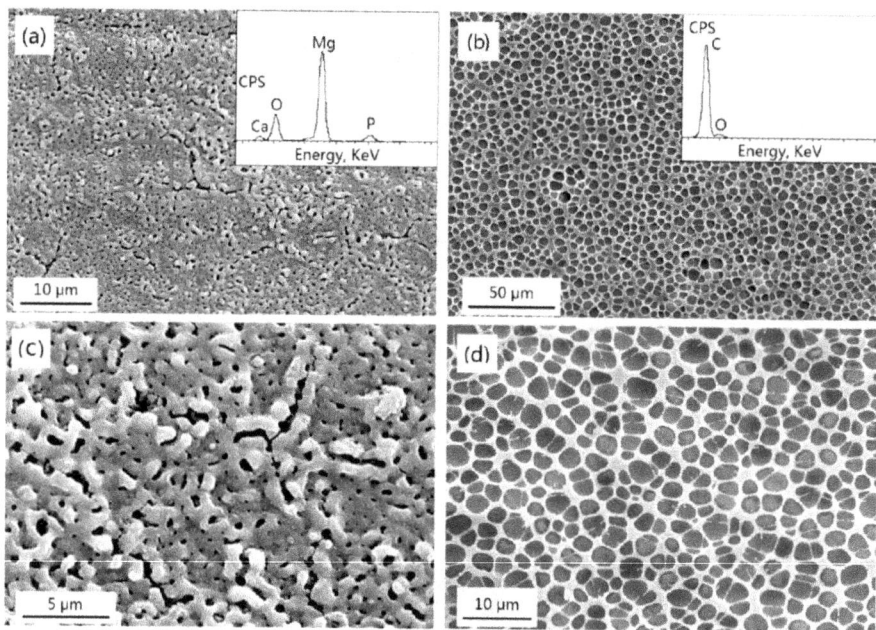

Figure 10.8 SEM morphologies of (a) the MAO coatings and (b) the MAO/PLLA composite coatings, (c and d) the high magnification of panels a and b, respectively. Reproduced from reference 87 with permission from American Chemical Society.

resulting in high solid, waterborne coatings with environmentally friendly nature. Recently, Alam *et al.* [94] reviewed in detail the vegetable oil based coating materials [94]. The review contained a brief description of role of VO as corrosion inhibitors and polymeric binders in coatings. Another review from Sharmin *et al.* [113] described in detail the modification of VO for the fabrication of eco-friendly hyperbranched, solvent free, water borne, high solids and UV curable coatings. Anand *et al.* [114] reported the preparation of polyurethane (PU) coatings using renewable material like sorbitol, diacids and 1,4-butanediol for the polyester polyol part of PU. Each component was selected based on their unique properties. The polyhydroxy structure provided high thermal, mechanical and solvent resistance properties to sorbitol. Renewable as well as non-renewable acids were selected based on their chain length and molecular weight. PU anti-corrosion coatings were prepared by mixing polyester polyol and 4,4'-methylenebis (phenylisocyanate) (MDI) in cyclohexanone keeping a 1.1: 1 ratio for NCO/OH and subsequently brushcoated on the mild steel substrate. Coatings prepared from long chain diacids like sebacic acid exhibited enhanced physical and chemical resistance to corrosion and better gloss and pencil hardness. Chaudhari *et al.* [115] fabricated an environment friendly polyurethane coating using neem oil polyetheramide. Initially, neem oil was allowed to react with diethanol amine to form fatty amide. The obtained fatty amide was modified to form polyetheramide by reacting it with bisphenol-A. Polyurethane coatings were prepared from the polyetheramide by treating it with methylene diphenyl diisocyanate. Nano-TiO_2 modified using silane coupling agents were also added from 0 to 4%. PU coatings with increased chemical resistance and thermal stability were obtained by the incorporation of nano-TiO_2 which increased the gloss and pencil hardness significantly. Neem seed oil is mainly used for pharmaceutical applications. Therefore, the current development in the use of need seed oil for the synthesis of high performing PU coatings confirms the potential utilization of renewable resources for the coatings suitable for industrial applications.

Petrovic and Javin [116] used epoxidized soybean oil with 4,4 -diphenylmethane diisocyanate (DMDI) polymer to obtain soybean oil based polyols which produced polyurethane coatings with strong adhesion and hardness. The authors introduced hydroxyl functional groups on the double bonds of the fatty acid and reported a hydroxyl number higher than 250 mg KOH/g. However, the polyols derived from vegetable oil had certain drawbacks in the coating applications because of their low glass transition temperature and low modulus. These resulted from the presence of pendant dangling chains as well as high molecular mobility of fatty acids. Therefore, it is required to develop means to reduce the content of pendant chains derived from fatty acids in the production of PU coatings. Kong *et al.* [117,118] attempted to prepare polyols having low

molecular weight and high hydroxyl number with low viscosity in order to prepare strong PU polymer suitable for coating and paint applications. The authors
prepared eco-friendly poly(ether ester) polyols using 1,2-propanediol/1,3-pro-
panediol and canola oil. The compounds were subjected to epoxidation and hy-
droxylation (esterification) reactions which led to the ring opening of epoxide
groups and transesterification of glycerides with diols, thus, resulting in a more
hydroxyl functionalized polyols. The PU generated from such polyols contained
less pendant chains and formed a highly crosslinked structure with a high T_g
and modulus. The authors also synthesized polyols (LiprolTM) with high hydroxyl numbers from canola (refined), sunflower (refined), camelina, linola flax
and NuLin flax for the production of high solid PU coatings [119]. The structure
and properties of various polyols depended on the degree of unsaturation, fatty
acid content of the starting oils and extent of oligomerization. PU coatings were
prepared by reacting the polyols with petrochemical derived diisocyanate (polymeric aromatic diphenylmethylene diisocyanate (pMDI)) and other additives.
PU coatings with high T_g, high contact angle, better corrosion resistance and
high cross-linking density were obtained which have strong potential to replace
the commercial PU materials derived from petroleum byproducts.

Smart polyurethane coatings encapsulated with corrosion inhibitor were
prepared by Marathe *et al.* [120] from neem oil. The authors incorporated quinoline as a corrosion inhibitor in the PU coating generated from neem oil acetylated polyester polyol (NAPP). The smart coatings containing 3% microcapsules exhibited a better corrosion protection in both aqueous and acid media.
This method of preparing smart coatings derived from bio-based polyols encapsulated with corrosion inhibitors is envisaged to provide a self-healing character to the polyurethane coatings and can be applied for upgrading other vegetable oil based coatings.

Rajput *et al.* [121] prepared polyols by substituting almost all components of
petroleum feed stock with renewable sources. The authors used dimer fatty
acid and palmitic acid as the renewable source for PU production. Initially, pal-
mitamide was prepared by amidation reaction of palmitic acid. It was then reacted with dimer fatty acid to form a polyesteramide resin (PEPAD). PU formulations were subsequently prepared by reacting a polymeric hexamethylene
diisocyanate (Desmodur N 75 BA/X) with a solution of 50% PEPAD in xylene at
different NCO/OH ratios (0.9, 1.1 and 1.3). Transparent PU coatings with high
thermal stability were obtained for the PU sample containing 1.1:1 NCO/OH
content. On increasing the mole ratio, enhanced properties such as corrosion
resistance, thermal stability and transparency was observed due to the formation of urea linkages.

In addition to the PU coatings, VO are also used for the development of alkyd resins which are used widely for the protective coatings since 1940 [122]. These resins are VO based polyesters and possess high thermal stability and good color retention. By changing the fatty acid chain length and suitable chemical modification, the range of applications of alkyd coatings can be enhanced. It has been reported that the physical drying and acid/alkali resistance of alkyds can be improved by alkylation reactions [123-125]. Linseed oil and soy oil are the commonly used VOs for preparing alkyd resins. These oils have a fast drying process and the unsaturation sites form polymeric networks by reaction with oxygen which enhances the properties suitable for protective coatings. The literature reports have confirmed the protection generated by linseed oil based alkyds to be better than soy oil. Araujo *et al.* [125] confirmed this behavior by conducting a comparison study on the anti-corrosion property of alkyd resin modified with linseed oil and soy oil in accelerated conditions and in marine or industrial atmospheric exposure. Linseed oil based coating provided good adhesion and corrosion resistance in the potentiodynamic and EIS analysis.

Poly(ester amide), consisting of amide (–NHCOR) and ester (–COOR) groups on the polymer backbone, is another class of systems derived from vegetable oil based renewable resources. Mesua ferrea seed oil, linseed oil, coconut oil, Pongamia glabra oil, and Ricinus communis are the commonly used VOs for the synthesis of linear poly(ester amide). Hyperbranched poly(ester amide) (HBPEA) derived from VO is a new class of materials having the properties of barrier coatings because of their unique features of high functionality, low melt and solution viscosity and the three dimensional non-entangled structure [126,127]. An ABx (x ≥ 2) type monomers were used for the synthesize of HBPEA polymer resins. Pramanik *et al.* [128] synthesized HBPEA for the first time using N,N -bis(2-hydroxy ethyl) castor oil fatty amide, isophthalic acid, phthalic anhydride and maleic anhydride, as A2 monomers, whereas a diethanol amine was used as B3 monomer. The degree of branching (DB) enhanced the thermal stability and flow behavior of the HBPEA polymers. The coatings exhibited high adhesion strength, mechanical properties, scratch hardness, impact strength and abrasion resistance desirable for the polymeric applications as surface coatings.

Palm oil is another VO used for the preparation of alkyd films. This non-drying oil, however, does not produce air oxidized coherent films due to its small iodine value. Ataei *et al.* [129] reported the palm oleic acid based barrier coatings with fast physical drying and high water and salt resistance. Palm oleic acid based alkyd were prepared by combining oleic acid and glycerol with phthalic anhydride (PA). The coatings were prepared by free-radical polymerization of alkyd with methylmethacrylate (MMA). The coatings exhibited a decreased drying time and increased alkali resistance with the increment of MMA content. Furthermore, the film hardness and adhesion strength of the coating improved significantly by the alkyd incorporation in the coating formulation.

In summary, it can be concluded that the vegetable oil based coatings have a good potential to be used as eco-friendly resins in surface coatings and can gradually substitute the petroleum based binders. Further developments will enhance the opportunities for making high performance coatings suitable for metal substrates exposed to aggressive environmental conditions at industrial areas.

References

1. Gonzalez-Garcia, Y., Gonzalez, S., and Souto, R. M. (2007) Electrochemical and structural properties of a polyurethane coating on steel substrates for corrosion protection. *Corrosion Science*, **49**, 3514-3526.
2. Leidheiser, H. (1982) Corrosion of painted metals - A review. *Corrosion,* **38**, 374-383.
3. Walter, G. W. (1986) A critical review of the protection of metals by paints. *Corrosion Science*, **26**, 27-38.
4. Chaudhari, A., Kuwar, A., Mahulikar, P., Hundiwale, D., Kulkarni, R., and Gite, V. (2014) Development of anticorrosive two pack polyurethane coatings based on modified fatty amide of Azadirachta indica Juss oil cured at room temperature – a sustainable resource. *RSC Advances,* **4**, 17866-17872.
5. Rajput, S. D., Mahulikar, P. P., and Gite, V. V. (2014) Biobased dimer fatty acid containing two pack polyurethane for wood finished coatings. *Progress in Organic Coatings,* **77**, 38-46.
6. Meshram, P. D., Puri, R. G., Patil, A. L., and Gite, V. V. (2013) Synthesis and characterization of modified cottonseed oil based polyesteramide for coating applications. *Progress in Organic Coatings,* **76**, 1144-1150.
7. Gandini, A., and Belgacem, M. N. (2002) Recent contributions to the preparation of polymers derived from renewable resources. *Journal of Polymers and Environment,* **10**, 105-114.
8. Derksen, J. T. P., Cuperus, F. P., and Kolster, P. (1996) Renewable resources in coatings technology: a review. *Progress in Organic Coatings,* **27**, 45-53.
9. Mahanta, A. K., Mittal, V., Singh, N., Dash, D., Malik, S., Kumar, M., and Maiti, P. (2015) Polyurethane-grafted chitosan as new biomaterials for controlled drug delivery. *Macromolecules,* **48**(8), 2654-2666.
10. Umoren, S. A., and Ekanem, U. F. (2010) Inhibition of mild steel corrosion in H_2SO_4 using exudate gum from *pachylobus edulis* and synergistic potassium halide additives. *Chemical Engineering Communications*, **197**, 1339-1356.
11. Umoren, S. A., and Eduok, U. M. (2016) Application of carbohydrate polymers as

corrosion inhibitors for metal substrates in different media: A review. *Carbohydrate Polymers*, **140**, 314-341.

12. Bayol, E., Gurten, A. A., Dursun, M., and Kayakirilmaz, K. (2008) Adsorption behavior and inhibition corrosion effect of sodium carboxymethyl cellulose on mild steel in acidic medium. *Acta Physico-Chimica Sinica*, **24**(12), 2236–2242.

13. Solomon, M. M., Umoren, S. A., Udosoro, I. I., and Udoh, A. P. (2010) Inhibitive and adsorption behaviour of carboxymethyl cellulose on mild steel corrosion in sulphuric acid solution. *Corrosion Science*, **52**(4), 1317-1325.

14. Rajeswari, V., Kesavan, D., Gopiraman, M., Viswanathamurthi, P. (2013) Physicochemical studies of glucose, gellan gum, and hydroxypropyl cellulose – Inhibition of cast iron corrosion. *Carbohydrate Polymers*, **95**(1), 288-294.

15. Arukalam, I. O., Madu, I. O., Ijomah, N. T., Ewulonu, C. M., and Onyeagore, G. N. (2014) Acid corrosion inhibition and adsorption behaviour of ethyl hydroxyethyl cellulose on mild steel corrosion. *Journal of Materials*, DOI:10.1155/2014/101709.

16. Arukalam, I. O., Madufo, I. C., Ogbobe, O., and Oguzie, E. E. (2014) Adsorption and inhibitive properties of hydroxypropyl Methylcellulose on the acid corrosion of mild steel. *International Journal of Applied Sciences and Engineering Research*, **3**(1), 241-256.

17. Arukalam, I. O., Madufor, I. C., Ogbobe, O., and Oguzie, E. E. (2015) Inhibition of mild steel corrosion in sulfuric acid medium by hydroxyethyl cellulose. *Chemical Engineering Communications*, **202**, 112-122.

18. Mobin, M., and Rizvi, M. (2017) Adsorption and corrosion inhibition behavior of hydroxyethyl cellulose and synergistic surfactants additives for carbon steel in 1 M HCl. *Carbohydrate Polymers*, **156**, 202-214.

19. Arukalam, I. O., Madufor, I. C., Ogbobe, O., and Oguzie, E. E. (2013) Adsorption and inhibitive properties of hydroxypropyl methylcellulose on the acid corrosion of mild steel. *International Journal of Applied Sciences and Engineering Research*, **2**(6), 613-629.

20. Rajeswari, V., Kesavan, D., Gopiraman, M., and Viswanathamurthi P. (2013) Physicochemical studies of glucose, gellan gum, and hydroxypropyl cellulose - Inhibition of cast iron corrosion. *Carbohydrate Polymers*, **95**, 288-294.

21. Umoren, S. A., Solomon, M. M., Udosoro, I. I., and Udoh, A. P. (2010) Synergistic and antagonistic effects between halide ions and carboxymethyl cellulose for the corrosion inhibition of mild steel in sulphuric acid solution. *Cellulose*, **17**(3), 635-648.

22. Manimaran, N., Rajendran, S., Manivannan, M., Thangakani, J. A., and Suriya Prabha A. (2013) Corrosion Inhibition by carboxymethyl cellulose. *European Chemical Bulletin*, **2**(7), 494-498.

23. Rajendran, S., Sridevi, S. P., Anthony, N., Amalraj, A. J., and Sundearavadivelu, M. (2005) Corrosion behavior of carbon steel in polyvinyl alcohol. *Anti-Corrosion Methods and Materials*, **52**(2), 102-107.

24. Li, M. M., Xu, Q. J., Han, J., Yun, H., and Min, Y. L. (2015) Inhibition action and adsorption behavior of green inhibitor sodium carboxymethyl cellulose on copper. *International Journal of Electrochemical Science*, **10**, 9028-9041.

25. Tian, H., Li, W., and Hou, B. (2011) Novel application of a hormone biosynthetic inhibitor for the corrosion resistance enhancement of copper in synthetic seawater. *Corrosion Science*, **53**, 3435-3445.

26. Yang, Q., Fukuzumi, H., Saito, T., Isogai, A., and Zhang, L. (2011) Transparent cellulose films with high gas barrier properties fabricated from aqueous alkali/urea solutions.

Biomacromolecules, **12**(7), 2766-2771.
27. Qi, Z. D., Saito, T., Fan, Y., and Isogai, A. (2012) Multifunctional coating films by layer-by-layer deposition of cellulose and chitin nanofibrils. *Biomacromolecules*, **13**(2), 553-558.
28. Wefers, K., Nitowski, G. A., and Wieserman, L. F. (1992) Phosphonic/Phosphinic Acid Bonded to Aluminum Hydroxide Layer, US Patent 5132181.
29. Arai, T., Shin, R. B., Yamamoto, T. (2008) Agents for the Surface Treatment of Zinc or Zinc Alloy Products, US Patent 20080113102.
30. Walters, D. N., Schneider, J. R. (2008) Coating Compositions Exhibiting Corrosion Resistance Properties and Related Coated Substrates, US Patent 20080090069.
31. Tamborim, S. M., Dias, S. L. P., Silva, S. N., Dick, L.F.P., and Azambuja, D.S. (2011) Preparation and electrochemical characterization of amoxicillin-doped cellulose acetate films for AA2024-T3 aluminum alloy coatings. *Corrosion Science*, **53**, 1571-1580.
32. Abdallah, M. (2004) Antibacterial drugs as corrosion inhibitors for corrosion of aluminium in hydrochloric solution. *Corrosion Science*, **46**, 1981-1996.
33. Yabuki, A., Kawashim, A., and. Fathon, I. W. (2014) Self-healing polymer coatings with cellulose nanofibers served as pathways for the release of a corrosion inhibitor. *Corrosion Science*, **85**, 141-146.
34. Yabuki, A., and Nishisaka, T. (2011) Self-healing capability of porous polymer film with corrosion inhibitor inserted for corrosion protection. *Corrosion Science*, **53**, 4118-4123.
35. Yabuki, A., Shiraiwa, T., Fathon, I. W. (2016) pH-controlled self-healing polymer coatings with cellulose nanofibers providing an effective release of corrosion inhibitor. *Corrosion Science*, **103**, 117-123.
36. Hirano, S., Inui, H., Kosaki, H., Uno, Y., and Toda, T. (1994) Chitin and chitosan: ecologically bioactive polymers. In: *Biotechnology and Bioactive Polymers*, Gebelein, C. G., and Carraher, Jr., C. E. (eds.), Plenum Press, New York, pp. 43-54.
37. Nishiyama, Y., Langan, P., and Chjanzy, H. (2002) Crystal structure and hydrogen-bonding system in cellulose Iß from synchrotron x-ray and neutron fiber diffraction. *Journal of the American Chemical Society*, **124**, 9074-9082.
38. Sandford, P. A., and Steinners, A. (1991). Biomedical applications of high-purity chitosan. In: *Water-soluble Polymers*, Shalaby, S. W., Cormick, C. L., and Butles, G. B. (eds.), ACS Symposium Series 467, Washington, USA, p. 430.
39. Eduok, U. M., and Khaled, M. M. (2014) Retraction Note to: Corrosion protection of steel sheets by chitosan from shrimp shells at acid pH. *Cellulose*, **21**, 3139-3143.
40. Vathsala, K., Venkatesha, T. V., Praveen, B. M., and Nayana, K. O. (2010) Electrochemical generation of Zn-chitosan composite coating on mild steel and its corrosion studies. *Engineering*, **2**, 580-584.
41. El-Haddad, M. N. (2013) Chitosan as a green inhibitor for copper corrosion in acidic medium. *International Journal of Biological Macromolecules*, **55**, 142-149.
42. Solmaz, R., Kardas, G., Yazıcı, B., and Erbil, M. (2008) Adsorption and corrosion inhibitive properties of 2-amino-5-mercapto-1,3,4-thiadiazole on mild steel in hydrochloric acid media. *Colloids and Surfaces A: Physicochemical and Engineering Aspects*, **312**, 7-17.
43. Fekry, A. M., and Mohamed, R. R. (2010) Acetyl thiourea chitosan as an eco-friendly inhibitor for mild steel in sulphuric acid medium. *Electrochimica Acta*, **55**, 1933-1939.

44. Cheng, S., Chen, S., Liu, T., Chang, X., and Yin, Y. (2007) Carboxymethylchitosan + Cu^{2+} mixture as an inhibitor used for mild steel in 1 M HCl. *Electrochima Acta*, **52**, 5932-5938.

45. Mohamed, R. R., and Fekry, A. M. (2011) Antimicrobial and anticorrosive activity of adsorbents based on chitosan Schiff's base. *International Journal of Electrochemical Science*, **6**, 2488-2508.

46. Goissis, G., Junior, E. M., Marcantonio, R. A. C., Lai, R. C. C., Cancian, D. C. J., and DeCaevallho, W. M. (1999) Biocompatibility studies of anionic collage membranes with different degree of glutaraldehyde cross-linking. *Biomaterials*, **20**, 27-34.

47. Huang, R. Y. M., Pal, R., and Moon, G. Y. (1999) Crosslinked chitosan composite membrane for the pervaporation dehydration of alcohol mixtures and enhancement of structural stability of chitosan/polysulfone composite membranes. *Journal of Membrane Science*, **160**, 17-30.

48. Wei, Y. C., Hudson, S. M., Mayer, J. M., and Kaplan, D. L. (1992) The crosslinking of chitosan fibers. *Journal of Polymer Science, Part A: Polymer Chemistry*, **30**, 2187-2193.

49. Schmidt, C. E., and Baier, J. M. (2000) Acellular vascular tissues: natural biomaterials for tissue repair and tissue engineering. *Biomaterials*, **21**, 2215-2231.

50. Luckachan, G. E., and Mittal, V. (2015) Anti-corrosion behavior of layer by layer coatings of cross-linked chitosan and poly(vinyl butyral) on carbon steel. *Cellulose*, **22**, 3275-3290.

51. Luckachan, G. E., and Mittal, V. Stable corrosion-resistant chitosan coatings: effect of crosslinking with silica, poly(vinyl butyral) over coating and graphene on the coating performance, in preparation.

52. Liu, Y., Zou, C., Yan, X., Xiao, R., Wang, T., and Li, M. (2015) β-Cyclodextrin modified natural chitosan as a green inhibitor for carbon steel in acid solutions. *Industrial & Engineering Chemistry Research*, **54**(21), 5664-5672.

53. Hoagland, P. D., and Parris, N. (1996) Chitosan/pectin laminated films. *Journal of Agricultural and Food Chemistry*, **44**, 1915-1919.

54. Szymanska, E., and Winnicka, K. (2015) Stability of chitosan – a challenge for pharmaceutical and biomedical applications. *Marine Drugs*, **13**, 1819-1846.

55. Tian, D., Dubois, P. H., Grandfile, C. H., Jermome, P., Viville, P., Lazzaroni, R., Bredas, J. L., and Leprince, P. (1997) A novel biodegradable and biocompatible ceramer prepared by the sol–gel process. *Chemistry of Materials*, **9**, 871-874.

56. Sugama, T., and Milian-Jimenez, S. (1999) Dextrine-modified chitosan marine polymer coatings. *Journal of Material Science*, **34**, 2003-2014.

57. Kumar, G., and Buchheit, R. G. (2006) Development and characterization of corrosion resistant coatings using the natural biopolymer chitosan. *ECS Transactions*, **1**(9), 101-117.

58. Sugama, T., and Cook, M. (2000) Poly(itaconic acid)-modified chitosan coatings for mitigating corrosion of aluminum substrates. *Progress in Organic Coatings*, **38**, 79-87.

59. Chui, V. W. D., Mok, K. W., Ng, C. Y., Luong, B. P., and Ma, K. K. (1996) Removal and recovery of copper(II), chromium(III), and nickel(II) from solutions using crude shrimp chitin packed in small columns. *Environment International*, **22**, 463-468.

60. Schlick, S. (1986) Binding sites of copper$_{2+}$ in chitin and chitosan. An electron spin resonance study. *Macromolecules*, **19**, 192-195.

61. Domard, A. (1987) pH and c.d. measurements on a fully deacetylated chitosan:

application to Cu^{II}—polymer interactions. *International Journal of Biological Macromolecules*, **9**, 98-104.

62. Nieto, J. M., Covas, C. P., and Del Bosque, J. (1992) Preparation and characterization of a chitosan-Fe(III) complex. *Carbohydrate Polymers*, **18**, 221-224.

63. Lundvall, O., Gulppi, M., Paez, M. A., Gonzalez, E., Zagal, J. H., Pavez, J., and Thompson, G. E. (2007) Copper modified chitosan for protection of AA-2024. *Surface & Coatings Technology*, **201**, 5973-5978.

64. Zheludkevich, M. L., Tedim, J., Freire, C. S. R., Fernandes, S. C. M., Kallip, S., Lisenkov, A., and. Ferreira, M. G. S. (2011) Self-healing protective coatings with "green" chitosan based pre-layer reservoir of corrosion inhibitor. *Journal of Material Chemistry*, **21**, 4805-4812.

65. Carneiro, J., Tedim, J., Fernandes, S. C. M., Freire, C. S.R., Gandini, A., Ferreira, M. G. S., and Zheludkevich, M. L. (2013) Functionalized chitosan-based coatings for active corrosion protection. *Surface & Coatings Technology*, **226**, 51-59.

66. Wang, Y., Dong, C., Zhang, D., Ren, P., Li, L., and Xiao-gang, L. (2015) Preparation and characterization of a chitosan-based low-pH-sensitive intelligent corrosion inhibitor. *International Journal of Minerals, Metallurgy and Materials*, **22**, 998-1004.

67. Carneiro, J., Tedim, J., Fernandes, S. C. M., Freire, C. S. R., Gandini, A., Ferreira, M. G. S., and Zheludkevich, M. L. (2013) Chitosan as a smart coating for controlled release of corrosion inhibitor 2-mercaptobenzothiazole. *ECS Electrochemistry Letters*, **2**(6), C19-C22.

68. Carneiro, J., Tedim, J., Fernandes, S. C. M., Freire, C. S. R., Silvestre, A. J. D., Gandini, A., Ferreira, M. G. S., and Zheludkevich, M. L. (2012) Chitosan-based self-healing protective coatings doped with cerium nitrate for corrosion protection of aluminum alloy 2024. *Progress in Organic Coatings*, **75**, 8-13.

69. Hassannejad, H., and Nouri, A. (2016) Synthesis and evaluation of self-healing cerium-doped chitosan nanocomposite coatings on AA5083-H321. *International Journal of Electrochemical Science*, **11**, 2106-2118.

70. Franchin, C., Muraglia, M., Corbo, F., Florio, M. A., Di Mola, A., Rosato, A., Matucci, R., Nesi, M., Bambeke, F. V., and Vitali, C. (2009) Synthesis and biological evaluation of 2-mercapto-1,3-benzothiazole derivatives with potential antimicrobial activity. *Archiv der Pharmazie*, **342**, 605-613.

71. Bao, Q., Zhang, D., and Wan, Y. (2011) 2-Mercaptobenzothiazole doped chitosan/11-alkanethiolate acid composite coating: Dual function for copper protection. *Applied Surface Science*, **257**, 10529-10534.

72. Fayyad, E. M., Sadasivuni, K. K., Ponnamma, D., and Al-Maadeed, M. A. A. (2016) Oleic acid-grafted chitosan/graphene oxide composite coating for corrosion protection of carbon steel. *Carbohydrate Polymers*, **151**, 871-878.

73. Sugama, T., and DuVall, J. E. (1996) Polyorganosiloxane-grafted potato starch coatings for protecting aluminum from corrosion. *Thin Solid Films*, **289**, 39-48.

74. Rosliza, R., and Wan Nik, W. B. (2010) Improvement of corrosion resistance of AA6061 alloy by tapioca starch in seawater. *Current Applied Physics*, **10**, 221-229.

75. Gao, B., Zhang, X., and Sheng, Y. (2008) Studies on preparing and corrosion inhibition behaviour of quaternized polyethyleneimine for low carbon steel in sulfuric acid. *Materials Chemistry and Physics*, **108**, 375-381.

76. Al-Juhni, A. A., and Newby, B. Z. (2006) Incorporation of benzoic acid and sodium

benzoate into silicone coatings and subsequent leaching of the compound from the incorporated coatings. *Progress in Organic Coatings,* **56**, 135-145.

77. Abd El Haleem, S. M. (1986) Anodic behaviour and pitting corrosion of plain carbon steel in NaOH solutions containing Cl- ions. *Surface and Coatings Technology,* **27**, 167-173.

78. Bello, M., Ochoa, N., Balsamo, V., López-Carrasquero, F., Coll, S., Monsalve, A., and Gonzalez, G. (2010) Modified cassava starches as corrosion inhibitors of carbon steel: An electrochemical and morphological approach. *Carbohydrate Polymers,* **82**, 561-568.

79. Ochoa, N., Bello, M., Sancristóbal, J., Balsamo, V., Albornoz, A., and Brito, J. L. (2013) Modified cassava starches as potential corrosion inhibitors for sustainable development. *Materials Research,* **16**(6), 1209-1219.

80. Carrasco, F., Pages, P., Gamez-Perez, J., Santana, O. O., and Maspoch, M. L. (2010) Kinetics of the thermal decomposition of processed poly(lactic acid). *Polymer Degradation and Stability,* **95**, 2508-2514.

81. Lasprilla, A. J. R., Martinez, G. A. R., Lunelli, B. H., Jardini, A. L., and Filho, R. M. (2012) Poly-lactic acid synthesis for application in biomedical devices - A review. *Biotechnology Advances,* **30**, 321-328.

82. Nampoothiri, K. M., Nair, N. R., and John, R. P. (2010) An overview of the recent developments in polylactide (PLA) research. *Bioresource Technology,* **101**, 8493–8501.

83. Gupta, B., Revagadea, N., and Hilborn, J. (2007) Poly(lactic acid) fiber: An overview. *Progress in Polymer Science,* **32**, 455-482.

84. Lim, L. T., Auras, R., and Rubino, M. (2008) Processing technologies for poly(lactic acid). *Progress in Polymer Science,* **33**, 820–852.

85. Carrasco, F., Pagesb, P., Gamez-Perez, J., Santana, O. O., and Maspoch, M. L. (2010) Processing of poly(lactic acid): Characterization of chemical structure, thermal stability and mechanical properties. *Polymer Degradation and Stability,* **95**, 116-125.

86. Svagan, A. J., Akesson, A., Cardenas, M., Bulut, S., Knudsen, J. C., Risbo, J., and Plackett, D. (2012) Transparent films based on PLA and montmorillonite with tunable oxygen barrier properties. *Biomacromolecules,* **13**(2), 397-405.

87. Zeng, R. C., Cui, L. Y., Jiang, K., Liu, R., Zhao, B. D., and Zheng, Y. F. (2016) In vitro corrosion and cytocompatibility of a microarc oxidation coating and poly(L-lactic acid) composite coating on Mg-1Li-1Ca alloy for orthopedic implants. *ACS Applied Materials & Interfaces,* **8**(15), 10014-10028.

88. Ocak, Y., Aysun, S., Funda, T., and Hasan B. (2009) Protection of marble surfaces by using biodegradable polymers as coating agent. *Progress in Organic Coatings,* **66**, 213-220.

89. Ocak, Y., Sofuoglu, A., Tihminlioglu, F., and Boke, H. (2015) Sustainable bio-nanocomposite coatings for the protection of marble surfaces. *Journal of Cultural Heritage,* **16**(3), 299-306.

90. Frediani, M., Rosi, L., Camaiti, M., Berti, D., Mariotti, A., Comucci, A., Vannucci, C., and Malesci, I. (2010) Polylactide/perfluoropolyether block copolymers: Potential candidates for protective and surface modifiers. *Macromolecular Chemistry and Physics,* **211**, 988-995.

91. Giuntoli, G., Rosi, L., Frediani, M., Sacchi, B., and Frediani, P. (2012) Fluoro-functionalized PLA polymers as potential water-repellent coating materials for protection of stone. *Journal of Applied Polymer Science,* **125**, 3125-3133.

92. Giuntoli, G., Frediani, M., Pedna, A., Rosi, L., and Frediani, P. (2012) New perspectives for the application of PLA in cultural heritage. In: *Polylactic Acid: Synthesis, Properties and Applications*, Nova Science Publishers, USA, pp.161-189.

93. Pedna, A., Pinho, L., Frediani, P., Mosquera, M. J. (2016) Obtaining SiO_2–fluorinated PLA bionanocomposites with application as reversible and highly-hydrophobic coatings of buildings. *Progress in Organic Coatings*, **90**, 91-100.

94. Alam, M., Akram, D., Sharmin, E., Zafar, F., and Ahmad, S. (2014) Vegetable oil based eco-friendly coating materials: A review article. *Arabian Journal of Chemistry*, **7**, 469-479.

95. Dutton, H. J., and Scholfield, C. R. (1963) Recent developments in the glyceride structure of vegetable oils. *Progress in the Chemistry of Fats and other Lipids*, **6**, 313-339.

96. Baumann, H., Buhler, M., Fochem, H., Hirsinger, F., Zoebelein, H., and Falbe, J. (1988) Natural fats and oils - Renewable raw materials for the chemical industry. *Angewandte Chemie International Edition*, **27**, 41-62.

97. Wisniak, J. (1977) Jojoba oil and derivatives. *Progress in the Chemistry of Fats and other Lipids*, **15**, 167-218.

98. Schuchardt, U., Sercheli, R., and Vargas, R. M. (1998) Transesterification of vegetable oils: A review. *Journal of the Brazilian Chemical Society*, **9**, 199-210.

99. Lu, Y., and Larock, R. C. (2009) Novel polymeric materials from vegetable oils and vinyl monomers: Preparation, properties, and applications. *ChemSusChem*, **2**, 136-147.

100. Xia, Y., and Larock, R. C. (2010) Vegetable oil-based polymeric materials: synthesis, properties, and applications. *Green Chemistry*, **12**, 1893-1909.

101. Salimon, J., Salih, N., and Yousif, E. (2012) Industrial development and applications of plant oils and their biobased oleochemicals. *Arabian Journal of Chemistry*, **5**, 135-145.

102. Odetoye, E., Ogunniyi, D. S, and Olatunji, G. A. (2010) Preparation and evaluation of Jatropha curcas Linneaus seed oil alkyd resins. *Industrial Crops and Products*, **32**(3), 225-230.

103. Shikha, D., Kamani, P. K., and Shukla, M. C. (2003) Studies on synthesis of water-borne epoxy ester based on RBO fatty acids. *Progress in Organic Coatings*, **47**(2), 87-94.

104. Ramasri, M., Srinivasa Rao, G. S., Sampathkumaran, P. S., and Shirsalkar, M. M. (1990) Water-soluble epoxy binders modified with boron ester for cathodic electrodeposition. *Progress in Organic Coatings*, **18**(1), 103-115.

105. Argyropoulos, J., Popa, P., Spilman, G., Bhattacharjee, D., and Koonce, W. (2009) Seed oil based polyester polyols for coatings. *Journal of Coating Technology and Research*, **6**(4), 501-508.

106. Laxmikanth Rao, J., Balakrishna, R. S., and Shirsalkar, M. M. (1992) Cathodically electrodepositable novel coating system from castor oil . *Journal of Applied Polymer Science*, **44**(11), 1873-1881.

107. Petrovic, Z. S. (2008) Polyurethanes from vegetable oils. *Polymer Reviews*, **48**(1), 109-155.

108. Lligadas, G., Ronda, J. C., Galia, M., and Cadiz, V. (2010) Plant oils as platform chemicals for polyurethane synthesis: Current state-of-the-art. *Biomacromolecules*, **11**, 2825-2835.

109. Meshram, P. D., Puri, R. G., Patil, A. L, and Gite, V. V. (2013) High performance moisture cured poly (ether–urethane) amide coatings based on renewable resource (cottonseed oil). *Journal of Coating Technology and Research*, **10**, 331-338.

110. Mahapatra, S. S., and Karak, N. (2004) Synthesis and characterization of polyesteramide resins from Nahar seed oil for surface coating applications. *Progress in Organic Coatings*, **51**, 103-108.
111. Alam, M., Sharmin, E., Ashraf, S. M., and Ahmad, S. (2004) Newly developed urethane modified polyetheramide-based anticorrosive coatings from a sustainable resource. *Progress in Organic Coatings*, **50**, 224-230.
112. Gaikwad, M. S., Gite, V. V., Mahulikar, P. P., Hundiwale, D. G., and Yemul, O. S. (2015) Eco-friendly polyurethane coatings from cottonseed and karanja oil. *Progress in Organic Coatings*, **86**, 164-172.
113. Sharmin, E., Zafar, F., Akram, D., Alam, M., and Ahmad, S. (2015) Recent advances in vegetable oils based environment friendly coatings: A review. *Industrial Crops and Products*, **76**, 215-229.
114. Anand, A., Kulkarni, R. D., Patila, C. K., and Gite, V. V. (2016) Utilization of renewable bio-based resources, *viz.* sorbitol, diol, and diacid, in the preparation of two pack PU anticorrosive coatings. *RSC Advances*, **6**, 9843-9850.
115. Chaudhari, A. B., Anand, A., Rajput, S. D, Kulkarni, R. D, and Gite, V. V. (2013) Synthesis, characterization and application of Azadirachta indica juss (neem oil) fatty amides (AIJFA) based polyurethanes coatings: A renewable novel approach. *Progress in Organic Coatings*, **76**, 1779-1785.
116. Petrovic, Z. S., and Javni, I. J. (2002) Process for the Synthesis of Epoxidized Natural Oil-based Isocyanate Prepolymers for Application in Polyurethanes, US Patent 6,399,698.
117. Kong, X. H., Liu, G. G., and Curtis, J. M. (2012) Novel polyurethane produced from canola oil based poly(ether ester) polyols: Synthesis, characterization and properties. *European Polymer Journal*, **48**, 2097-2106.
118. Anuar, S. T, Zhao, Y. Y., Mugo, S. M., and Curtis, J. M. (2012) Monitoring the epoxidation of canola oil by non-aqueous reversed phase liquid chromatography/mass spectrometry for process optimization and control. *Journal of American Oil Chemists Society*, **89**, 1951-1960.
119. Kong, X., Liu, G., Qi, H., and Curtis, J. M. (2013) Preparation and characterization of high-solid polyurethane coating systems based on vegetable oil derived polyols. *Progress in Organic Coatings*, **76**, 1151-1160.
120. Marathe, R., Tatiya, P., Chaudhari, A., Lee, J., Mahulikar, P., Sohn, D., and Gite, V. (2015) Neem acetylated polyester polyol - Renewable source based smart PU coatings containing quinoline (corrosion inhibitor) encapsulated polyurea microcapsules for enhance anticorrosive property. *Industrial Crops and Products*, **77**, 239-250.
121. Rajput, S. D., Hundiwale, D. G., Mahulikar, P. P., and Gite, V. V. (2014) Fatty acids based transparent polyurethane films Dilip G and coatings. *Progress in Organic Coatings*, **77**, 1360-1368.
122. Heiskanen, N., Jamsa, S., Paajanen, L., and Koshimies, S. (2010) Synthesis and performance of alkyd–acrylic hybrid binders. *Progress in Organic Coatings*, **67**, 329-338.
123. Akbarinezhad, E., Ebrahimi, M., Kassiriha, S. M., and Khorasani, M. (2009) Synthesis and evaluation of water-reducible acrylic–alkyd resins with high hydrolytic stability. *Progress in Organic Coatings*, **65**, 217–221.
124. Akintayo, C. O., and Adebowale, K. O. (2004) Synthesis and characterization of acrylated Albizia benth medium oil alkyds. *Progress in Organic Coatings*, **50**, 207-212.

125. Araujo, W. S., Margarit, I. C. P., Mattos, O. R., Fragata, F. L., and Lima-Netoe, P. (2010) Corrosion aspects of alkyd paints modified with linseed and soy oils. *Electrochimica Acta*, **55**, 6204-6211.
126. Mezzenga, R., and Manson, J. A. E. (2001) Thermo-mechanical properties of hyperbranched polymer modified epoxies. *Journal of Material Sciences*, **36**, 4883-4891.
127. Pilla, S., Kramschuster, A., Lee, J., Clemons, C., Gong, S., and Turng, LS. (2010) Microcellular processing of polylactide-hyperbranched polyester-nanoclay composites. *Journal of Material Science*, **45**, 2732-2746.
128. Pramanik, S., Konwarh, R., Sagar, K., Konwar, B. K., and Karak, N. (2013) Bio-degradable vegetable oil based hyperbranched poly(ester amide) as an advanced surface coating material. *Progress in Organic Coatings*, **76**, 689-697.
129. Ataei, S., Yahya, R., and Gan, S. N. (2011) Fast drying, high water and salt resistance coatings from non-drying vegetable oil. *Progress in Organic Coatings*, **72**, 703-708.

11

Epoxy Composite Coatings for Enhanced Corrosion Resistance

11.1 Introduction

High performance polymers are widely used as resistive coatings for the corrosion protection of metallic structures in aggressive environments. The main role of these polymeric coatings is to protect the metal from external corrosive agents (oxygen, water etc.) by acting as an effective barrier [1,2]. In fact, most polymers are not absolute barriers against water vapor, gases and organic substances. Recent studies indicate that addition of nano-fillers enhance the barrier properties of the coatings [3-7]. Nano-fillers improve adhesion strength between the polymer and metal as well as the gas impermeability and mechanical strength of the polymer, thus, enhancing the corrosion resistance of the coatings [5,6,8]. Graphene and graphene derived materials like graphene oxide (GO) and reduced graphene oxide (RGO) have been used recently as nano-filler for corrosion inhibition in reinforced polymer coatings because of the unique properties, such as exceptional chemical and thermal resistance, mechanical strength as well as high electrical and thermal conductivities [9-16]. Nano-layered graphene platelets provide an extraordinarily zigzag torturous diffusion path that lead to enhanced barrier performance for gas, moisture and oxygen transmissions. In addition to the diffusion, the solubility of the corrosive species in polymer nanocomposites (PGNs) can be markedly influenced by the presence of graphene.

Several studies have been reported on the effect of graphene on the protective nature of polymer composites. Liao et al. [17] synthesized polyurethane acrylate/graphene composite (PUAGC) via in-situ polymerization. The results exhibited enhanced mechanical properties and improved electrical conductivity with increasing graphene loading. Fabbri et al. [18] synthesized polybutylene terephthalate (PBT)/graphene composite and concluded that the composite could function as a conductive coating. The optimum oxidative degradation temperature was reported at 0.75% graphene loading, while the melting temperature of the composite decreased with increasing graphene loading. Martin-Gallego et al. [19] incorporated functionalized graphene in an epoxy matrix. The UV cured coating had higher glass transition temperature, higher stiffness and

Gisha E. Luckachan, Khaled Hassan and Vikas Mittal*, The Petroleum Institute (part of Khalifa University of Science and Technology), Abu Dhabi, UAE
*Current address: Bletchington, Wellington County, Australia

higher storage modulus relative to the pure epoxy matrix. In another study, Zaman *et al.* [20] studied the toughening effect of graphene nano-sheets on the epoxy matrix bisphenol A diglycidyl ether (BADGE). Graphene was produced by mechanical exfoliation and chemical modification. A 14.7% increase in T_g was reported (relative to pure BADGE resin) and the fracture energy release rate increased to 613.4 Jm^{-2} as compared to 204.2 Jm^{-2} for the pure resin.

In the present work, the effect of graphene derivatives on the anti-corrosion performance of epoxy coatings has been studied by reinforcing the BADGE polymer with graphene oxide and reduced graphene oxide. The anti-corrosion performance was analyzed using electrochemical impedance spectroscopy (EIS). Other analytical characterization techniques such as thermogravimetric analysis (TGA), differential scanning calorimetry (DSC) and infra-red (IR) spectroscopy were also used for analyzing the physical properties of the epoxy composite coatings.

11.2 Experimental Details

11.2.1 Materials

The uncured epoxy prepolymer BADGE D.E.R. 332 was purchased from Sigma Aldrich with an EEW in the range of 172-176 g/eq. The polyetherdiamine curing agent Jeffamine ED-600 (0,0'-bis(2-aminopropyl) polypropylene glycol-block-polyethylene glycol-block-polypropylene glycol) with an amine hydrogen equivalent weight (AHEW) of 132 g/eq was also purchased from Sigma Aldrich and used as received. RGO was purchased from Angstron Materials, USA (grade N008-100-N). Also to mention that the terms graphene and reduced graphene oxide will be used interchangeably throughout this paper. GO was synthesized in the laboratory via the modified Hummers method. Carbon steel coupons were obtained from a local supplier and used as substrates to test the formulated coatings. Fuming hydrochloric acid (35%) reagent grade and acetone were supplied by Merck for treating the steel coupons before coating.

11.2.2 Substrate Preparation

Carbon steel coupons of 5 cm x 2 cm x 2 cm were first pickled with hydrochloric acid for 2 h in order to remove the oxide layer from the surface. After acid treatment, the coupons were polished with sandpaper (60, 150 and 180 grits) followed by water and acetone rinsing. Finally, the coupons were sonicated in acetone for 10 min and dried in oven at 90 °C for 1 h.

11.2.3 BADGE Composite Coatings

For the pure BADGE coating, 2.5 g of the polymer was added to a beaker and heated at 40 °C to allow melting. The stoichiometric amount of Jeffamine was added to the beaker after cooling to room temperature and the mixture was stirred for 20 min. The mixture was applied to the polished steel substrates by brush, left at room temperature for 12 h and subsequently kept in an atmospheric oven overnight at 50 °C. For RGO/GO coatings, the procedure required the use of a solvent to disperse the nano-sheets. Acetone was selected as a suitable solvent as it can be removed easily at the final stage. The procedure followed for this purpose was adapted from Liu *et al.* [21]. Graphene was stirred in 100 mL acetone for 3 h and the dispersion was probe sonicated (40% amplitude) for 1.5 h in an ice bath. The mixture was stirred further for 1.5 h and 2.5 g of the epoxy prepolymer was added to the mixture while stirring. After the epoxy prepolymer was dissolved, the mixture was bath sonicated for 1.5 h. Finally to remove acetone, the mixture was heated at 70 °C in an oil bath with gentle stirring, and followed by heating in a vacuum oven. The stoichiometric amount of curing agent was added to the epoxy/filler dispersion after cooling to room temperature and stirred for 20 min. The mixture was degassed using a vacuum pump and applied by brush to the polished steel substrates. The brush coated steel substrates were left overnight in an atmospheric oven at 60-70 °C to obtain fully cured touch dry coating films. It was noticed that a higher temperature was needed to cure the composite coatings as compared to the pure polymer coatings. BADGE coatings containing 1% and 5% each of RGO and GO fillers were prepared using this method. Free-standing films of all samples were also generated by casting the coating formulations onto Teflon molds for carrying other physical property testing.

11.2.4 Characterization Techniques

The thickness of each coating was measured using PosiTector 6000 coating thickness gauge from DeFelsko Corporation, USA. Tape adhesion tests were carried out on the coated substrates as per the ASTM D-3359 standard. The adhesion toolkit was supplied by GARDCO® where a carbide knife was used to scribe the coatings. A cross-cut was created (test method B) and a pressure sensitive tape was applied over the scribed area. The grid pattern which results on the tape was used to evaluate the adhesion quality as per the standard. The electrical conductivity of the coatings was measured using the Prostat PRS-812 meter.

Thermogravimetric analysis of the prepared films was carried out using Discovery TGA supplied by TA Instruments. The synthesized films of epoxy and composites were heated from 25 °C to 500 °C at 10 °C/min heating rate under N_2 atmosphere. The Discovery DSC differential scanning calorimetry from TA Instruments was used to study the thermal transitions associated with the polymer composites. The samples were tested using two heating and cooling cycles (25 °C ↔ 250 °C) at a 10 °C/min heating/cooling rate under N_2 atmosphere. The Fourier transform infrared spectroscopy (FTIR) of the samples was performed using a Bruker VERTEX 70 FTIR spectrometer attached with a DRIFT accessory. IR acquisition was achieved by 120 scans at a resolution of 4 cm^{-1} in the frequency range of 370 cm^{-1} to 4000 cm^{-1} using OPUS software. Wide-angle X-ray diffraction was used to evaluate the degree of dispersion of the nano-fillers in epoxy.

Edges of the coated coupons were sealed using Nippon epoxy primer and dried for 24 h at room temperature. The coated substrates were immersed covering up to half of their height in 3.5 wt% NaCl solution at room temperature. The immersion tests were carried out at room temperature and were stopped upon coating degradation.

A three electrode cell of 250 mL volume with a platinum gauze counter electrode and a saturated calomel reference electrode (SCE) with bridge tube was used to perform electrochemical tests on the flat coated coupons. The coated coupon was the working electrode and an area of 1 cm^2 was exposed to 3.5 wt% NaCl solution. All tests were carried out at room temperature by connecting the corrosion cell to the BioLogic VMP-300 multipotentiostat (controlled by a computer running EC-Lab 10.40 software). Ultra-low current cables connected to the potentiostat were used for the accurate measurement of the current. This option included current ranges from 100 nA down to 100 pA with additional gains extending the current ranges to 10 pA and 1 pA. The resolution on the lowest range was 76 aA. The open circuit potential (OCP) was measured for 5 min in order to allow the potential to stabilize before the electrochemical impedance and potentiodynamic polarization test. The impedance measurements were performed at amplitude of 20 mV over the frequency range from 10^5 Hz to 10^{-2} Hz. After each measurement, the samples were kept in sodium chloride solution outside the corrosion cell. The impedance behavior of the samples was simulated using the same software. Polarization measurements were conducted by polarizing the working electrode from an initial potential of -250 mV up to a final potential of +250 mV as a function of open circuit potential using a scan rate of 1.66 mV/s. The corrosion rate was calculated in milli-inches per year (mpy) assuming an equivalent weight of 27.92 g/eq and density of 7.87

g/cm³. The surface area exposed to the test solution was 1 cm². To study the reproducibility of the measurements, each set of experiments was repeated three times on newly coated samples. The choice of the coated samples was made after visual inspection so as to ensure the absence of any macro-bubbles or voids in the coatings.

11.3 Results and Discussion

The FTIR spectra of the BADGE prepolymer and the crosslinked polymer were recorded to confirm crosslinking (Figure 11.1). The FTIR spectrum of uncross-linked BADGE exhibited the typical absorption bands for methylene groups

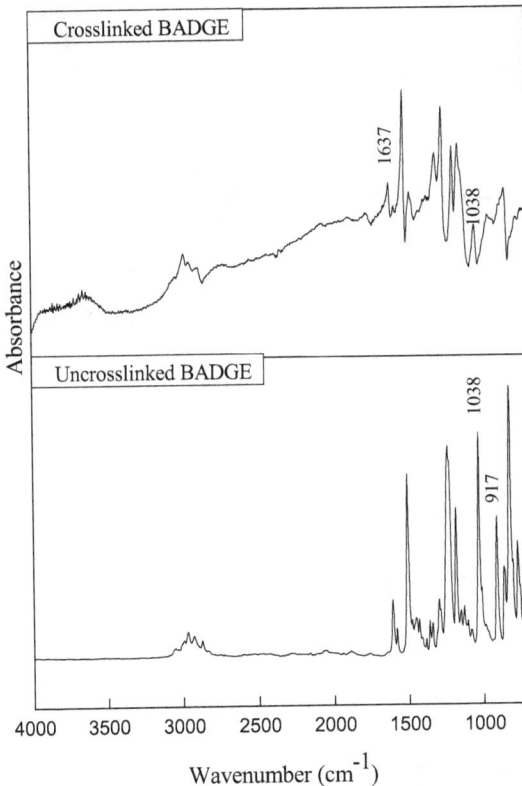

Figure 11.1 FTIR spectra of uncrosslinked and crosslinked BADGE polymer.

bending, methyl asymmetrical and symmetrical stretching and Ar-C=C-H stretching (1610 and 1583 cm^{-1}). The strong band around 830 cm^{-1} corresponded to the para substitution of the aromatic rings [22-24]. The strong band at 917 cm^{-1} was attributed to the oxirane CH$_2$-O-CH- group present in the structure of the uncrosslinked prepolymer [23]. Another strong absorption band was observed at ~1038 cm^{-1} which corresponded to -C-O-C-stretching vibrations. The FTIR spectrum of the crosslinked polymer indicated a band at ~1637 cm^{-1} reflecting the formation of hydroxyl groups [23]. The absence of the CH$_2$-O-CH- epoxy band at 917 cm^{-1} confirmed the crosslinking of the prepolymer via oxirane ring opening reactions [23]. Moreover, there was a decrease in the intensity of the ~1038 cm^{-1} band which indicated that the number of -C-O-C- present in the structure was reduced after the crosslinking reaction. Further, the appearance of more absorption bands (medium to weak) in the region 1360-1080 cm^{-1} was characteristic of C-N bond stretching vibrations [23,25,26]. The FTIR spectra of the BADGE composites with RGO and GO were inconclusive and did not reflect the formation of new bonds. This is, however, expected as the solvent blending method did not lead to the formation of chemical bonds between the filler and the host polymer.

Figure 11.2a demonstrates the TGA thermograms of the epoxy composites synthesized with GO. The onset of thermal degradation temperature (T_{onset}) of BADGE was observed to be 383 °C. The onset temperature value decreased by 7 °C with the introduction of 1 & 5% GO (Table 11.1). C-N and O-CH$_2$ are the weak sites in the cured BADGE resin according to the bond enthalpy data [27]. However, the presence of a large number of hydroxyl groups in the structure results in the formation of intramolecular hydrogen bonding between the neighboring chains. Thus, as also corroborated from the literature [27], dehydration was the most energetically favorable reaction. The difference in amount of char produced between neat BADGE and BADGE/5% GO was almost 1%. The amount of char produced from BADGE was generally high due to the presence of aromatic rings in the chemical structure of BADGE which yielded unsaturated moieties and cross-linked polyaromatics [21]. In the case of BADGE/5% GO composite, slightly more char was produced as a result of the increased GO loading. The decrease in T_{onset} with the addition of GO could be attributed to the arrangement of the GO sheets in the thermoset structure. The addition of the curing agent to the mixture was the last step i.e. after the introduction and sonication of GO in BADGE. This could have led to spatial obstruction caused by the GO sheets which prevented the development of a fully cross-linked epoxy matrix. The differences in onset mass % were not large, however, in the case of 5% GO composite, a slight increase was observed and attributed to the adsorbed

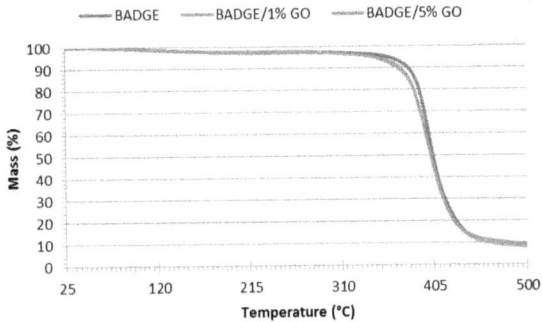

(a)

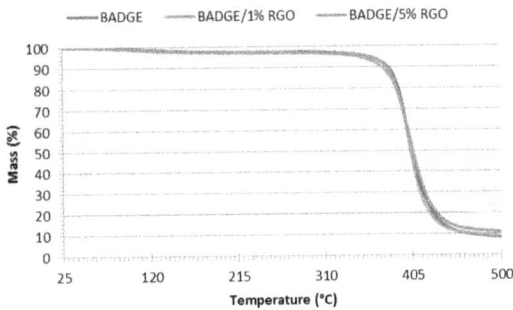

(b)

Figure 11.2 TGA thermograms of (a) BADGE/GO and (b) BADGE/RGO composites.

moisture due to the hydrophilic nature of GO. The thermal performance of RGO incorporated BADGE was similar to the case of BADGE/GO, as T_{onset} and onset mass % followed the same trend (as shown in Figure 11.2b, Table 11.1), though lesser decrease in the onset degradation temperature was observed in RGO composites as compared to GO composites. A noticeable difference in the amount of produced char was also observed with RGO as filler. On comparing the results, RGO appeared to promote the formation of char residue. There was a 1.6% difference in the amount of char produced between the neat BADGE and the 5% RGO composite. This increase in char yield could have resulted due to an alteration in the mechanism of char formation caused by the presence of a high RGO loading, which however, was not prominent in the GO composites. A more likely explanation of this effect could be the role of RGO as a catalyst for the char producing reactions. The formation of char was attributed to an enhanced flame resistance [21,28]. In addition to these findings, it was clear that the peak degradation temperature of the composites remained largely unaffected on the addition of the filler (Figure 11.2). As the degradation temperature of epoxy polymer itself is significantly high, thus, the composites still represent functional coatings systems for high temperature operation.

Table 11.1 Thermal data of BADGE/GO and BADGE/RGO composites

Sample	T_{onset} (°C)	Onset mass (%)	T_{end} (°C)	End mass (%)
BADGE	383	95.97	418	11.99
BADGE/1%GO	376	95.85	420	11.60
BADGE5%GO	376	96.52	419	12.92
BADGE/1%RGO	383	95.79	416	12.25
BADGE/5%RGO	380	95.91	422	13.62

DSC analysis of the synthesized BADGE composites (Figure 11.3) did not reveal a value for T_g, which was attributed to a highly crosslinked polymer system [29]. Crosslinking affects segmental mobility and at a very high degree of crosslinking as a result of covalent bonds between the individual polymer backbones, T_g is not observable [29-31]. Further, as can be observed from the thermograms, filler addition with varying loadings did not alter this behavior. In addition, the crosslinked polymer did not yield any thermal transition in the range exhibited by its uncrosslinked counterpart. The thermal transition of the uncrosslinked prepolymer was found to be 43 °C which agrees with values reported in the literature [30-33]. Absence of this transition in the crosslinked polymer concluded that a high degree of crosslinking was achieved and that filler addition did not impact the behavior. Combining these findings with the TGA analysis, it can be suggested that the addition of filler did not affect the curing behavior, thus, the decrease in the onset degradation temperature, especially in the case of GO containing composites, would have resulted from other factors. The presence of structural defects or impurities in the filler structure may also have contributed to the observed decrease in the onset degradation temperature. The measured electrical conductivity of the neat BADGE polymer was similar to that reported in the literature (approx. 10^{-13} S/cm). Upon addition of 1% RGO, no noticeable enhancement was observed in conductivity. 5% RGO composites manifested a 2 orders magnitude drop in the surface resistivity, indicating significant increase in electrical conductivity (Table 11.2).

Adding GO to BADGE did not alter its conductivity with no noticeable increase observed at both filler loadings.

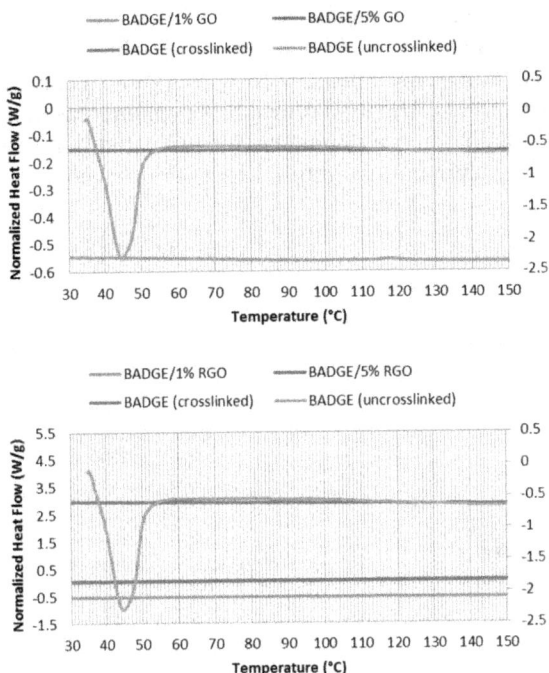

Figure 11.3 DSC thermograms of BADE/GO and BADGE/RGO composites.

Table 11.2 Electrical conductivity of BADGE/nanofiller composites

Sample	Surface resistivity (Ω/sq)	Surface conductivity (S cm^{-1})
Neat BADGE	7.90×10^{10}	1.27×10^{-12}
BADGE+1% RGO	1.60×10^{11}	6.25×10^{-13}
BADGE+5%RGO	1.71×10^{9}	5.85×10^{-11}
BADGE+1% GO	9.71×10^{10}	1.03×10^{-12}
BADGE+5% GO	2.70×10^{10}	3.70×10^{-12}

The XRD patterns of BADGE and its GO composites presented in Figure 11.4 indicated the generation of an amorphous material. The amorphous nature was

Polymers in Oil and Gas Industry

not altered with increased GO loading. GO was also dispersed well in the epoxy polymer since its crystalline structure was lost. This may be attributed to polar-polar chemical interactions between the oxygen functionalities of GO and BADGE. The observation was also similar to other reported studies [22,34,35]. The XRD patterns of BADGE/RGO composites exhibited the broad amorphous peak of BADGE and a sharp peak at 2θ ≈ 26.5° which corresponded to the layered structure of RGO. The RGO diffraction peak increased in intensity with increased RGO loading in the polymer. This indicated that RGO may have re-aggregated in the polymer and did not disperse homogeneously. Thus, despite

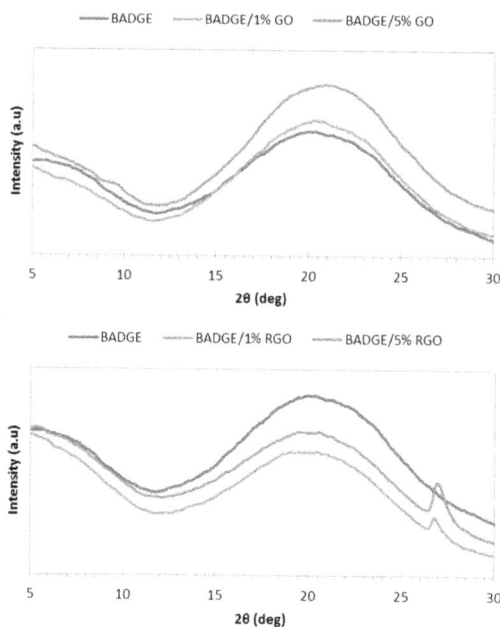

Figure 11.4 XRD patterns of BADGE/GO and BADGE/RGO composites.

mixing and sonication, delamination of some of the RGO sheets was not fully achieved and these sheets existed in the polymer in an aggregated arrangement. Therefore, RGO maintained its crystalline structure, which would also have resulted due to relatively weak interactions of RGO with the epoxy polymer. This was expected due to the absence of oxygen functionalities in RGO. Although π − π interactions existed between the aromatic regions of RGO and the para substituted aromatic rings of BADGE, the interactions were not sufficient to for complete exfoliation of the RGO sheets. Similar findings were also observed in other studies reporting about BADGE systems [21,35]. It is also possible that $\pi - \pi$ interactions were sterically hindered due to the presence of methyl groups protruding from the plane of the aromatic rings in BADGE.

Corrosion resistance behavior of BADGE and nano-fillers reinforced BADGE was studied by EIS measurements. The Bode plot of pure BADGE shown in Figure 11.5 indicated the initial impedance modulus in the low frequency region as $10^{5.80}$ Ωcm^2 at the beginning of the 1 day analysis. This value was maintained for 6 h, which is similar to the findings reported in literature [26]. The impedance modulus dropped to $10^{4.84}$ Ωcm^2 by the end of the 1 day measurement period. This value did not correspond to strong anti-corrosion protection as the impedance modulus of bare uncoated steel is in the order of $10^{3.50}$ Ωcm^2. The decrease in impedance with immersion time corresponded to ongoing corrosion initiated by the ingress of electrolyte and interaction with the metal substrate. Despite the high degree of crosslinking of the polymer (as shown by DSC results), the coating was not totally impermeable to aggressive species such as Cl^-.

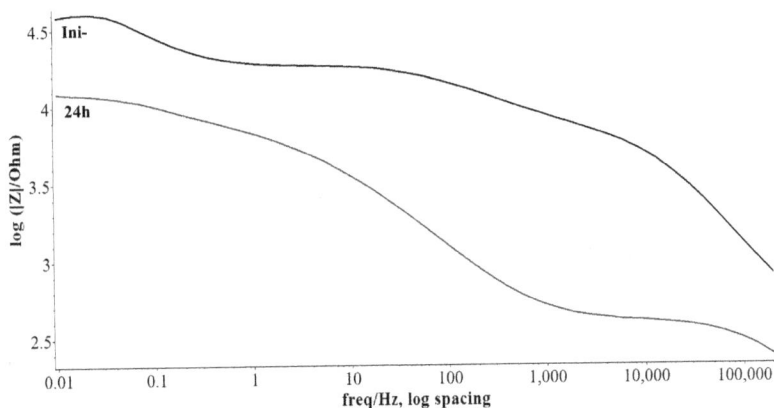

Figure 11.5 Bode plots of pure BADGE coating in 3.5% sodium chloride solution.

The addition of RGO significantly enhanced the anti-corrosion performance of the BADGE coating (Figure 11.6). The addition of 1% RGO filler to the epoxy thermoset resin yielded an initial impedance modulus of $10^{8.21}$ Ω cm^2 at the beginning of the 1 day analysis, which indicated an increase of almost 2.5 orders of magnitude over neat BADGE. The impedance in the low frequency region of the Bode plot was $10^{6.83}$ Ω cm^2 at the end of the 1 day analysis which was an improvement of 2 orders of magnitude over the neat epoxy system. The enhancement in the anti-corrosion performance with the addition of RGO was maintained over the whole immersion cycle. Thus, the addition of RGO imparted barrier properties and made the thermoset coating more impermeable to oxygen and other aggressive species [26,36]. Further addition of RGO to the epoxy coating did not enhance the anti-corrosion performance further. The impedance modulus in the low frequency region of BADGE/5%RGO varied between $10^{5.56}$ Ω cm^2 to $10^{4.8}$ Ω cm^2 over the course of 1 day measurement (Figure 11.6). This represented almost 3 orders of magnitude reduction in the impedance modulus as compared to BADGE/1% RGO. Nevertheless, based on the impedance moduli over the course of the 1 day measurement it was clear that BADGE/1% RGO exhibited better anti-corrosion performance as compared to the 5% RGO sample.

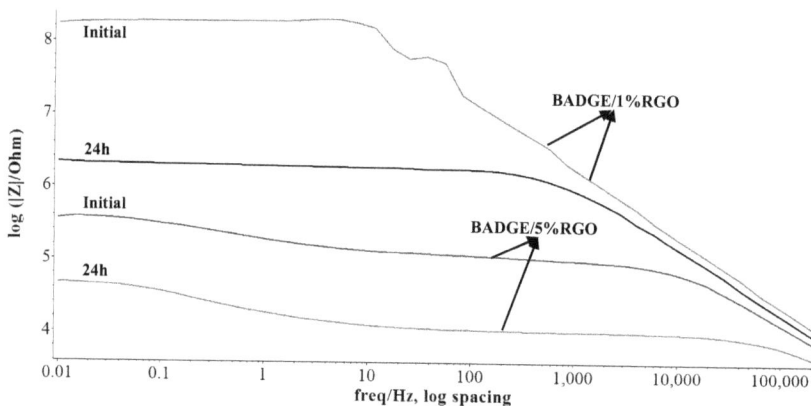

Figure 11.6 Bode plots of BADGE/RGO composite coatings in 3.5% sodium chloride solution.

The addition of GO to BADGE resulted in a better anti-corrosion performance than the RGO samples. The EIS results of BADGE/1% GO and BADGE/5% GO were very similar with only a small difference in the low frequency region impedance moduli at the end of the immersion cycle. The impedance modulus in the low frequency region of BADGE/5% GO exhibited a value of $10^{8.90}$ Ωcm^2 at the beginning of the 1 day measurement and $10^{8.42}$ Ωcm^2 at the end (Figure 11.7). These values were the highest among all BADGE samples. The value was changed to $10^{9.00}$ Ωcm^2 after 144 h of immersion. Therefore, only BADGE/5% GO was chosen for further corrosion analysis and discussed herein. From the results, it was evident that BADGE/5% GO exhibited the highest initial impedance moduli which was also stabilized over the course of the immersion cycle.

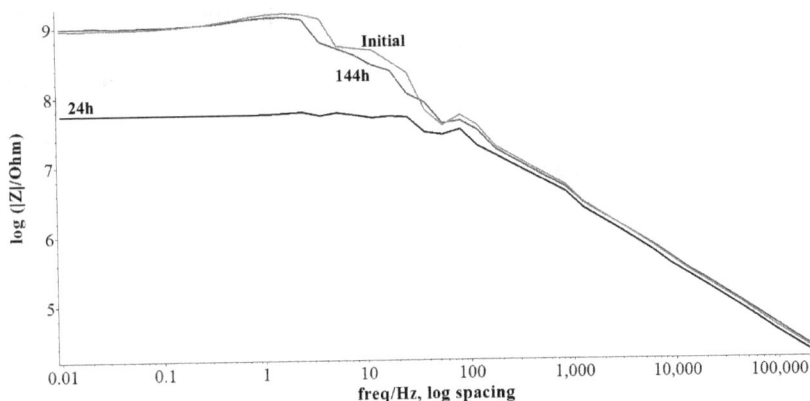

Figure 11.7 Bode plots of BADGE/5% GO after immersion.

The above analyses indicated that the addition of GO to the epoxy polymer resulted in more protective coating formulations as compared to RGO. Generally, the drop in the impedance modulus in the low frequency region occurred less rapidly for the GO based coatings. The visual inspection of the immersed specimens further validated the conclusions drawn from the EIS analyses. The better anti-corrosion performance of the GO composite coatings as compared to the RGO coatings can be attributed to the component interactions present in the composites. As also observed earlier in the X-ray analysis, RGO was observed to have re-aggregated in the composite and did not disperse homogeneously which negatively affected the barrier properties of the final composite.

To further evaluate the anti-corrosion performance and investigate the processes occurring at the metal-coating interface, the experimental data were fitted to a circuit model shown in Figure 11.8. The equivalent circuit in Figure 11.8 fitted the experimental data well and the resulting errors were less than 3%. In the circuit, R_s is the solution resistance accounting for the resistance of the electrolyte and the position of the reference electrode. C_{coat} is the capacitance of the nanocomposite coating and is an indication of water or electrolyte uptake by the film. As the coating pores fill up with the electrolyte, the dielectric constant increases in magnitude (the dielectric constant of water is around 80 compared to 4-8 for the organic coating) [37]. R_{pore} is the charge transfer resistance within the coating pores. It represents all electrical resistances to charge transfer through the organic medium due to the presence of defects or voids. The decrease in R_{pore} corresponds to the permeation of the electrolyte solution through the coating pores [38]. The model circuit obtained in this study was also similar to that reported in other literature studies on BADGE composite coatings systems [26,39]. Table 11.3 also shows the variation of parameter values with immersion time for the BADGE/5%CO composite coating.

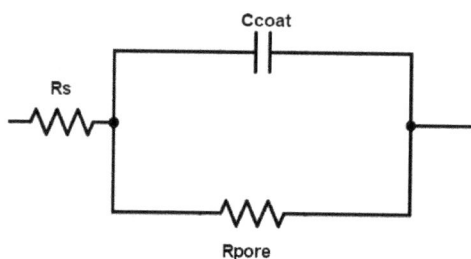

Figure 11.8 Equivalent circuit model used to fit the Bode plots of BADGE/5%GO composite coatings.

Table 11.3 Fitted parameters of Bode plots of BADGE/5%GO in the equivalent circuit model as function of immersion time

Model parameters	Immersion Time			
	Initial	6 h	12 h	18 h
log ($|Z|$) (Ω cm^2)	8.90	8.85	8.80	8.65
C_{coat} (Ω^{-1} cm^{-2})	0.17$\times$10^{-9}	0.21$\times$10^{-9}	0.21$\times$10^{-9}	0.24$\times$10^{-9}
R_{pore} (Ω cm^2)	5.42$\times$10^{15}	6.23$\times$10^8	5.34$\times$10^8	4.11$\times$10^8

Model parameters	Immersion Time			
	24 h	72 h	120 h	144 h
log ($|Z|$) (Ω cm^2)	8.42	8.71	8.91	9.00
C_{coat} (Ω^{-1} cm^{-2})	0.28$\times$10^{-9}	0.18$\times$10^{-9}	0.18$\times$10^{-9}	0.19$\times$10^{-9}
R_{pore} (Ω cm^2)	3.68$\times$10^8	5.45$\times$10^8	5.64$\times$10^8	5.79$\times$10^8

For the first 24 h of immersion, a slight increase in the coating capacitance was observed, which was caused by electrolyte uptake by the coating. However, after 24 h, the increase in C_{coat} stopped and a smaller value was maintained for the rest of the immersion cycle indicating reduced amount of electrolyte uptake after 24 h. The initial increase in C_{coat} was accompanied by a decrease in R_{pore} which was attributed to the permeation of the electrolyte through the coating pores and voids. Towards the end of the 24 h, the C_{coat} stabilized indicating that the water entering the coating did not reach the metal surface for corrosion and the water was dispersed inside the coating. At the same time, the increased R_{pore} indicated the barrier protection imparted by the GO sheets as a result of the increased tortuous diffusion path of water through the epoxy matrix. Table 11.4 presents the results obtained from the polarization analysis. The corrosion potential E_{corr} (mV), corrosion current I_{corr} (μA) and corrosion rate CR (mpy) were measured before and after 24 h immersion in 3.5 wt% NaCl solution. E_{corr} of the coated steel indicated the protection of the underlying metal as it remained more positive than that of bare steel before and after immersion. The corrosion current of the coated specimen was significantly lower than that of bare steel which suggested the isolation of the anodic and cathodic regions on the steel surface and, hence, high resistance to corrosion.

Table 11.4 Tafel analysis of BADGE/5% GO coating before and after immersion

	BADGE/ 5% GO			Bare steel		
	E_{corr} **(mV)**	I_{corr} **(μA)**	**CR (mpy)**	E_{corr} **(mV)**	I_{corr} **(μA)**	**CR (mpy)**
Before	128.39	0.61×10^{-9}	2.81×10^{-10}	-613.53	10.90	5.03
After	-578.11	1.42×10^{-6}	6.55×10^{-7}	-827.28	56.32	25.97

11.4 Conclusion

In this study, the effect of GO and RGO on the anti-corrosion behavior of epoxy polymer (BADGE) was studied by preparing BADGE/GO and BADGE/RGO nanocomposite coatings on carbon steel substrates. Thermal stability of BADGE was not affected significantly by the nano-fillers. Surface conductivity increased by an order of two with the incorporation of 5% RGO, however, it was not impacted significantly by the GO addition. The weak interaction of RGO with epoxy polymer because of the absence of oxygen functionalities led to an agglomerated arrangement of the sheets in the composite. However, a well dispersed arrangement of GO in BADGE polymer was observed in the XRD pattern that also correspondingly led to an enhanced corrosion resistance of BADGE/GO nanocomposite coatings. Both RGO and GO nano-fillers improved the low frequency impedance of BADGE coating significantly. However,

BADGE/5%GO nanocomposite coatings provided a long-term protection by exhibiting a high low frequency impedance of $10^{9.00}$ Ωcm^2 over 144 h of immersion in the corrosive media as compared with BADGE/RGO nanocomposite coatings. It was also confirmed that the uniform dispersion of GO in the epoxy matrix increased the tortuous path of water dispersion through the coating, which in turn imparted an enhanced corrosion resistance to the BADGE coatings. The epoxy/GO coatings, thus, exhibited functional coatings systems for effective corrosion protection of metal surfaces, which makes these coatings systems relevant in many oil and gas industry applications.

References

1. Wicks, Jr. Z. W., Jones, F. N., Pappas, S. P., and Wicks, D. A. (2007) *Organic Coatings: Science and Technology*, John Wiley & Sons, USA.
2. Ahmad, Z. (2006) Principles of Corrosion Engineering and Corrosion Control, Butterworth-Heinemann, UK.
3. Makhlouf, A. S. H. (2014) Handbook of Smart Coatings for Materials Protection, Woodhead Publishing, UK.
4. Behzadnasab, M., Mirabedini, S., and Esfandeh, M. (2013) Corrosion protection of steel by epoxy nanocomposite coatings containing various combinations of clay and nanoparticulate zirconia. *Corrosion Science*, **75**, 134-141.
5. Zhang, Y., Shao, Y., Zhang, T., Meng, G., and Wang, F. (2013) High corrosion protection of a polyaniline/organophilic montmorillonite coating for magnesium alloys. *Progress in Organic Coatings*, **76**, 804-811.
6. Piromruen, P., Kongparakul, S., and Prasassarakich, P. (2014) Synthesis of polyaniline/montmorillonite nanocomposites with an enhanced anticorrosive performance. *Progress in Organic Coatings*, **77**, 691-700.
7. Radhakrishnan, S., Siju, C., Mahanta, D., Patil, S., and Madras, G. (2009) Conducting polyaniline–nano-TiO$_2$ composites for smart corrosion resistant coatings. *Electrochimica Acta*, **54**, 1249-1254.
8. Bagherzadeh, M., and Mousavinejad, T. (2012) Preparation and investigation of anticorrosion properties of the water-based epoxy-clay nanocoating modified by Na+-MMT and Cloisite 30B. *Progress in Organic Coatings*, **74**, 589-595.
9. Dong, Y., Liu, Q., and Zhou, Q. (2015) Time-dependent protection of ground and polished Cu using graphene film. *Corrosion Science*, **90**, 69-75.
10. Sun, W., Wang, L., Wu, T., Wang, M., Yang, Z., Pan, Y., and Liu, G. (2015) Inhibiting the corrosion-promotion activity of graphene. *Chemistry of Materials*, **27**, 2367-2373.
11. Miskovic-Stankovic, V., Jevremovic, I., Jung, I., and Rhee, K. (2014) Electrochemical study of corrosion behavior of graphene coatings on copper and aluminum in a chloride solution. *Carbon*, **75**, 335-344.
12. Liu, J., Hua, L., Li, S., and Yu, M. (2015) Graphene dip coatings: An effective anticorrosion barrier on aluminum. *Applied Surface Science*, **327**, 241-245.

13. Singh, B. P., Nayak, S., Nanda, K. K., Jena, B. K., Bhattacharjee, S., and Besra, L. (2013) The production of a corrosion resistant graphene reinforced composite coating on copper by electrophoretic deposition. *Carbon*, **61**, 47-56.
14. Ming, H., Wang, J., Zhang, Z., Wang, S., Han, E. H., and Ke, W. (2014) Multilayer graphene: A potential anti-oxidation barrier in simulated primary water. *Journal of Materials Science & Technology*, **30**, 1084-1087.
15. Kumar, S., Sun, L., Caceres, S., Li, B., Wood, W., Perugini, A., Maguire, R., and Zhong, W. (2010) Dynamic synergy of graphitic nanoplatelets and multi-walled carbon nanotubes in polyetherimide nanocomposites. *Nanotechnology*, **21**, 105702.
16. Li, J., Cui, J., Yang, J., Li, Y., Qiu, H., and Yang, J. (2016) Reinforcement of graphene and its derivatives on the anticorrosive properties of waterborne polyurethane coatings. *Composites Science and Technology*, **129**, 30-37.
17. Liao, K.-H., Qian, Y., and Macosko, C. W. (2012) Ultralow percolation graphene/polyurethane acrylate nanocomposites. *Polymer*, **53**, 3756-3761.
18. Fabbri, P., Bassoli, E., Bon, S. B., and Valentini, L. (2012) Preparation and characterization of poly (butylene terephthalate)/graphene composites by in-situ polymerization of cyclic butylene terephthalate. *Polymer*, **53**, 897-902.
19. Martin-Gallego, M. Verdego, R., Lopez-Manchado, M. A., and Sangermano, M. (2011) Epoxy-graphene UV-cured nanocomposites. *Polymer*, **52**, 4664-4669.
20. Zaman, I., Phan, T. T., Kuan, H.-C., Meng, Q., La L. T. B., Luong L., Youssf, O., and Ma, J. (2011) Epoxy/graphene platelets nanocomposites with two levels of interface strength. *Polymer*, 52, 1603-1611.
21. Liu, S., Yan, H., Fang, Z., and Wang, H. (2014) Effect of graphene nanosheets on morphology, thermal stability and flame retardancy of epoxy resin. *Composites Science and Technology*, **90**, 40-47.
22. Hassan, K. (2016) Development of Graphene-based Anticorrosion Coatings. MSc Thesis, The Petroleum Institute, UAE
23. Nikolic, G., Zlatkovic, S., Cakic, M., Cakic, S., Lacnjevac, C., and Rajic, Z. (2010) Fast Fourier transform IR characterization of epoxy GY systems crosslinked with aliphatic and cycloaliphatic EH polyamine adducts. *Sensors*, **10**, 684-96.
24. Gonzalez, M. G., Cabanelas, J. C., and Baselga, J. (2012) Applications of FTIR on epoxy resins –identification, monitoring the curing process, phase separation and water uptake. In: Infrared Spectroscopy - Materials Science, Engineering and Technology, Theophanides, T. (ed.), Intechopen, Croatia. Online: http://www.intechopen.com/books/infrared-spectroscopy-materials-science-engineering-and-technology/applications-of-ftir-on-epoxy-resins-identification-monitoring-the-curing-process-phase-separatio (assessed 22nd February 2017)
25. Krishnamoorthy, K., Jeyasubramanian, K., Premanathan, M., Subbiah, G., Shin, H. S., and Kim, S. J. (2014) Graphene oxide nanopaint. *Carbon*, **72**, 328-337.
26. Chang, K.-C., Hsu, M.-H., Lu, H.-I., Lai, M.-C., Liu, P.-J., Hsu, C.-H., Ji, W. F., Chuang, T. L., Wei, Y., Yeh, J. M., Liu, W. R., Yeh, J. M. (2014) Room-temperature cured hydrophobic epoxy/graphene composites as corrosion inhibitor for cold-rolled steel. *Carbon*, **66**, 144-153.
27. Grassie, N., and Guy, M. I. (1986) Degradation of epoxy polymers: Part 4 – Thermal degradation of bisphenol-A diglycidyl ether cured with ethylene diamine. *Polymer Degradation and Stability*, **14**, 125-137.

Polymers in Oil and Gas Industry

28. Pielichowski, K., Leszcynska, A., and Njuguna, J. (2010) Mechanisms of thermal stability enhancement in polymer nanocomposites. In: *Optimization of Polymer Nanocomposite Properties*, Mittal, V. (ed.), Wiley VCH, Germany, pp. 195-210.
29. Daniels, C. A. *Polymers: Structure and Properties*, CRC Press, USA.
30. Arab, B., Shokuhfar, A., and Ebrahimi-Nejad, S. (2012) Glass Transition Temperature of Cross-Linked Epoxy Polymers: a Molecular Dynamics Study. International Conference Nanomaterials: Applications and Properties, Crimea. Online: http://essuir.sumdu.edu.ua/handle/123456789/35102 (assessed 22nd February 2017).
31. Askadskii, A. A. (1992) Analysis of the Structure and Properties Of High Crosslinked Polymer Networks, Volume 16, Hardwood Academic Publishers, UK.
32. White, S. R., Mather, P. T., and Smith, M. J. (2002) Characterization of the cure-state of DGEBA-DDS epoxy using ultrasonic, dynamic mechanical, and thermal probes. *Polymer Engineering and Science*, **42**, 51-67.
33. Acton, Q. A. (2013) Amines - Advances in Research and Application, Scholarly Editions, USA.
34. Wang, X., Jin, J., and Song M. (2013) An investigation of the mechanism of graphene toughening epoxy. *Carbon*, **65**, 324-333.
35. Laachachi, A., Burger, N., Apaydin, K., Sonnier, R., and Ferriol, M. (2015) Is expanded graphite acting as flame retardant in epoxy resin? *Polymer Degradation and Stability*, **117**, 22-29.
36. Zomorodian, A., Garcia, M. P., Moura e Silva, T., Fernandes, J. C., Fernandes, M. H., and Montemor, M. F. (2013) Corrosion resistance of a composite polymeric coating applied on biodegradable AZ31 magnesium alloy. *Acta Biomaterialia*, **9**, 8660-70.
37. Brondel, D., Edwards, R., Hayman, A., Hill, D., and Mehta, S. (1994) Corrosion in the oil industry. *Oilfield Review*, **6(2)**, 4-18.
38. Yu, B., Wang, X., Xing, W., Yang, H., Song, L., and Hu, Y. (2012) UV-curable functionalized graphene oxide/polyurethane acrylate nanocomposite coatings with enhanced thermal stability and mechanical properties. *Industrial & Engineering Chemistry Research*, **51**, 14629-14636.
39. Mohammadi, S., Taromi, F. A., Shariatpanahi, H., Neshati, J., and Hemmati, M. (2014) Electrochemical and anticorrosion behavior of functionalized graphite nanoplatelets epoxy coating. *Journal of Industrial and Engineering Chemistry*, **20**, 4124-4139.

12

Thermally Conducting Polymer Nanocomposites: Synthesis, Properties and Applications

12.1 Introduction

Polymers are best known for their insulating behavior and relatively higher specific heat capacities. As most of the polymers (e.g. polyethylene, polypropylene, epoxy) have low thermal conductivities in the range of 0.11-0.44 W.m^{-1}.K^{-1}, optimization of their thermal conductivities is vital for melt processing and specific applications. During polymer melt processing, thermal conductivity plays a key role due to the requirement of efficient heat transfer for processing cycle times. Similarly, certain applications of polymers also require enhanced thermal conduction such as electronic packaging, assembly, batteries, solar cells, satellite devices, etc. Generally, in perfect solids, atoms are in constant modes of coupled and coordinated vibrations with high frequency and comparatively small amplitudes (shorter wavelengths) which can be assumed as elastic waves. Analogous to photon, quantum of vibrational energy is called phonon. Contrary to metals, the dominant heat conduction mechanism in most of the polymers is through phonons [1]. Specifically, heat conduction in polymers is accomplished through combination of phonons and motion of chains molecules i.e. vibrational and rotational motion. Heat transfer in polymers mainly depends on the extent of crystallinity of the material, thus, the polymer with high crystallinity has higher conductivity due to ordered structure as compared to the polymer with amorphous or disordered structure. Thermal conductivity can be defined as analogous to Fick's first law:

$$q = kA\frac{\Delta T}{L}$$

where q is the heat flux or rate of heat transfer, A is the cross-sectional area, k is the thermal conductivity, ΔT is the temperature difference and L is the conduction path length [2]. Enhancement in the thermal conductivity of polymeric

Ali U. Chaudhry* and Vikas Mittal**, The Petroleum Institute (part of Khalifa University of Science and Technology), Abu Dhabi, UAE
*Current address: Texas A&M University, Qatar; **Current address: Bletchington, Wellington County, Australia

materials can be achieved by fabricating polymer composites and alloys by the addition of organic or inorganic fillers.

Nano-scale materials have at least one external dimension or internal feature in the size range of 1-100 nanometers [3]. Nanocomposites generated using nano-fillers such as layered silicates, graphene nano-sheets, graphite nanoparticles, carbon nanotubes (CNTs) and metal oxide nanoparticles have attracted much research attention [4]. Using nano-fillers in polymer matrices is advantageous due to high surface to volume ratio and improved properties at much lower filler loading such as low density, mechanical strength as well as enhanced electrical, thermal, optical, and rheological properties. Same approach has also been adopted to improve the thermal conductivity of polymers by the addition of heat conductive fillers in the matrices. In literature, various thermally conductive fillers such as aluminum nitride, aluminum oxide (Al_2O_3), silica (SiO_2), graphite, boron nitride (BN), graphene, metallic particles (Cu), diamond, etc. have been reported. In conducting polymer composites, the improvement of thermal conductivity is thus mainly attributed to the incorporated fillers due to very low heat conduction ability of polymers. Filler properties like shape, size, concentration, interaction with polymer, alignment and orientation have strong effect on the conductivity of the composite systems (Figure 12.1) [1,5]. This review discusses recent developments in enhancing thermal conductivity of polymer based nanocomposites. Various polymer and filler

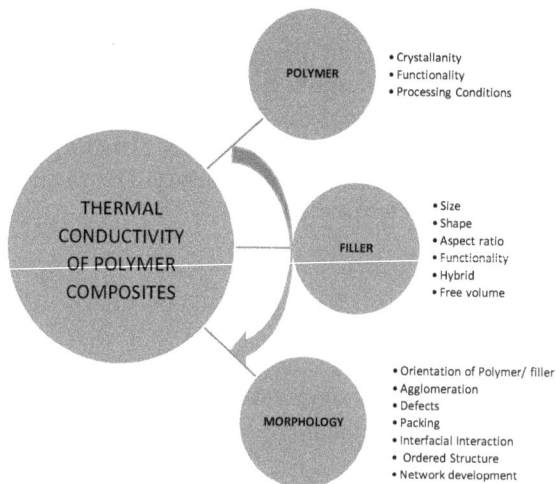

POLYMER
• Crystallanity
• Functionality
• Processing Conditions

THERMAL
CONDUCTIVITY
OF POLYMER
COMPOSITES

FILLER
• Size
• Shape
• Aspect ratio
• Functionality
• Hybrid
• Free volume

MORPHOLOGY
• Orientation of Polymer/ filler
• Agglomeration
• Defects
• Packing
• Interfacial Interaction
• Ordered Structure
• Network development

Figure 12.1 Design parameters for thermal conductivity of polymer composites.

types as well as techniques used to evaluate the thermal conductivity and microstructure of thermally conducting polymer nanocomposites are discussed, followed by the applications of these composites.

12.2 Modeling of Thermal Conductivity

Designing of polymer composites for improved thermal conductivity depends on many factors such as type of polymer and filler, concentration of filler (loading), processing conditions, combinations of different reinforcements, etc. Prediction of thermal conductivity through modeling is vital for designing effective polymer composites with optimum set of properties. Among various available approaches, two mainly employed models for thermal conductivity prediction are analytical micro-mechanical models (MME) (constitutive equations) and finite element (FC) simulations. Modeling of pure material is comparatively easier than the composite system due to evolution of complex morphology during processing i.e. varying degrees of filler dispersion, network formation, and interfacial resistance. Thus, the predicted thermal conductivity values for polymer composites from these models are generally qualitative or semi-quantitative. MME models are very useful for rapid evaluation of thermal conductivity for composite system on the basis of composition and properties of the host polymers and fillers. However, these models are often limited to certain composite systems like homogenous matrix incorporated with mono-disperse sphere-shaped or perfectly aligned ellipsoidal fillers. In this regard, two main approaches are adopted to predict the thermal conductivity of the composites i.e. upper bound (parallel model, linear mixing rule) and lower bound (series model, inverse mixing rule). In basic parallel model, the overall conductivity is predicted based on the assumption that the heat flux is the weighted sum of heat fluxes through domain of each composite phase, whereas temperature gradient is uniform. Parallel model also assumes that the particles are in complete contact forming percolating network. According to the parallel model, thermal conductivity can be calculated in terms of volume fractions of each phase as

$$k_c = k_p \Phi_p + k_m \Phi_m$$

where k_c, k_p and k_m are the thermal conductivities of the composite, polymer phase and filler phase, whereas ϕ_p and ϕ_m are the volume fractions of the polymer and filler respectively. On the contrary, the basic series model assumes that the overall temperature gradient is based on the weighted sum of temperature gradients through the domain of each phase, whereas heat flux is uniform. It also assumes that filler particles are completely out of contact with each other and percolation is impossible to achieve. According to the series model, the thermal conductivity can be calculated in terms of volume fractions as [1,6]

$$k_c = \frac{1}{(k_p + \Phi_p) + (k_m + \Phi_m)}$$

Among these models, inverse mixing rule (series model) is more common due to the ability to make predictions matching the experimental data. Series model has also formed the basis of many models built on complex weighted averages of k_c, k_p, ϕ_p and ϕ_m. These complex models frequently consider semi-theoretical fitting parameters and depend on effective medium approximations (EMA) or effective medium theories (EMT). Based on EMA approach, many common models such as Maxwell model, Bruggeman model, Hasselman-Johnson model, Rayleigh models, Lewis-Nielsen model and percolation model have been developed [7,8]. Maxwell model considers spherical non-interacting particles embedded in a continuous matrix, where effective thermal conductivity of the composite k_{eff} is given as

$$k_{eff} = k_m \left[1 + \frac{3\Phi_f}{\dfrac{(k_f + 2k_m)}{k_f - k_m} - \Phi_f} \right]$$

where k_f and ϕ_f represent the thermal conductivity and volume fraction of the filler phase. The validity of Maxwell model is traditionally for composite systems under 25% of filler fraction, which led to other modifications for particle shape and different phases of filler particles [9,10]. Similarly, Rayleigh introduced another model based on the thermal interactions between the filler particles. The model considers the filler particles as cubically arranged spherical shape inclusions in the matrix and can be used for higher volume fraction of filler [11]. According to Rayleigh model, k_{eff} can be calculated as

$$k_{eff} = k_m \left[1 + \frac{3\Phi_f}{\left(\dfrac{k_f + 2k_m}{k_f - k_m}\right) - \Phi_f + 1.569 \left(\dfrac{k_f - k_m}{3k_f - 4k_m}\right) \Phi_f^{\frac{10}{3}} + \dots} \right]$$

Using Maxwell and Rayleigh models, Hasselman-Johnson developed another model considering filler volume fraction, interfacial gaps, thermal resistance and particle size in order to develop the relation for k_{eff}. According to Hasselman-Johnson model, equations of k_{eff} for spherical, cylindrical, and flat plate filler geometry are given as

$$k_{eff} = k_m \left[\frac{2\left(\dfrac{k_f}{k_m} - \dfrac{k_f}{ah_c}\right)\Phi_f + \dfrac{k_f}{k_m} + \dfrac{2k_f}{ah_c} + 2}{\left(1 - \dfrac{k_f}{k_m} + \dfrac{k_f}{ah_c}\right)\Phi_f + \dfrac{k_f}{k_m} + \dfrac{2k_f}{ah_c} + 2} \right] \quad ----Spherical$$

$$k_{eff} = k_m \left[\frac{\left(\dfrac{k_f}{k_m} - \dfrac{k_f}{ah_c} - 1\right)\Phi_f + \left(1 + \dfrac{k_f}{k_m} + \dfrac{2k_f}{ah_c}\right)}{\left(1 - \dfrac{k_f}{k_m} + \dfrac{k_f}{ah_c}\right)\Phi_f + \left(1 - \dfrac{k_f}{k_m} + \dfrac{k_f}{ah_c}\right)} \right] \quad ----Cylindrical$$

$$k_{eff} = \left[\frac{k_f}{\left(1 - \dfrac{k_f}{k_m} + \dfrac{2k_f}{ah_c}\right)\Phi_f + \left(\dfrac{k_f}{ah_c}\right)} \right] \quad ----Flat\ plate$$

where a and h_c represent the particle radius and boundary conductivity respectively [12,13]. Further, Bruggeman used differential equations to calculate infinitesimal changes in incrementally constructed composite system. This approach is usually called differential effective medium theory (DEM) and can be used for different systems and high filler volume fractions. Using Bruggeman approach, many research studies obtained k_{eff} for different systems given as

$$(1-\Phi_f)^3 = \left(\frac{k_m}{k_{eff}}\right)^{(1+2\alpha)/(1-\alpha)} \left[\frac{k_{eff} - k_f(1-\alpha)}{k_m - k_f(1-\alpha)}\right]^{\frac{3}{(1-\alpha)}} \quad ----Particulate-Composite$$

$$\frac{k_f}{k_m} = \frac{1}{(1-\Phi)^{3(1-\alpha)/(1+2\alpha)}} \quad ----ZnS-Diamond$$

$$(1-\Phi_f) = \left(\frac{k_m}{k_{eff}}\right)^{1/3} \left[\frac{k_{eff}k_f R_{int} + ak_{eff} - \alpha k_1)}{k_m k_f R_{int} + ak_m - \alpha k_1}\right] \quad ----Composite$$

where α is a dimensionless parameter, which depends on the interfacial thermal resistance (R_{int}) between the filler and matrix and $\alpha = a_k/a$, where a is the particle size, a_k is the Kapitaz radius ($a_k = R_{int}k_m$) [14,15]. Lewis-Nielsen model was another simple and popular model reported in literature for moderate filler fractions (<40%). The benefits of this model are its applicability for a broad range of particle shapes and arrangements. The k_{eff} of a composite according to this model is given as

$$k_{eff} = \frac{1 + AB\Phi_f}{1 - B\psi\Phi_f}, \quad B = \frac{\left(\frac{k_f}{k_m}\right) - 1}{\left(\frac{k_f}{k_m}\right) + A}, \quad \psi = 1 + \left(\frac{1 - \Phi_m}{\Phi_m^2}\right)$$

where ϕ_m and A represent maximum filler volume fraction and shape coefficient for the filler particles respectively [8,16,17].

12.3 Measurement Techniques

Many techniques with different precision levels have been introduced for determining the thermal conductivity of materials. In general, the measurement techniques can be divided into two basic groups i.e. steady state and transient methods. The techniques in the former group perform the measurement when system has attained stability, whereas the techniques in the latter group take the measurement during heating or cooling process. The ability of the measuring techniques in both groups primarily depends on factors such as temperature range, material type and range of thermal conductivity [18]. Steady state techniques deal with the measurements of heat flux and temperature gradient through the materials and generally take longer time to achieve equilibrium. Steady state techniques include methods like guarded hot plate, axial flow, heat flow meter and pipe method [1]. Guarded hot plate method is usually employed for materials with low thermal conductivity and requires large sample size. The advantage of this method is its high accuracy (2%) and its usability in broad range of temperature of 80-1500 K. In this method, the sample is placed between the cold and hot plates attached with heaters and thermal insulation. The heat is allowed to pass through the specimen and the temperature is measured at both sides when the system attains steady state. Thermal conductivity is calculated on the basis of heat flux and temperature gradient across the sample, thickness and surface area of specimen [19]. Axial flow method is a longitudinal method to measure wide range of thermal conductivity of the materials using low temperature. In this method, the conductivity is usually measured by creating a temperature difference across the sample which is placed between two suitable references of known conductivity and attached with heater and heat sink [20]. Heat flow meter is similar to guarded hot plate method except the presence of heat flux sensor instead of main heater. It is simple in construction and has the ability to operate between -173-473 K, 90-1300 K, 298-2600 K for normal, axial and radial heat flow respectively [21]. Pipe method is another kind of radial heat flow method with wide range of operating temperature (293-2700 K) and targeted thermal conductivity range (0.02-2 W/(m.K)). It uses the principle of temperature gradient where cylindrical sample contains a core heater surrounded by a heat sink [18,22]. As mentioned earlier, transient methods perform the thermal conductivity measurements during transient addition of heat to the sample. Various transient methods include laser flash method, transient hot wire method and transient plane source method. In these methods, the thermal diffusivity is recorded as a function of time which is further used to calculate the thermal conductivity. In laser flash method, a sample disk is heated by laser pulse and the temperature is monitored by a detector. Transient hot wire method is similar to the pipe method where a hot wire is embedded in the test sample which acts as a sensor for the heater. The heating in transient plane source method is achieved with a known electrical current pulse. In this method, the heating element is placed between two test samples of same material, while temperature resistance is measured as a function of time [23-26].

12.4 Polymer and Fillers for Thermally Conducting Nanocomposites

As mentioned earlier, most of the polymers are insulator at macro-scale and usually have thermal conductivity below 0.5 W/(m.K). Generally, the polymers with higher degree of crystalline domains have higher thermal conductivity than the amorphous polymers. Similarly, presence of side chains in the main chain also result in lower thermal conductivity of the polymer. Specifically, in thermoplastic polymers, thermal conductivity is affected by the factors such as chain structure and orientation, crystallinity and interchain interactions. On the contrary, the factors which affect the thermal conductivity in thermoset polymers are mainly liquid crystal domain size and content, curing conditions and orientation [1].

Recent studies have indicated that the polymers with stiff backbone, such as conducting polymers, can have improved thermal conductivity due to suppressed chain conformation. Similarly, polymers with strong interchain interactions (e.g. Teflon) can also suppress the chain conformation [27,28]. In similar manner, polymers containing more crystalline phases have higher thermal conductivity than the amorphous polymers due to the fact that amorphous phases cause phonon scattering, which is not observed in crystalline phase (e.g. poly (ether ether ketone)). In crystalline phases, thermal conductivity further depends on the extent of extended crystal structure and thickness of lamellae in the polymer morphology [29]. Moreover, the thermal conductivity is also affected with the direction of chains i.e. transverse direction can have significantly higher thermal conductivity. Similarly, orientation of polymer chains through mechanical stretching can also enhance thermal conductivity [30]. It was also reported that the thermal conductivity of a H-bond-accepting amorphous polymer ((poly(N-acryloyl piperidine)) was improved due to engineered interchain interaction by blending with H-bond donating polymers such as (poly(acrylic acid), poly(vinyl alcohol), or poly(4-vinyl phenol)) [31].

Thermosets matrices are formed by crosslinking of either liquid monomers (epoxy resin) or solid polymers (rubber, polyethylene). In most of the cases, the mechanical and thermal properties of thermosets are superior to those of thermoplastic polymers. Similar to thermoplastics, thermosets also have amorphous (epoxy) and crystalline phases (liquid crystals) with low and high thermal conductivity, respectively. It was reported that the thermal conductivity of thermosets from liquid crystals strongly depends on the liquid crystal domain size and content. It usually increases with the volume percentage of anisotropic structure and liquid crystal domain size. As mentioned earlier, other factors like curing conditions have also significant great influence on the thermal conductivity of the polymer. For instance, in case of liquid crystal processing, higher temperature and slow cooling reduce the crystalline phase by changing from highly oriented smectic state, then to weakly oriented nematic state and finally to isotropic amorphous state [32-34]. Similar to thermoplastics, the thermal conductivity also varies with the orientation of polymer chains and significantly increases in the orientation direction, e. g. in highly oriented liquid crystalline thermosets [35].

It is also vital to discuss the factors associated with various fillers which significantly determine the thermal conductivity of the whole composite system due to low thermal conductivity of the polymer matrices. Different types of filler materials based on ceramic (BeO, Al_2O_3), metal (Cu, Al) and carbon (graphite) categories have been used with polymers for improved thermal conductivity. Generally, the materials which conduct heat through phonons are less efficient than those through electrons due to the resistance to scattering and higher speed of electrons [36]. Metallic fillers are quite effective in this process and are widely used in order to improve the thermal conductivity of polymers. The major disadvantage associated with metallic fillers is their high density and inherited electrical conductivity which can increase the weight and reduce the dielectric breakdown voltage, respectively [37]. On the other hand, carbon based materials belong to the class of conductive materials which have very high thermal conductivity and comparatively lower weight. As a result, materials like graphite, exfoliated graphite, carbon nanotubes, carbon fibers, and conductive carbon black have been extensively considered as conductive fillers for improved thermal conductivity [38]. Contrary to metal and carbon bases fillers, ceramic fillers are intrinsically electrically insulating materials. Transport of heat through ceramic fillers is mostly through phonons and this category usually comprises of both oxides (BeO, SiO_2) and non-oxides (AlN, BN, Si_3N_4) based fillers. Some shortcomings associated with ceramic fillers were also reported in the literature such as anisotropic behavior, crystal structure, crystallinity, presence of surface impurities, grain boundaries, etc. [39,40]. In addition to the inherited properties of fillers, the thermal conductivity of the system also depends on other factors such as filler loading level, shape and size of particles and interfacial adhesion. Usually, the thermal conductivity of the composites exhibits non-linear behavior with increasing filler loading and requires higher loading of fillers to achieve filler-to-filler connections [41]. Moreover, fillers with low aspect ratio exhibit a lower thermal conductivity as compared to the fillers with high aspect ratio (fibers, tubes). Usually, one dimensional fillers (1-D) develop well connected long conductive pathways in composite systems which result in significantly improved thermal conductivity. Examples of 1-D fillers are carbon nanotubes, carbon fibers, Si_3N_4 nanowires, boron nitride nanotubes, silver nanowires, copper nanowires, etc. [42-44]. Similarly, two dimensional (2-D) fillers, such as platelet fillers, are also considered as high aspect ratio fillers and can exhibit very high in-plane thermal conductivity, examples of such fillers with plate like morphology are boron nitride, graphene, Al_2O_3, TiB_2 and SiC [45]. Further, filler size is also vital for improving the thermal conductivity of the polymer composite systems as smaller size filler particles provide more interfacial area which results in more phonon scattering. Larger sized filler particles create effective network in the polymer composite system which reduces the thermal interfacial resistance. It has also been reported that the thermal conductivity does not depend on the particle size as nano-sized fillers were also observed to exhibit similar enhancement in thermal conductivity as the micro-sized fillers. Overall, the dependence of the thermal conductivity is generally linked to network formation by conductive fillers regardless of the size [46-48]. Further, the thermal conductivity can also be improved by using other techniques such as the use of hybrid fillers, different filler surface treatments, etc. [49-51]. Microstructure of the composite also affects the thermal conductivity of the system. Development of the composite micro-structure occurs during processing of the composite and depends on many factors such as processing conditions and ability of filler to orientate, agglomerate and form network in the polymer. During processing, filler particles can also be oriented using externally applied fields and shear or extensional flow. Similarly, formation of network by filler particles in polymers can be achieved by self-assembly of the filler particles in the polymer, molding of the mixture of filler and polymer powders, in-situ polymerization or double percolation. Agglomeration of filler particles is also occasionally required in order to achieve composite systems with isotropic thermal conductivity [52-56].

Before moving to the applications of various thermally conducting polymer nanocomposite systems, few general examples of thermally conducting nanocomposites are presented here in order to underline the above-mentioned factors. Shahil and Balandin reported epoxy nanocomposites using graphene - multilayer graphene hybrid filler [57]. Significant increase in the thermal conductivity was of the composites was observed as a function of filler volume fraction, as shown in Figure 12.2. As observed in Figure 12.2a, an enhancement of nearly 25 times in the thermal conductivity of the nanocomposites was achieved, which was significantly higher than corresponding composites with graphite and carbon black. In another study, Yao *et al.* [58] reported the nanocomposites of polyaniline with single-wall nanotubes. The nanocomposites exhibited small increase in thermal conductivity, however, the enhancement was

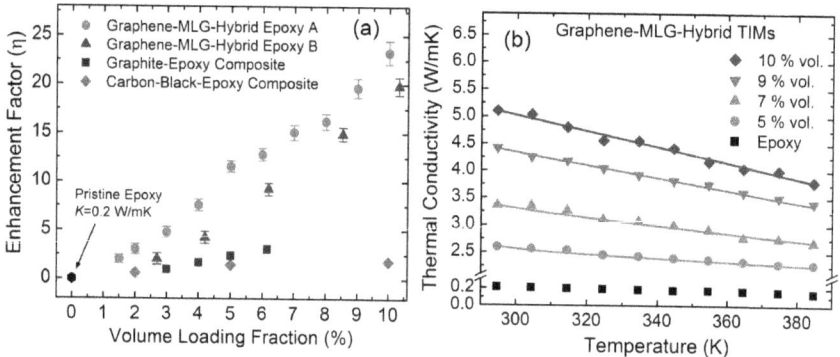

Figure 12.2 (a) Thermal conductivity enhancement factor as a fraction of the filler volume fraction; (b) thermal conductivity of the thermal interface materials as a function of filler fraction and temperature. Reproduced from Reference 57 with permission of American Chemical Society.

auch less as compared to the predictions, as shown in Figure 12.3. Even with filler fraction of 41 wt%, the thermal conductivity of the nanocomposite was 1.5 W/(mK). The authors attributed this phenomenon to the phonon scattering effect of the nano-interfaces generated by polyaniline/nanotubes nanocomposite structure. Thus, varying degrees of thermal conductivity enhancements have been achieved in literature.

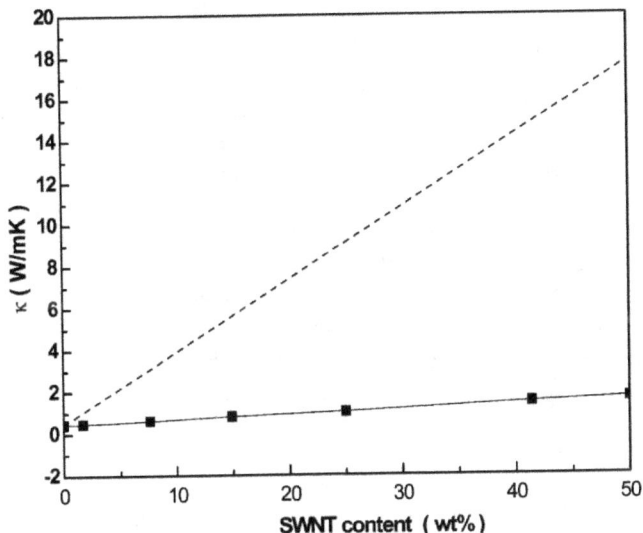

Figure 12.3 Thermal conductivity of the polyaniline-single walled nanotubes nanocomposites as a function of filler content. Dotted line indicates the prediction of thermal conductivity development with filler amount. Reproduced from Reference 58 with permission from American Chemical Society.

12.5 Applications

As mentioned earlier, polymer nanocomposites have advantages over conventional composite systems due to significant improvements in mechanical properties, gas barrier, thermal stability, chemical resistance, electrical and thermal conductivity. The improved properties of polymer nanocomposites at low volume fraction are due to the incorporated fillers with high surface to volume ratio and large interfacial area between the polymer and filler phases. With increasing demand in thermal management applications, where heat removal has become tremendously significant, the field of thermally conductive nanocomposites is expanding rapidly. These materials represent new potential for replacing metal, ceramic and pure polymer components in numerous applications owing to recyclability, electrical insulation, manufacturability, compactness, easy processing, light weight, anti-corrosion performance, etc. [59]. The following sections discuss selected evolving application areas of thermally conducting nanocomposites.

12.5.1 Heat Exchange Materials

Heat exchangers plays an important role in refrigeration, air conditioning as well as energy recovery applications in industries, especially oil and gas industries. The thermal performance in heat exchangers is driven by efficiency, cost, size, and quality of the heat exchanging system. Use of conventional materials such as metals has encountered limitations in many applications. The advantages of using polymers as heat exchange materials over conventional metallic alloys are ease of processing and handling, vast range of mechanical properties, high volume to weight ratio, light weight, low cost, resistance against fouling and absence of corrosion. Examples of such polymers useful as heat exchange materials after modification of thermal conductivity by the addition of thermally conducting fillers are fluoro-plastics, epoxy liquid crystal polymer, phenolic resins, polyamide, polyester resins, polyethylene terephthalate, polycarbonate, low density polyethylene, polyetheretherketone, polypropylene, polysulfone, polyvinyl chloride, polyvinylidene difluoride, polyphenylsulfone, etc. [60]. For instance, in a study underlining the impact of synergism between the polymer and filler phases to enhance thermal conductivity, Xu and Buehler [61] studied heat transfer performance at interfaces of carbon nanotubes. By employing the technique of polymer wrapping, interfacial thermal conductivity of carbon nanotube junctions was significantly improved. As shown in Figure 12.4, optimum polymer density led to more than 40% increase in the interfacial thermal conductivity.

Fouling of heat exchanger surfaces in process industries is one of the major problems which considerably increases the process costs. This problem may be overcome by introducing hydrophobicity on the internal surfaces of the heat exchanger. Rungraeng *et al.* [62] introduced thermally conducting hydrophobic anti-fouling coating comprising of carbon nanotubes (MWCNTs) and polytetrafluoroethylene on the heat exchanger internal surface. The nanocomposite coating exhibited a reduction of 97% in the surface energy as compared to the stainless steel surface. Superhydrophobic nature of the nanocomposite coating

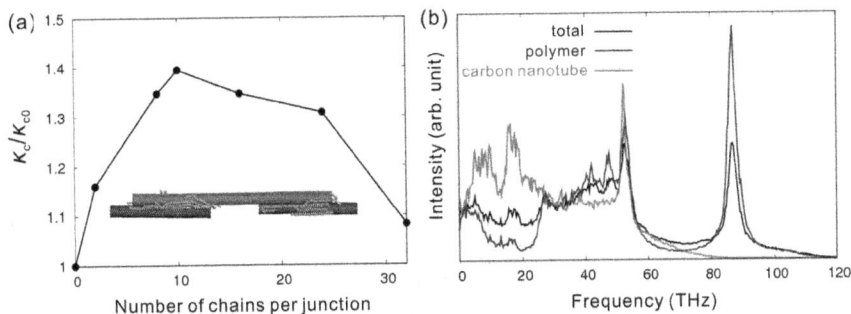

Figure 12.4 (a) Enhancement of the interfacial thermal conductivity by polyethylene chain wrapping as a function of number of chains per junction; (b) phonon spectrum of the hybrid system containing carbon nanotubes and polyethylene chains. Reproduced from Reference 61 with permission from American Chemical Society.

(5 wt% MWCNTs) was revealed by increased contact angle from 71.2° to 141.1° due to increased micro/nanoscale surface roughness. The thermal conductivity of the nanocomposite coating was observed to be 10.25 W/m.K, which was close to the thermal conductivity of stainless steel (16.3 W/m.K) [62]. Similar technique was adopted by Kananeh *et al.* [63] where thermoset polymers (epoxy, polyurethane and polyamide) were reinforced with nano-powders of Al_2O_3, SiO_2, etc., and coated on stainless steel. Cleaning in place (CIP) time was reduced by 70% in the case of plates coated with thermally conducting nano-composite [63].

Scale deposition in heat exchanger tubes of process plants is also a key concern which causes fouling, corrosion and erosion, thus, using a tube surface with properties like high thermal conductivity and super hydrophobicity may reduce the cleaning and maintenance efforts. Hwang *et al.* [64] used adherent and protective coatings consisting of conducting nanocomposites of polytetra-fluoroethylene (PTFE) and polyphenylenesulfide (PPS) blend with different concentrations of carbon nanotubes (MWNTs) for high-performance heat exchanger used in geothermal binary cycle power plants. It was reported that using a blend of PTFE/PPS (3:2) helped to overcome the drawbacks associated with PTFE and PPS. The blend components also complemented each other as the poor thermal conductivity, abrasion resistance and adhesion of PTFE were improved by the addition of PPS, whereas low glass transition temperature (T_g), low strength and brittleness of PPS were improved by blending with PTFE. Addition of surface modified MCWNTs (up to 5 wt%) to the blend further improved the hydrophobicity, electrical, thermal and mechanical properties. The contact angle measurements revealed that the addition of MWCNTs (5 wt%) to the PTFE/PPS blend increased the water contact angle of blend from 131° to 171° due to surface roughness. It was also observed that the addition of functionalized MWCNTs changed the morphology of the nanocomposites from smooth to nano-pin-like [64].

In another study, Marconnet *et al.* [65] developed epoxy nanocomposites reinforced with aligned carbon nanotubes. At 1 vol% of filler fraction in epoxy, the thermal conductivity of the polymer increased by 100%. Increasing the filler content to 17 vol% led to an increase in the thermal conductivity by a factor of 18, as shown in Figure 12.5. The developed nanocomposites exhibit high potential for heat exchange applications. Similarly, for heat exchanger material, the thermal conductivity of poly(vinylidene fluoride) (PVDF) was improved with the help of reduced graphene oxide (rGO) [66]. The nanocomposites were prepared by solvent casting in the absence as well as presence of an external

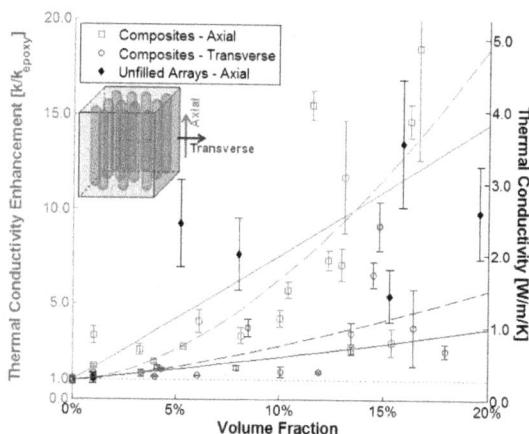

Figure 12.5 The axial and transverse thermal conductivity of CNT nanocomposites and unfilled arrays as a function of volume fraction. Reproduced from Reference 65 with permission from American Chemical Society.

electric field. It was observed that with the addition of up to 10 wt% rGO to PVDF, the thermal conductivity of the nanocomposites increased due to uniform filler dispersion. Beyond 10 wt% rGO in PVDF, the conductivity increased at a faster rate due to effective contact and connection between the rGO sheets. At 20 wt% filler content, the thermal conductivity of the system was observed to be maximum i.e. 0.562 W/m.K, which was 212% as compared to PVDF. Beyond 20 wt% of rGO content in PVDF, the thermal conductivity of the composite decreased owing to filler agglomeration, which led to scattering of phonons and increased ITR between PVDF and agglomerated graphene. The study also demonstrated that the application of external electric field during solution processing of nanocomposites was advantageous to align the rGO sheets in the polymer. In case of applied electric field, the thermal conductivity of the composite with 7 wt% filler content was significantly higher due to the alignment of the rGO sheets in the composite. Application of electric field reduced the filler percolation threshold owing to the better dispersion in the composite, thus, leading to enhanced thermal conductivity at lower filler concentration. As a comparison, rGO at 1 wt% level exhibited 23% increased thermal conductivity as compared to the composite without filler alignment [66].

Hagenmueller *et al.* [67] studied the thermal conductivity of the high density and low density polyethylene nanocomposites with single wall carbon nanotubes. The evolution of thermal conductivity with filler loading, degree of polymer crystallinity as well as polyethylene alignment. High density polyethylene composite with a filler fraction of 0.2 had twice the thermal conductivity as low density polyethylene composite. It indicated that the polymer with high crystallinity had lower interfacial thermal resistance. The thermal conductivity along the alignment direction was observed to increase with polyethylene alignment irrespective of the filler loading (Figure 12.6). The thermal conductivity of the composite was observed to evolve from both polymer and filler phases, thus, underlining the importance of polymer matrix in defining the thermal conductivity and corresponding applications of the composites. As expected, the electrical conductivity of the composites was also increased with addition of carbon nanotubes.

12.5.2 Materials for Harnessing Solar Energy

The importance of energy from alternative resources is becoming progressively vital owing to shrinking crude oil reserves and environmental hazards of liberated gases. Bringing renewable energy into the main stream is also a significant challenge of the current century. Alternative or renewable energy refers to the energy which is obtained from the sources other than earth's natural resources and which do not damage the environment. In this category, main technologies can be wind, hydro, solar, geothermal, biomass, etc. Among these different technologies providing renewable and sustainable energy, usage of solar energy has increased rapidly in recent years. Usually harnessing of solar

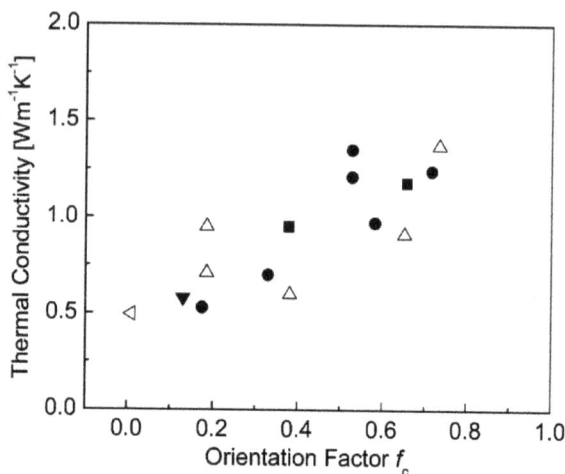

Figure 12.6 Thermal conductivity of various HDPE nanocomposites as a function of the chain orientation; (◁) isotropic HDPE, (▼) nominally isotropic SWNT/HDPE with φ ~ 0.006, (△) aligned HDPE fibers, (•) aligned SWNT/HDPE with φ ~ 0.006, and (■) aligned SWNT/ HDPE composites with φ ~ 0.012. Reproduced from Reference 67 with permission from American Chemical Society.

energy is achieved through photovoltaic cells, solar heating, molten salt power plants, solar architecture, and artificial photosynthesis. Photovoltaic cells have the characteristic of zero emission and represent clean and efficient route to convert sunlight into electric energy. However, inefficient heat dissipation from the cells is one of the major drawbacks in these cells. It was also reported that with an increase of 1 °C in the operating temperature leads to approximately 0.4–0.5% and 0.25% decline of the relative conversion efficiency for crystalline and amorphous silicon based cells, respectively [68]. In order to improve the heat dissipation ability of photovoltaic cells, the thermal conductivity of electrically insulating ethylene-vinyl acetate (EVA) layer was improved using thermally conductive fillers [69,70]. Cross-linked EVA composites were fabricated with different thermally conductive fillers such as Al_2O_3, SiC, ZnO, BN, AlN or MgO. Thermal conductivity of the composites containing SiC, ZnO and BN system was observed to be 2.85, 2.26 and 2.08 W/m·K respectively at filler content of 60 vol%. Thermally conductive layers produced from these composites were able to successfully improve the heat dissipation ability of the photovoltaic cells [69]. In another attempt, the thermal conductivity of EVA was improved by incorporating BN particles in varying concentrations from 0-60 wt% [71]. The thermal conductivity increased from 0.24W/m·K to 0.80W/m·K for the 60 wt% composite. It was also observed that the performance of the solar cell was increased by 0.3% owing to 6% more cooling than the standard laminate [71].

12.5.3 Batteries

Similar to solar cells, thermal management in batteries has pivotal role in battery's performance and service life. Higher temperature (>30°C) affects the battery life, whereas slow heat dissipation rate may result in internal pressure accumulation which could lead to catastrophic destruction of the cell. In order to overcome the overheating and safety challenges related to battery operation, highly thermally conductive materials are desirable in batteries which are capable to increase the rate of heat dissipation. In a related study, composites of graphite with polyvinylidene fluoride and carbon-black were studied for generating negative electrodes [72]. Introduction of a phase change material (PCM) in Li-ion battery packs for thermal management has also become a common practice. PCM takes excessive heat from the battery pack and stores it as latent heat, followed by phase change over a small temperature range. Usually PCMs have very low thermal conductivity which can be further improved by the addition of thermally conductive fillers. Paraffin wax is extensively applied as thermal PCM owing to its availability, stability, and heat storing and phase change ability. Shirazi *et al.* [73] conducted theoretical studies on nanocomposites of paraffin with fullerene, graphene, and CNTs with different fractions as phase change materials for lithium ion batteries. Computational investigations indicated that with increasing the concentration of fillers in the paraffin matrix, the rate of heat dissipation of the battery pack to the environment increased based on charging/discharging rates (C-rates). Similarly, Goli *et al.* [74] used nanocomposites comprising of paraffinic hydrocarbons and graphene sheets of different dimensions. The thermal conductivity of the Li-ion batteries was improved dramatically on incorporating graphene sheets, thus, leading to significant decrease in the temperature rise. The improved thermal conductivity of the nanocomposites was attributed to the attachment of hydrocarbon molecules to the graphene sheets which reduced the thermal interface resistance between the paraffin and graphene, as shown by Raman spectroscopy. In another study, graphene was used as an additive in paraffin wax from 0% up to 15%, where graphene/paraffin showed an enhanced thermal performance due to the graphene additive [75]. Similar experimental studies using different fillers such as carbon nano-fibers, SWCNTs, MWCNTs, SiO_2, Al_2O_3, Fe_2O_3, ZnO with paraffin have also been reported [76,77].

12.5.4 Light-Emitting Diode Devices

Light-emitting diode (LED) is a semiconductor p-n junction diode which is used to convert electrical energy into light energy, commonly known as electroluminescence phenomenon. Some of the significant applications of LEDs are illuminated displays, lighting, signaling, data communication, etc. Thermal management in LED devices is important as generated heat can affect the performance and service life. Consequently, efficient thermal dissipation system for LEDs is vital to maintain the luminous efficiency and lifetime of a LED. Generally, for thermal interface materials (TIM), PCMs like grease, gels (low conductivity), etc., are used whereas Cu and Al alloys (high weight and cost) are used as heat spreader and heat sink. Thermally conducting nanocomposites with improved thermal conductivity can be used as alternative materials for TIM, heat spreader and heat sink purposes in LEDs. Hu *et al.* [78] studied nanocomposite of silicone matrix with varying volume fractions of CNTs and micro-scale metallic (Ni) particles/spheres for TIM. The inclusion of CNTs (2%) helped to improve the thermal conductivity of the composite by 2.25 times (at 40 vol% Ni). From the percolation theory, improved thermal conductivity of the composite was attributed to ultra-high conductivity of CNTs and formation of conductive paths through interconnection of CNTs and nickel particles [78]. Similarly, nanocomposites of epoxy resin with functionalized SiC particles, MWNTs and hexagonal boron nitride (h-BN) were fabricated and applied as TIM [79, 80].

12.5.5 Electronic Packaging

Heat dissipation in electronic packaging is a challenge due to continuous development of higher power density and miniaturization. The purpose of thermal management in electronic packaging is to guarantee the performance and service life of devices by maintaining the temperature in the working range. Similar to LEDs and batteries, applications of electronic packaging such as single and stack chip packages and electronic control units (ECU) also need TIM and the heat sink with improved thermal conductivity [81]. Along with the improved thermal conductivity for TIM for chip packages, the material should also have good adhesion and rheological properties. Recently, Sun *et al.* [82] reported novel nanocomposites as TIM for integrated circuits and electronic packaging. The nanocomposite consisted of Ag coated polyimide network and indium matrix. Nanocomposites were fabricated in three steps. During the first step, electro-spinning technique was used to produce polyimide nano-fibers followed by surface modification for imide-ring cleavage. The infiltration of Ag was achieved by placing the fibrous network in $AgNO_3$ aqueous solution for ion exchange reaction. Further, liquid indium was infiltrated in the nanocomposite by using an infiltration device. Thermal cycling under extreme environments (−40 to 115 °C at 2 cycles/h) indicated slight increase of thermal resistance after 1000 cycles. The TIM nanocomposite had good adhesion, comparably unchanging reliability and consistency due to indium [82]. Similarly, nano-structured polyurethane and indium tin bismuth (InSnBi, melting point of 60°C) composite was fabricated by injecting metal into the nano-fibers. Mechanical testing indicated that shear strength of the sample decreased to 50% after 500 cycles due to overflow of the alloy [83]. Similarly, Xu *et al.* [84] prepared nanocomposites from Ag nanowires embedded in polycarbonate templates as TIM for electronic cooling. Nanocomposites exhibited improved intrinsic thermal conductivity of 30.3 W/m.K. Further, as a replacement of metals for TIM, high-performance TIM was prepared from composites of few-layer graphene (FLG) and epoxy resin [85]. In this research, FLG was prepared by interlayer catalytic exfoliation which is a cost-effective method, whereas the nanocomposites were prepared by solution mixing followed by molding (Figure 12.7). Nanocomposites exhibited improved thermal and mechanical reliability within ranges of

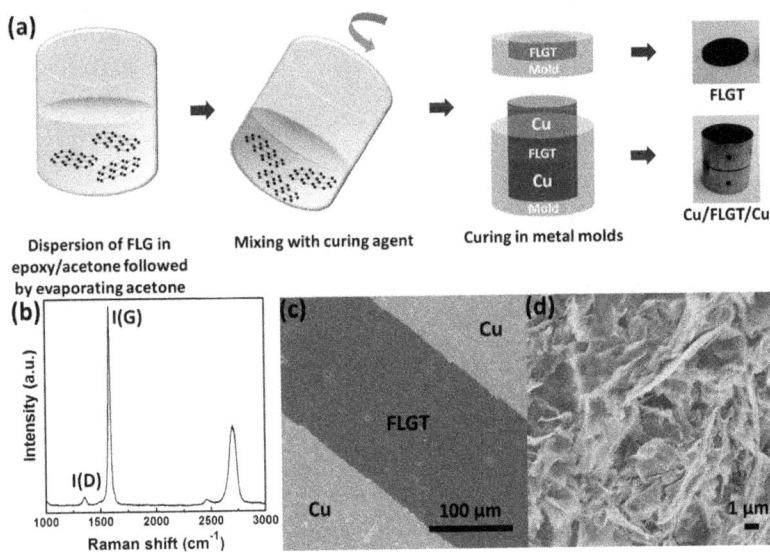

Figure 12.7 (a) Schematic of the FLG-TIM and Cu/FLG-TIM/Cu preparation; (b) Raman spectrum of a FLG flake; and (c) SEM image of the interface. Reproduced from Reference 85 with permission from American Chemical Society.

operating temperature and pressure. Thermal interface resistance between FLG TIM and copper was observed in the range of 3.2 and 4.3 mm²K/W for 5 vol% and 10 vol% at 330 K respectively, which was lower than that of commercially available TIMs. Similar studies were also reported where nanocomposites were fabricated using graphene nano-platelets and epoxy resin [86,87].

Further, in order to generate synergist effect of graphene and Ag for TIM, nanocomposites from hybrid fillers of thiophenol modified silver nanowires (mAgNWs) and acrylamide-modified reduced graphene oxide (AA-RGO) were fabricated with silicone rubber [88]. Improved thermal conductivity (1.152 W/m.K) of the composite was observed owing to synergetic effects and development of three-dimensional network structure in silicone rubber due to π-π interactions. It was observed that mAgNWs aided the graphene sheets to avoid re-stacking and aggregation in the composites, thus, leading to new heat conduction networks [88]. The full use of superior thermal properties of graphene is sometimes restricted due to high thermal resistance at the polymer-filler interface [89]. Interfacial thermal resistance mainly depends on factors like covalent and noncovalent functionalization, weak interactions between polymer chains and graphene surface, graphene doping, acetylenic linkage in graphene, etc. Molecular dynamics simulations on graphene-epoxy nanocomposites indicated that previously mentioned factors reduce the interfacial thermal resistance, as shown in Figure 12.8 for covalent functional groups [89].

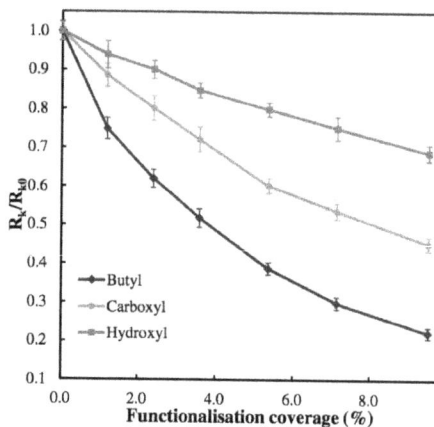

Figure 12.8 Relative interfacial thermal resistance with respect to the coverage of different covalent functional groups. Reproduced from Reference 89 with permission from American Chemical Society.

Electrically insulated hexagonal boron nitride (h-BN) has excellent in-plane ($\sim$600 W/m.K) and poor out-of-plane thermal conductivity. Nanocomposites prepared from randomly oriented h-BN nanoparticles generate materials with low thermal conductivity and limit its application for electronic packaging materials. Research studies have exhibited that directing the orientation of h-BN nanoparticles in the polymer matrix can produce nanocomposites with enhanced thermal conductivity. For this purpose, magnetically activated h-BN particles were synthesized by surface modification of h-BN nanoparticles with super paramagnetic Fe_2O_3 nanoparticles [90]. The surface modification was achieved by electrostatic interaction of $(Fe_2O_3)^+$ nanoparticles with (h-BN)$^-$ nano-platelets, which generated magnetically reactive h-BN nanoparticles (mh-BN). Nanocomposites were generated for electronic encapsulation by solution mixing of epoxy resin and mh-BN followed by placing between two rare-earth magnets, whereas curing was performed in the presence of magnets. The thermal conductivity ($\kappa_{composite}/\kappa_{matrix}$ = 5.7) of mh-BN/epoxy nanocomposites at 20 wt% filler content was significantly enhanced by vertically aligned filler as compared to the thermal conductivity ($\kappa_{composite}/\kappa_{matrix}$=2.8) of randomly oriented rh-BN/epoxy nanocomposites. Finite element analysis of nanocomposites as underfill material in a flip-chip package exhibited promising material performance for advanced micro-electronic packaging [90].

12.5.6 Energy Storage

Depleting fossil fuel resources and rising awareness of environmental concerns indicate that there is a need to reassess our energy needs by discovering alternative sources of energy and effectively utilizing the existing resources. Use of PCMs is a convenient way to store latent heat through melting (store) and solidifying (emit). For energy storage purposes, mainly paraffins or paraffinic hydrocarbons are employed due to their abundance from natural and synthetic resources, chemically inert nature, high latent heat, etc. Various methods of energy storage reported in the literature are storage through mechanical, electrical, thermal and thermochemical processes [91]. Many energy storage applications using PCMs have been discussed in literature such as thermal energy storage, conditioning of buildings [92], solar cooking, solar power plants [93], cold energy battery [94], medical applications [95], turbine inlet chilling with thermal energy storage, cooling of heat and electrical engines, waste heat recovery [96], spacecraft thermal systems [97], textiles [98], etc. [99]. A great deal of research focus has been devoted to improve the thermal conductivity of thermoplastics, thermosets and paraffin waxes for energy storage by the incorporation of different inorganic and organic thermally conductive fillers. Some of the thermal management applications require shape stabilization and structure integrity during energy storage operation. Polymers with rigid backbone can provide support and protection for the PCM materials from the leakage issues. For this purpose, grafting of PCM on to the rigid chain can be beneficial owing to distinctive thermal management. In this way, the attached PCM molecules can act as energy storage units whereas rigid chains perform the supporting role.

In order to achieve the unique characteristics of PCM materials, Pielichowska *et al.* [101] developed shape stabilized PCM by grafting octadecanol on poly(styrene-co-maleic anhydride) and graphene oxide nano-platelets (GONPs) were added to the grafted polymer through solution blending method [100]. The purpose of adding GONPs was to improve thermal stability of the PCM at higher temperature due to adsorption and shielding behavior of GONPs sheet-like structure. Similarly, in order to improve the leakage and low thermal conductivity of PCM, thermosets incorporated with thermally conductive fillers were also used as carrier of PCM. Poly(ethylene glycol) (PEG) is a semi-crystalline promising PCM due to its properties such as non-toxicity, inertness, low super cooling and cost effectiveness. In order to avoid the leakage above its melting point, chemical linking with polyisocyanates was performed to produce polyurethane using one-step bulk polymerization. Moreover, graphite nano-platelets GNPs were also incorporated into PU/PEG to improve thermal conductivity and shape stability [102]. Three-dimensional (3-D) microcrystalline cellulose aerogels (CAG) is a light weight and porous material and can be used as support for PCM for thermal energy storage applications. Use of PEG with 3-D CAG can also be beneficial owing to inter-molecular interactions between PEG and cellulose molecules. It was reported that thermal conductivity of PEG/3-DCAG blends was improved by 463% with the incorporation of 5.3 wt% of GNPs, along with retaining of the shape upon compressing at melting point of PEG [103]. Nano-fiber nanocomposites produced from electro-spinning of nylon (polyamide 6, PA6), PEG and SiO_2, Fe_2O_3, ZnO, Al_2O_3 nanoparticles were also reported as PCM with improved thermal stability and stability [77,104].

Similarly, direct nanocomposites of PCM materials with different thermally conductive fillers were reported for energy storage applications using solar energy. In a related study, nanocomposites of different waxes like paraffin wax and microcrystalline wax with α nano-alumina (α-Al_2O_3) (0.5, 1.0 and 2.0 wt%) were prepared using melt mixing and intensive sonication. The results exhibited that the thermal conductivity of the nanocomposite improved, thus, leading to rise in latent heat and reduction in melting temperature [105]. Xia *et al.* [106] prepared nanocomposites of acetamide and expanded graphite (EG) (10 wt%) for energy storage application in the active solar systems. It was reported that with the addition of EG, the thermal conductivity of the nanocomposites increased by five-fold. Along with the thermal conductivity, other thermal properties of were also improved such as shifting of melting/freezing points 66.95/42.46 °C to 65.91/65.52 °C and latent heat from 194.92 to 163.71 kJ.kg^{-1} for pure acetamide and acetamide/EG respectively. The use of acetamide/EG in a latent thermal energy storage unit performed better by reducing heat storage and retrieval periods to 45% and 78% than that of acetamide alone [106]. In a similar work, nanocomposites of paraffin/HDPE with 3 wt% of EG were generated for enhanced thermal conductivity. The material exhibited improved thermal conductivity by 24% due to thermally conductive networks and was used as PCM for thermal energy storage [107]. In another study, attempt to improve the shape stability and thermal storage performance of PCM was made by generating nanocomposites of paraffin/halloysite nanotube (P-HNT) by absorbing P into the pores of HNT and applied for solar energy storage. The nanocomposite materials were proved to be effective for solar energy storage after 50 melt–freeze cycles [108]. For medium range temperature solar energy storage applications, sebacic acid/expanded graphite nanocomposites were synthesized in order to avoid the low extent of sub-cooling of sebacic acid. The PCM exhibited negligible sub-cooling along with excellent thermal properties, thermal reliability, stability, and formability [109]. Use of cellulosic matrix loaded with stable microcapsules of PCM (25 and 50 wt%) was reported for cold storage of perishable products. Using approaches of experiments and computations, it was observed that PCM/cellulose composite served as effective cold storage material [110].

Recently, it was estimated that 40% of entire energy is usually spent in industrial and residential building activities like lighting, heating and cooling. In accordance with the global policies of effective and efficient consumption of energy, it is necessary to pay attention to the thermal management of buildings. Research on energy efficiency in buildings has been expanding rapidly to generate improved systems and design approaches to ensure efficient use of building activities, especially in many energy-intensive industries. Fang *et al.* [111] introduced a novel method of incorporating organic PCMs into building materials. Firstly, nanocomposites were prepared by adding organically modified montmorillonite (OMM) in organic PCMs followed by incorporation in gypsum. The aim of OMM addition in PCM was to improve the thermal conductivity and generate compatibility with gypsum. Nanocomposites generated with 20 wt% of OMM performed much better in decreasing the building energy consumption due to higher latent heat, appropriate phase change temperature and solidity

[111]. Similar studies have also been reported on the generation of nanocomposites with PCMs [112,113]. Moreover, the direct incorporation of octadecane (C18) and tetracosane (C24) (eutectic mixture) as PCM into gypsum was also reported. The composite as PCM exhibited outstanding thermal regulation properties even after 1000 melting and freezing cycles. In addition, the composite did not lose its chemical structure and performed as favorable material to regulate room temperature [114]. Warzoha *et al.* [115] studied the effect of different volume fractions of randomly oriented herringbone style graphite nanofibers (HGNF) on thermal properties of paraffin used as PCM. It was observed that as the percolation approached, the thermal conductivity of the nanocomposites increased exponentially. Also, the thermal boundary resistance analysis demonstrated that the thermal resistance was higher at HGNF–PCM interface as compared to HGNF–HGNF interfaces in solid and liquid phases [115].

12.6 Conclusion

In this chapter, different factors affecting the thermal conductivity of the polymers and their nanocomposites have been discussed. Theoretical calculations and modeling of nanocomposites may help to select the polymer, filler and parameters of nanocomposites fabrication so as to achieve suitable thermally conducting polymer nanocomposite. Generally, the thermal conductivity of the polymer nanocomposites increases with increasing filler loading. Improvement in the thermal conductivity can be achieved at lower loading of fillers by using crystalline and oriented polymers. The interfacial thermal resistance can be decreased using fillers with large sized particles. Further, the use of hybrid fillers and modification of filler surface are other techniques to develop conducting networks between different fillers in a matrix. Finally, different applications of thermally conducting polymer nanocomposites in alternative and sustainable energy resources along with efficient use of existing energy have been discussed so as to achieve effective use of these materials in a large number of industries, especially in energy intensive oil and gas industries.

References

1. Chen, H., Ginzburg, V. V., Yang, J., Yang, Y., Liu, W., Huanf, Y., Du, L., and Chen, B. (2016) Thermal conductivity of polymer-based composites: Fundamentals and applications. *Progress in Polymer Science*, **59**, 41-85.

2. Callister, W. D., and Rethwisch, D. G. (2014) *Thermal Properties*, in *Materials Science and Engineering: An Introduction*, Wiley, USA.
3. Lovestam, G., Rauscher, H., Roebben, G., Kluttgn, B. S., Gibson, N., Putaud, J.-P., and Stamm, H. (2010) *Considerations on a definition of nanomaterial for regulatory purposes*. JRC Reference Reports. Online: https://ec.europa.eu/jrc/sites/jrcsh/files/jrc_reference_report_201007_nanomateri als.pdf (assessed 24th February 2017).
4. Mittal, V. (2014) Functional polymer nanocomposites with graphene: A review. *Macromolecular Materials and Engineering*, **299**, 906-931.
5. Idumah C. I., and Hassan, A. (2016) Recently emerging trends in thermal conductivity of polymer nanocomposites. *Reviews in Chemical Engineering*, **32**, 413-457.
6. Ebadi-Dehaghani, H., and Nazempour, M. (2012) Thermal conductivity of nanoparticles filled polymers. In: *Smart Nanopraticles Technology*, Hashim, A. (ed.), INTECH Open. Online: http://www.intechopen.com/books/smart-nanoparticles-technology/thermal-conductivity-of-nanoparticles-filled-polymers- (assessed 10th February 2017).
7. Kumar, P. M., Kumar, J., Tamilarasan, R., Sendhilnathan, S., and Suresh, S. (2015) Review on nanofluids theoretical thermal conductivity models. *Engineering Journal*, **19**(1). Online: http://citeseerx.ist.psu.edu/viewdoc/download;jsessionid=2F215B408291AC43A1 47C40DF7276C83?doi=10.1.1.676.9543&rep=rep1&type=pdf (assessed 31st January 2017).
8. Pietrak, K., and Wisniewski, T. S. (2015) A review of models for effective thermal conductivity of composite materials. *Journal of Power Technologies*, **95**(1), 14-24.
9. Burger, H. (1915) Das lertvermogen verdummter mischristallfreier lonsungen. *Phys. Zs.*, **20**, 73-76.
10. Hamilton, R., and Crosser, O. (1962) Thermal conductivity of heterogeneous two-component systems. *Industrial & Engineering Chemistry Fundamentals*, **1**(3), 187-191.
11. Rayleigh, L. (1892) LVI. On the influence of obstacles arranged in rectangular order upon the properties of a medium. *Philosophical Magazine Series 5*, **34**(211), 481-502.
12. Swartz, E. T., and Pohl, R. O. (1989) Thermal boundary resistance. *Reviews of Modern Physics*, **61**(3), 605-668.
13. Powell, Jr., B. R., Youngblood, G. E., Hasselman, D. P., and Bentsen, L. D. (1980) Effect of thermal expansion mismatch on the thermal diffusivity of glass-Ni composites. *Journal of the American Ceramic Society*, **63**(9-10), 581-586.
14. Benveniste, Y. (1987) Effective thermal conductivity of composites with a thermal contact resistance between the constituents: Nondilute case. *Journal of Applied Physics*, **61**(8), 2840-2843.
15. Landauer, R. (1952) The electrical resistance of binary metallic mixtures. *Journal of Applied Physics*, **23**(7), 779-784.
16. Conway, J., and Sloane, N. J. A. (2013) *Sphere Packings, Lattices and Groups*, Springer, USA.

17. Devpura, A., Phelan, P. E., and Prasher, R. S. (2000) Percolation Theory Applied to the Analysis of Thermal Interface Materials in Flip-Chip Technology. In: *ITHERM 2000. The Seventh Intersociety Conference on Thermal and Thermomechanical Phenomena in Electronic Systems,* Cat. No.00CH37069.

18. Yuksel, N. (2016) The review of some commonly used methods and techniques to measure the thermal conductivity of insulation materials. In: Insulation Materials in Context of Sustainability, Almusard, A., and Almassad, A. (eds.), INTECH Open. Online: http://www.intechopen.com/books/insulation-materials-in-context-of-sustainability/the-review-of-some-commonly-used-methods-and-techniques-to-measure-the-thermal-conductivity-of-insul (assessed 25th February 2017).

19. Tong, X.C. (2011) Characterization methodologies of thermal management materials. In: *Advanced Materials for Thermal Management of Electronic Packaging,* Springer, USA, pp. 59-129.

20. Corsan, J. M. (1992) Axial heat flow methods of thermal conductivity measurement for good conducting materials. In: *Compendium of Thermophysical Property Measurement Methods: Volume 2, Recommended Measurement Techniques and Practices,* Maglic, K. D., Cezairliyan, A., and Peletsky, V. E. (eds.), Springer, USA, pp. 3-31.

21. Czichos, H., Saito, T., and Smith, L. (2006) *Springer Handbook of Materials Measurement Methods,* vol. 978, Springer, Germany.

22. Yesilata, B., and Turgut, P. (2007) A simple dynamic measurement technique for comparing thermal insulation performances of anisotropic building materials. *Energy and Buildings,* **39**(9), 1027-1034.

23. Vozar, L. (1996) A computer-controlled apparatus for thermal conductivity measurement by the transient hot wire method. *Journal of Thermal Analysis and Calorimetry,* **46**(2), 495-505.

24. Kwon, S. Y., and Lee, S. (2012) Precise measurement of thermal conductivity of liquid over a wide temperature range using a transient hot-wire technique by uncertainty analysis. *Thermochimica Acta,* **542**, 18-23.

25. Solorzano, E., Rodriguez-Perez, M. A., and de Saja, J. A. (2008) Thermal conductivity of cellular metals measured by the transient plane sour method. *Advanced Engineering Materials,* **10**(4), 371-377.

26. Min, S., Blumm, J., and Lindemann, A. (2007) A new laser flash system for measurement of the thermophysical properties. *Thermochimica Acta,* **455**(1–2), 46-49.

27. Zhang, T., Wu, X., and Luo, T. (2014) Polymer nanofibers with outstanding thermal conductivity and thermal stability: fundamental linkage between molecular characteristics and macroscopic thermal properties. *The Journal of Physical Chemistry C,* **118**(36), 21148-21159.

28. Zhang, T., and Luo, T. (2012) Morphology-influenced thermal conductivity of polyethylene single chains and crystalline fibers. *Journal of Applied Physics,* **112**, 094304.

29. Hansen, D., and Bernier, G. A. (1972) Thermal conductivity of polyethylene: The effects of crystal size, density and orientation on the thermal conductivity. *Polymer Engineering & Science,* **12**(3), 204-208.

30. Liu, J., and Yang, R. (2010) Tuning the thermal conductivity of polymers with mechanical strains. *Physical Review B*, **81**(17), 174122.
31. Kim, G.-H., Le, D., Shanker, A., Shao, L., Kwon, M. S., Gidley, D., Kim, J., and Pipe, K. P. (2015) High thermal conductivity in amorphous polymer blends by engineered interchain interactions. *Nature Materials*, **14**(3), 295-300.
32. Giang, T., Park, J., Cho, I., Ko, Y., and Kim, J. (2013) Effect of backbone moiety in epoxies on thermal conductivity of epoxy/alumina composite. *Polymer Composites*, **34**(4), 468-476.
33. Harada, M., Hamaura, N., Ochi, M., and Agari, Y. Thermal conductivity of liquid crystalline epoxy/BN filler composites having ordered network structure. *Composites, Part B: Engineering*, **55**, 306-313.
34. Akatsuka, M., Takazawa, Y., and Sugawara, K. (2006) Thermosetting Resin Compounds, US Patent 20060276568 A1.
35. Hammerschmidt, A., Geibel, K., and Strohmer, F. (1993 In situ photopolymerized, oriented liquid-crystalline diacrylates with high thermal conductivities. *Advanced Materials*, **5**(2), 107-109.
36. Wypych, G. (2000) *Handbook of Fillers*. Chem Tech Publishing, Canada.
37. Mamunya, Y. P., Davydenko, V. V., Pissis, P., and Lebedev, E. V. (2002) Electrical and thermal conductivity of polymers filled with metal powders. *European Polymer Journal*, **38**, 1887-1897.
38. Stoller, M. D., Park, S., Zhu, Y., An, J., and Ruoff, R. S. (2008) Graphene-based ultracapacitors. *Nano Letters*, **8**(10), 3498-3502.
39. Liu, Z., Wu, B., and Gu, M. (2007) Effect of hydrolysis of AlN particulates on corrosion behavior of Al/AlNp composite in neutral chloride solution. *Composites, Part A: Applied Science and Manufacturing*, **38**(1), 94-99.
40. Yokota, H., and Ibukiyama, M. (2003) Microstructure tailoring for high thermal conductivity of β-Si₃N₄ ceramics. *Journal of the American Ceramic Society*, **86**(1), 197-199.
41. Bhattacharya, S. K. (1986) *Metal Filled Polymers*, Taylor & Francis, USA.
42. Tavman, I. H., and Akinci, H. (2000) Transverse thermal conductivity of fiber reinforced polymer composites. *International Communications in Heat and Mass Transfer*, **27**(2), 253-261.
43. Kusunose, T., Yagi, T., Firoz, S. H., and Sekino, T. (2013) Fabrication of epoxy/silicon nitride nanowire composites and evaluation of their thermal conductivity. *Journal of Materials Chemistry A*, **1**(10), 3440-3445.
44. Yu, J., Chen, Y., Wuhrer, R., Liu, Z., and Ringer, S. P. (2005) In situ formation of BN nanotubes during nitriding reactions. *Chemistry of Materials*, **17**(20), 5172-5176.
45. Hill, R. F., and Supancic, P. H. (2002) Thermal conductivity of platelet-filled polymer composites. *Journal of the American Ceramic Society*, **85**(4), 851-857.
46. Pashayi, K., Fard, H. R., Lai, F., Iruvanti, S., Plawsky, J., and Borca-Tasciuc, T. (2012) High thermal conductivity epoxy-silver composites based on self-constructed nanostructured metallic networks. *Journal of Applied Physics*, **111**(10), 104310.
47. Li, T.-L., and Hsu, S.L.-C. (2010) Enhanced thermal conductivity of polyimide films via a hybrid of micro- and nano-sized boron nitride. *The Journal of Physical Chemistry B*, **114**(20), 6825-6829.

48. Zhou, W., Qi, S., Tu, C., Zhao, H., Wang, C., and Kou, J. (2007) Effect of the particle size of Al2O3 on the properties of filled heat-conductive silicone rubber. *Journal of Applied Polymer Science*, **104**(2), 1312-1318.
49. Leung, S. N., Khan. M. O., Chan, E., Naguib, H. E., Dawson, F., Adinkrah, V., and Lakatos-Hayward, L. (2013) Synergistic effects of hybrid fillers on the development of thermally conductive polyphenylene sulfide composites. *Journal of Applied Polymer Science*, **127**(5), 3293-3301.
50. Pak, S. Y., Kim, H. M., Kim, S. Y., and Youn, J. R. (2012) Synergistic improvement of thermal conductivity of thermoplastic composites with mixed boron nitride and multi-walled carbon nanotube fillers. *Carbon*, **50**(13), 4830-4838.
51. Zhu, B. L., Zheng, H., Wang, J., Ma, J., Wu, J., and Wu, R. (2014) Tailoring of thermal and dielectric properties of LDPE-matrix composites by the volume fraction, density, and surface modification of hollow glass microsphere filler. *Composites, Part B: Engineering*, **58**, 91-102.
52. Raman, C. (2008) Boron Nitride in Thermoplastics: Effect of Loading, Particle Morphology and Processing Conditions. in *Proceedings of the NATAS Annual Conference on Thermal Analysis and Applications*, USA.
53. Choy, C. L., Leung, W. P., Kowk, K. W., and Lau, F. P. (1992) Elastic moduli and thermal conductivity of injection-molded short-fiber–reinforced thermoplastics. *Polymer Composites*, **13**(2), 69-80.
54. Choy, C. L., Wong, Y. W., Yang, G. W., and Kanamoto, T. (1999) Elastic modulus and thermal conductivity of ultradrawn polyethylene. *Journal of Polymer Science, Part B: Polymer Physics*, **37**(23), 3359-3367.
55. Choy, C. L., Luk, W. H., and Chen, F. C. (1978) Thermal conductivity of highly oriented polyethylene. *Polymer*, **19**, 155-162.
56. Han, S., Lin, J. T., Yamada, Y., and Chung, D. D. L. (2008) Enhancing the thermal conductivity and compressive modulus of carbon fiber polymer–matrix composites in the through-thickness direction by nanostructuring the interlaminar interface with carbon black. *Carbon*, **46**(7), 1060-1071.
57. Shahil, K. M. F., and Balandin, A. A. (2012) Graphene–multilayer graphene nanocomposites as highly efficient thermal interface materials. *Nano Letters*, **12**(2), 861-867.
58. Yao, Q., Chen, L., Zhang, W., Liufu, S., and Chen, X. (2010) Enhanced thermoelectric performance of single-walled carbon nanotubes/polyaniline hybrid nanocomposites. *ACS Nano*, 4(4), 2445-2451.
59. Han, Z., and Fina, A. (2011) Thermal conductivity of carbon nanotubes and their polymer nanocomposites: A review. *Progress in Polymer Science*, **36**(7), 914-944.
60. T'Joen, C., Park, Y., Wang, Q., Sommers, A., Han, X., and Jacobi, A. (2009) A review on polymer heat exchangers for HVAC&R applications. *International Journal of Refrigeration*, **32**(5), 763-779.
61. Xu, Z., and Buehler, M. J. (2009) Nanoengineering heat transfer performance at carbon nanotube interfaces. *ACS Nano*, 3(9), 2767-2775.
62. Rungraeng, N., Cho, Y.-C., Yoon, S. H., and Jun, S. (2012) Carbon nanotube-polytetrafluoroethylene nanocomposite coating for milk fouling reduction in plate heat exchanger. *Journal of Food Engineering*, **111**(2), 218-224.

63. Kananeh, A. B., Scharnbeck, E., Kuck, U. D., and Rabiger, N. (2010) Reduction of milk fouling inside gasketed plate heat exchanger using nano-coatings. *Food and Bioproducts Processing*, **88**(4), 349-356.

64. Yang, E., Hwang, T., Kumar, A., and Kim, K. J. (2016) Anti-biofouling, thermal, and electrical performance of nanocomposite coating with multiwall carbon nanotube and polytetrafluoroethylene-blended polyphenylenesulfide. *Advances in Polymer Technology*, DOI: 10.1002/adv.21728.

65. Marconnet, A. M., Yamamoto, N., Panzer, M. A., Wardle, B. L., and Goodson, K. E. (2011) Thermal conduction in aligned carbon nanotube-polymer nanocomposites with high packing density. *ACS Nano*, **5**(6), 4818-4825.

66. Guo, H., Li, X., Li, B., Wang, J., and Wang, S. (2017) Thermal conductivity of graphene/poly(vinylidene fluoride) nanocomposite membrane. *Materials & Design*, **114**:, 355-363.

67. Haggenmueller, R., Guthy, C., Lukes, J. R., Fischer, J. E., and Winey, K. I. (2007) Single wall carbon nanotune/polyethylene nanocomposites: thermal and electrical conductivity. *Macromolecules*, **40**, 2417-2421.

68. Krauter, S., Araujo, R. G., Schroer, S., Hanitsch, R., Salhi, M. J., Triebel, C., and Lemoine, R. (1999) Combined photovoltaic and solar thermal systems for facade integration and building insulation. *Solar Energy*, **67**(4–6), 239-248.

69. Lee, B., Liu, J. Z., Sun, B., Shn, C. Y., and Dai, G. C. (2008) Thermally conductive and electrically insulating EVA composite encapsulants for solar photovoltaic (PV) cell. *eXPRESS Polymer Letters*, **2**(5), 357-363.

70. Kim, N., Kim, D., Kang, H., and Park, Y.-G. (2016) Improved heat dissipation in a crystalline silicon PV module for better performance by using a highly thermal conducting backsheet. *Energy*, **113**, 515-520.

71. Allan, J., Pinder, H., and Dehouche, Z. (2016) Enhancing the thermal conductivity of ethylene-vinyl acetate (EVA) in a photovoltaic thermal collector. *AIP Advances*, **6**(3), 035011.

72. Maleki, H., Selman, J. R., Dinwiddie, R. B., and Wang, H. (2001) High thermal conductivity negative electrode material for lithium-ion batteries. *Journal of Power Sources*, **94**(1), 26-35.

73. Shirazi, A. H. N., Mohebbi, F., Kakavand, M. R. A., He, B., and Rabczuk, T. (2016) Paraffin nanocomposites for heat management of lithium-ion batteries: A computational investigation. *Journal of Nanomaterials*, 2131946.

74. Goli, P., Legedza, S., Dhar, A., Salgado, R., Renteria, J., and Balandin, A. A. (2014) Graphene-enhanced hybrid phase change materials for thermal management of Li-ion batteries. *Journal of Power Sources*, **248**, 37-43.

75. Zhang, Y., Yue, W., Zhang, S., Huang, S., and Liu, J. (2016) Experimental Investigation of Paraffin Wax with Graphene Enhancement as Thermal Management Materials for Batteries. In: *17th International Conference on Electronic Packaging Technology (ICEPT)*, China, DOI: 10.1109/ICEPT.2016.7583385.

76. Shaikh, S., Lafdi, K., and Hallinan, K. (2008) Carbon nanoadditives to enhance latent energy storage of phase change materials. *Journal of Applied Physics*, **103**(9), 094302.

77. Babapoor, A., Karimi, G., and Sabbaghi, S. (2016) Thermal characteristic of nanocomposite phase change materials during solidification process. *Journal of Energy Storage*, **7**, 74-81.

78. Xuejiao, H., Linan, J., and Goodson. K. E. (204) Thermal Conductance Enhancement of Particle-Filled Thermal Interface Materials Using Carbon Nanotube Inclusions. *The Ninth Intersociety Conference on Thermal and Thermomechanical Phenomena In Electronic Systems*, USA, IEEE Cat. No. 04CH37543.

79. Liang, Q., Xiu, Y., Lin, W., Moon, K.-S., and Wong, C. P. (2009) Epoxy/h-BN Composites for Thermally Conductive Underfill Material. *59th Electronic Components and Technology Conference*, USA.

80. Liang, Q., Moon, K.-S., Jiang, H., and Wong, C. P. (2012) Thermal conductivity enhancement of epoxy composites by interfacial covalent bonding for underfill and thermal interfacial materials in Cu/low-K application. *IEEE Transactions on Components, Packaging and Manufacturing Technology*, **2**(10), 1571-1579.

81. Prasher, R. (2006) Thermal interface materials: historical perspective, status, and future directions. *Proceedings of the IEEE*, **94**(8), 1571-1586.

82. Sun, S., Chen, S., Luo, X., Fu, Y., Ye, L., and Liu, J. (2016) Mechanical and thermal characterization of a novel nanocomposite thermal interface material for electronic packaging. *Microelectronics Reliability*, **56**, 129-135.

83. Lu, X., Zhuang, M., Zhang, L., Ye, L., and Liu, J. (2012) Environmental Reliability of Nano-structured Polymer-Metal Composite Thermal Interface Material. *13th International Conference on Electronic Packaging Technology & High Density Packaging*, China.

84. Xu, J., Munari, A., Dalton, E., Mathewson, A., and Razeeb, K. M. (2009) Silver nanowire array-polymer composite as thermal interface material. *Journal of Applied Physics*, **106**(12), 124310.

85. Park, W., Guo, Y., Li, X., Hu, J., Liu, L., Ruan, X., and Chen, Y. P. (2015) High-performance thermal interface material based on few-layer graphene composite. *The Journal of Physical Chemistry C*, **119**(47), 26753-26759.

86. Jimenez-Suarez, A., Moriche, R., Prolongo, S. G., and Urena, A. (2016) GNPs reinforced epoxy nanocomposites used as thermal interface materials. *Journal of Nano Research*, **38**, 18-25.

87. Gu, J., Yang, X., Lv, Z., Li, N., Liang, C., anf Zhang, Q. (2016) Functionalized graphite nanoplatelets/epoxy resin nanocomposites with high thermal conductivity. *International Journal of Heat and Mass Transfer*, **92**, 15-22.

88. Lin, S.-C., Ma, C.-C. M., Liao, W.-H., Wang, J.-A., Zeng, S.-J., Hsu, S.-Y., Chen, Y.-H., Hsiao, S.-T., Cheng, T.-Y., Lin, C.-W., and Hsiao, P.-Y. (2016) Preparation of a graphene–silver nanowire hybrid/silicone rubber composite for thermal interface materials. *Journal of the Taiwan Institute of Chemical Engineers*, **68**, 396-406.

89. Wang, Y., Yang, C., Pei, Q.-X., and Zhang, Y. (2016) Some aspects of thermal transport across the interface between graphene and epoxy in nanocomposites. *ACS Applied Materials & Interfaces*, **8**(12), 8272-8279.

90. Lin, Z., Liu, Y., Raghavan, S., Moon, K.-s., Sitaraman, S. K., and Wong, C.-p. (2013) Magnetic alignment of hexagonal boron nitride platelets in polymer matrix: toward high performance anisotropic polymer composites for electronic encapsulation. *ACS Applied Materials & Interfaces*, **5**(15), 7633-7640.

91. Sharma, A., Tyagi, V. V., Chen, C. R., and Buddhi, D. (2009) Review on thermal energy storage with phase change materials and applications. *Renewable and Sustainable Energy Reviews,* **13**(2), 318-345.
92. Liu, L., and Khodadadi, J. M. (2016) Thermal conductivity enhancement of phase change materials for thermal energy storage: A review. *Renewable and Sustainable Energy Reviews,* **62**, 305-317.
93. Kenisarin, M., and Mahkamov, K. (2007) Solar energy storage using phase change materials. *Renewable and Sustainable Energy Reviews,* **11**(9), 1913-1965.
94. Huo, Y., and Rao, Z. (2017) Investigation of phase change material based battery thermal management at cold temperature using lattice Boltzmann method. *Energy Conversion and Management,* **133**, 204-215.
95. Grim, T. E., and Haines, J. R. (1990) Orthopedic Device Having Gel Pad with Phase Change Material, US Patent 4964402 A.
96. Kaizawa, A., Maruoka, N., Kawai, A., Kamano, H., Jozuka, T., Senda, T., and Akiyama, T. (2008) Thermophysical and heat transfer properties of phase change material candidate for waste heat transportation system. *Heat and Mass Transfer,* **44**(7), 763-769.
97. Mulligan, J. C., Colvin, D. P., and Bryant, Y. G. (1996) Microencapsulated phase-change material suspensions for heat transfer in spacecraft thermal systems. *Journal of Spacecraft and Rockets,* **33**(2), 278-284.
98. Mondal, S. (2008) Phase change materials for smart textiles – An overview. *Applied Thermal Engineering,* **28**(11-12), 1536-1550.
99. Farid, M. M., Khudhair, A. M., Razack, S. A. K., and Al-Hallaj, S. (2004) A review on phase change energy storage: materials and applications. *Energy Conversion and Management,* **45**(9–10), 1597-1615.
100. Pielichowska, K., Nowak, M., Szatkowski, P., and Macherzynska, B. (2016) The influence of chain extender on properties of polyurethane-based phase change materials modified with graphene. *Applied Energy,* **162**, 1024-1033.
101. Liu, L., Kong, L., Wang, H., Niu, R., and Shi, H. (2016) Effect of graphene oxide nanoplatelets on the thermal characteristics and shape-stabilized performance of poly(styrene-co-maleic anhydride)-g-octadecanol comb-like polymeric phase change materials. *Solar Energy Materials and Solar Cells,* **149**, 40-48.
102. Pielichowska, K., Bieda, J., and Szatkowski, P. (2016) Polyurethane/graphite nano-platelet composites for thermal energy storage. *Renewable Energy,* **91**, 456-465.
103. Yang, J., Zhang, E., Li, X., Zhang, Y., Qu, J., and Yu, Z.-Z. (2016) Cellulose/graphene aerogel supported phase change composites with high thermal conductivity and good shape stability for thermal energy storage. *Carbon,* **98**, 50-57.
104. Babapoor, A., Karimi, G., and Khorram, M. (2016) Fabrication and characterization of nanofiber-nanoparticle-composites with phase change materials by electrospinning. *Applied Thermal Engineering,* **99**, 1225-1235.
105. Mohamed, N. H., Soliman, F. S., El Maghraby, H., and Moustfa, Y. M. (2017) Thermal conductivity enhancement of treated petroleum waxes, as phase change material, by α nano alumina: Energy storage. *Renewable and Sustainable Energy Reviews,* **70**, 1052-1058.
106. Xia, L., and Zhang, P. (2011) Thermal property measurement and heat transfer analysis of acetamide and acetamide/expanded graphite composite phase change

material for solar heat storage. *Solar Energy Materials and Solar Cells*, **95**(8), 2246-2254.
107. Zhang, R., Moon, K.-s., Lin, W., and Wong, C. P. (2010) Preparation of highly conductive polymer nanocomposites by low temperature sintering of silver nanoparticles. Journal of Materials Chemistry, **20**, 2018-2023.
108. Zhang, J., Zhang, X., Wan, Y., Mei, D., and Zhang, B. (2012) Preparation and thermal energy properties of paraffin/halloysite nanotube composite as form-stable phase change material. *Solar Energy*, **86**(5), 1142-1148.
109. Wang, S., Qin, P., Fang, X., Zhang, Z., Wang, S., and Liu, X. (2014) A novel sebacic acid/expanded graphite composite phase change material for solar thermal medium-temperature applications. *Solar Energy*, **99**, 283-290.
110. Melone, L., Altomare, L., Cigada, A., and De Nardo, L. (2012) Phase change material cellulosic composites for the cold storage of perishable products: From material preparation to computational evaluation. *Applied Energy*, **89**(1), 339-346.
111. Fang, X., Zhang, Z., and Chen, Z. (2008) Study on preparation of montmorillonite-based composite phase change materials and their applications in thermal storage building materials. *Energy Conversion and Management*, **49**(4), 718-723.
112. Fang, X., and Zhang, Z. (2006) A novel montmorillonite-based composite phase change material and its applications in thermal storage building materials. *Energy and Buildings*, **38**(4), 377-380.
113. Zhang, Z., and Fang, X. (2006) Study on paraffin/expanded graphite composite phase change thermal energy storage material. *Energy Conversion and Management*, **47**(3), 303-310.
114. Karaipekli, A., Sarı, A., and Bicer, A. (2016) Thermal regulating performance of gypsum/(C18–C24) composite phase change material (CPCM) for building energy storage applications. *Applied Thermal Engineering*, **107**, 55-62.
115. Warzoha, R. J., Weigand, R. M., and Fleischer, A. S. (2015) Temperature-dependent thermal properties of a paraffin phase change material embedded with herringbone style graphite nanofibers. *Applied Energy*, **137**, 716-725.

Index

i